Micro- and Nanopatterning Polymers

ACS SYMPOSIUM SERIES **706**

Micro- and Nanopatterning Polymers

Hiroshi Ito, EDITOR
IBM Almaden Research Center

Elsa Reichmanis, EDITOR
Bell Laboratories, Lucent Technologies

Omkaram Nalamasu, EDITOR
Bell Laboratories, Lucent Technologies

Takumi Ueno, EDITOR
Hitachi Ltd.

Developed from a symposium sponsored by the
Division of Polymeric Materials: Science and Engineering
at the 214th ACS National Meeting
Las Vegas, Nevada,
September 7–11, 1997

American Chemical Society, Washington, DC

Library of Congress Cataloging-in-Publication Data

Micro- and nanopatterning polymers / Hiroshi Ito, editor ... [et al.].

p. cm.—(ACS symposium series, ISSN 0097–6156 ; 706)

"Developed from a symposium sponsored by the Division of Polymeric Materials: Science and Engineering at the 214th ACS National Meeting, Las Vegas, Nevada, September 7–11, 1997."

Includes bibliographical references and index.

ISBN 0–8412–3581–3

1. Microelectronics—Materials. 2. Polymeric composites. 3. Photoresists—Materials.

I. Ito, Hiroshi. II. American Chemical Society. Division of Polymeric Materials: Science and Engineering. III. American Chemical Society. Meeting (214th : 1997 : Las Vegas, Nev.) IV. Series.

TK7871.15.P6M517 1998
621.381—dc21

98–25955
CIP

The paper used in this publication meets the minimum requirements of American National Standard for Information Sciences—Permanence of Paper for Printed Library Materials, ANSI Z39.48–1984.

Distributed by Oxford University Press

PRINTED IN THE UNITED STATES OF AMERICA

Advisory Board

ACS Symposium Series

Foreword

THE ACS SYMPOSIUM SERIES was first published in 1974 to provide a mechanism for publishing symposia quickly in book form. The purpose of the series is to publish timely, comprehensive books developed from ACS sponsored symposia based on current scientific research. Occasionally, books are developed from symposia sponsored by other organizations when the topic is of keen interest to the chemistry audience.

Before agreeing to publish a book, the proposed table of contents is reviewed for appropriate and comprehensive coverage and for interest to the audience. Some papers may be excluded in order to better focus the book; others may be added to provide comprehensiveness. When appropriate, overview or introductory chapters are added. Drafts of chapters are peer-reviewed prior to final acceptance or rejection, and manuscripts are prepared in camera-ready format.

As a rule, only original research papers and original review papers are included in the volumes. Verbatim reproductions of previously published papers are not accepted.

ACS BOOKS DEPARTMENT

Contents

ADVANCED MICROLITHOGRAPHIC MATERIALS CHEMISTRY AND PROCESSING

LITHOGRAPHIC MATERIALS AND PROCESSES

Preface

Polymers have played an important role in the electronics industry as component structures and in manufacturing of microelectronics devices such as microprocessors and memory chips. The wide-spread use of polymeric materials arises from its film-forming properties, ease of fabrication, and especially the ability to synthesize and modify such materials for specific functionalities. Furthermore, patterning of such polymeric materials in a μm–nm scale is a key technology in electronic device manufacturing.

In the last decade, the performance advancement and price reduction of consumer electronic devices have been astounding, which was made possible by shrinking the minimum feature size on integrated circuits in microlithographic imaging technology, where radiation-sensitive polymeric materials called "resists" play a key role. A major shift in the technology has occurred recently—a shift from 365 to 248 nm and an accompanied shift from novolac–diazoquinone resists to chemically amplified resist systems. The chemically amplified 248-nm resists designed to resolve 250-nm features for manufacture of 256-megabit random access memory and related logic devices are now expected to support 1-gigabit (175-nm) and even 4-gigabit (150-nm) generations in the near future. Thus, refinement of chemically amplified 248-nm resists is critical, which could be accomplished through fundamental understanding of chemistry and physics of resist materials and imaging processes.

Further reduction of the wavelength to 193 nm (ArF excimer laser) has become the major thrust in the last several years, which has necessitated a development of new chemically amplified resists, spawning a massive research effort and providing enormous challenges and opportunities to resist chemists and engineers. X-ray and electron-beam lithographic technologies are also expected to come into the scene in the not-far future.

In parallel to the fast progress of the "conventional" lithographic technology, new polymeric materials must also be developed for new patterning technologies for process simplication, environmental consideration, and so on. Furthermore, the patterning technology is not confined in the electronics industry but finds its use in other areas such as biology and medicine.

We hope that this volume provides valuable information on the current status, future directions, challenges, and opportunities in the area of polymeric materials for use in μm-and nm-scale patterning.

Acknowledgments

This volume is based on the symposium "Polymeric Materials for Micro- and Nano-Patterning Science and Technology," which was sponsored by the American Chemical Society (ACS) Division of Polymeric Materials: Science and Engineering, Inc. (PMSE), and held in Las Vegas in September 1997. We thank the PMSE Division for sponsoring the symposium, the authors for their valuable contributions to both the symposium and this volume, and the referees for carefully reviewing the manuscripts. Financial support for the symposium was provided by Olin Microelectronic Materials, Inc., Hoechst-Celanese, Inc., SEMATECH, Lucent Technologies Bell Laboratories, IBM, the PMSE Division, and the Petroleum Research Fund administered by ACS, which is gratefully acknowledged.

Our sincere thanks are also extended to Anne Wilson, Tracie Barnes, and the production staff of the ACS Books Department for their efforts in assembling the volume.

HIROSHI ITO
IBM Almaden Research Center
650 Harry Road
San Jose, CA 95120

ELSA REICHMANIS
Bell Laboratories, Lucent Technologies
600 Mountain Avenue
Murray Hill, NJ 07974

OMKARAM NALAMASU
Bell Laboratories, Lucent Technologies
600 Mountain Avenue
Murray Hill, NJ 07974

TAKUMI UENO
Hitachi, Ltd.
Hitachi Research Laboratory
1-1 Omika-cho 7-Chome
Hitachi, Ibaraki 319-12, Japan

Novel Patterning Chemistry and Processing

Elsa Reichmanis

The whole field of electronic materials, and materials for microlithography in particular, continues to be a major area of research and the contribution made by polymers is substantial. Polymeric materials have found widespread use in the electronics industry in both the manufacturing patterning processes used to generate today's integrated circuits and as component structures in the completed device. The broad applicability of polymers arises from the ability to design and synthesize such materials with the precise functionalities and properties required for a given application. Notably, polymeric materials have been used as lithographic imaging materials, dielectric, passivation and insulating materials.

Recently, new materials and process techniques have been expanding the boundaries of patterning science and lithographic technology. Nanometer scale patterning is finding application in a wide range of materials systems. Of particular interest are functional nanostructures that can be prepared via self-assembly. Block copolymers prove to be particularly useful in that regard. Such materials with narrow polydispersity are known to microphase separate into regular domains while macrophase separation is precluded. While the microstructure morphologies do not possess the necessary control of pattern geometry for microelectronic device applications, many related applications could be envisaged. These may include the fabrication of quantum dots or anti-dots, the synthesis of DNA electrophoresis media, nanometer pore size filters, etc. A microphase separated, diblock copolymer could provide an almost ideal lithographic template for such applications. Use of such materials for the deposition of metallic species is the subject of advanced research efforts.

Alternatively, direct deposition of metal oxides and other dielectric and conducting materials is a growing area of interest as a means to decrease the number of processes required for fabrication of today's complex devices. Such approaches include direct photoinduced deposition of metal oxides onto device substrates, structurally designed organic conductors for advanced interconnect technologies, and image-wise chemical plating of Teflon induced by excimer laser irradiation.

Novel materials patterning technologies mentioned above and micro injection molding are being pursued for a vast array of device applications. Many of these provide avenues for further exploration of scientific phenomena, while others will see more immediate technological application. In all, they demonstrate the interdisciplinary nature of the field.

Chapter 1

Lithography with a Pattern of Block Copolymer Microdomains as a Positive or Negative Resist

Christopher Harrison[1], Miri Park[1], Paul M. Chaikin[1], Richard A. Register[2], and Douglas H. Adamson[3]

[1]Department of Physics, [2]Department of Chemical Engineering, and [3]Princeton Materials Institute, Princeton University, Princeton, NJ 08544

Dense, periodic arrays of holes and troughs have been fabricated in silicon, silicon nitride, and germanium, at a length scale inaccessible by conventional lithographic techniques. The holes are approximately 20 nanometers (nm) wide, 20 nm deep, spaced 40 nm apart, and uniformly patterned with 3×10^{12} holes on a three inch silicon wafer. To access this length scale, self-assembling resists were synthesized to produce either a layer of hexagonally ordered polyisoprene (PI) spheres or polybutadiene (PB) cylinders in a polystyrene (PS) matrix. The PI spheres or PB cylinders were then chemically modified by either degradation or stained with metal compounds to produce a useful mask for pattern transfer by fluorine-based reactive ion etching (RIE). A mask of spherical microdomains was used to fabricate a lattice of holes or posts and a mask of cylindrical voids was used to produce parallel troughs. This technique accesses a length scale difficult to produce by conventional lithography and opens a route for the patterning of surfaces via self-assembly.

Recent advances in photolithography have pushed the feature size down to 150 nm in production processes(*1*), and even smaller feature sizes have been reported in experimental research(*2*,*3*,*4*). However, dramatic improvements in the circuit density with photolithographic processes are not anticipated because the minimum feature size is limited by the wavelength of light, typically 193 or 248 nm in current processes. As an alternative to photolithography, self-assembled structures, such as monolayers of spheres or cylinders(*5*,*6*), have been advanced by researchers due to their nanoscopic feature sizes and the control demonstrated in uniformly coating substrates. Though these morphologies do not allow one the high degree of pattern control necessary for microelectronic circuits, there are a variety of applications in which regular patterning or texturing of a surface at the 10 nm lengthscale is ideal: for example, the periodic patterning of an electric potential on a two-dimensional electron gas system(*7,8,9,10,11*), fabrication of quantum dots or anti-dots(*12*), synthesis of DNA electrophoresis media(*13*), filters with nanometer pore sizes, and creation of quantum confinements for light emission. For these applications, the self-assembled structures observed in ordered diblock copolymer thin films would seem ideal as lithography templates because of their self-assembly and the ability to parallel-process wafer-sized areas.

Block copolymers with a narrow polydispersity (the chains are of near uniform length) and with $\chi N>10$ microphase separate above their glass transition temperature, where χ is the Flory-Huggins interaction parameter and N is the degree of polymerization(*14*). Macroscopic phase separation of the components of the block copolymer is prevented by a covalent bond which connects the unlike blocks. The resulting morphology depends largely on the relative volume fraction of the components. Some of the more commonly seen microdomain morphologies are lamellae, cylinders, and spheres (Figure 1). The periodicity of these structures is determined by the length of the polymer chains and is typically on the order of 20-100 nm.

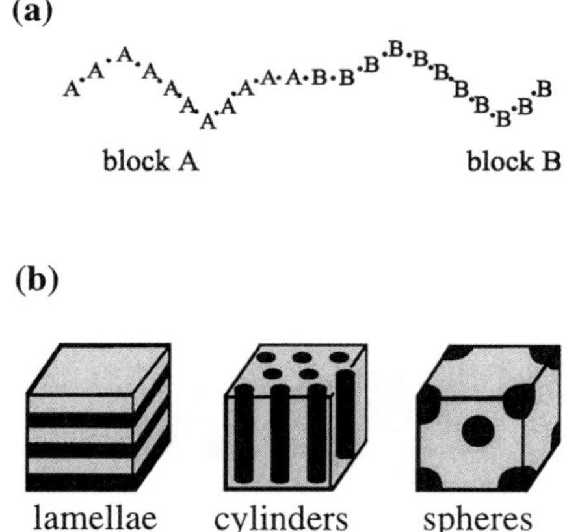

Figure 1: (a) A block copolymer consists of two or more homogeneous blocks, drawn here as blocks A and B. (b) Microphase separation typically produces a lamellar morphology for equal lengths of blocks A (light) and B (dark). As block B is shortened with respect to block A, hexagonally packed cylinders are typically observed. For an even shorter block B, packed spheres on a body centered-cubic lattice are observed.

Researchers have recently studied thin films of block copolymers because of both the rich set of phenomena that have been observed(*15,16,17*) and because of their application to lithography(*18,19,20*). Earlier work with poly(styrene-*b*-butadiene), PS-PB, and subsequent work with poly(styrene-*b*-isoprene), PS-PI, block copolymers showed that by choosing the appropriate film thickness, a single layer of spherical or cylindrical microdomains (a *monolayer* of spheres or cylinders, see Figure 2) could be produced as a lithographic template. Brush-like wetting layers of PB or PI (collectively referred to as the diene component) on the free and confined surfaces were shown to sufficiently decouple the microdomains from the substrate to allow the microdomains to order with a grain size of up to 25 by 25 unit cells, or about 1 square micron. Subsequent dynamic secondary ion mass spectrometry analysis confirmed the existence of wetting layers of PB on the free and confined surface of films spin coated onto silicon wafers(*21*). The diene component preferentially wets the free surface due to its lower surface tension(*22*) while the confined surface is wet by the diene component due to a combination of a lower interfacial tension and a possible silica-polydiene chemical bonding(*23,24*).

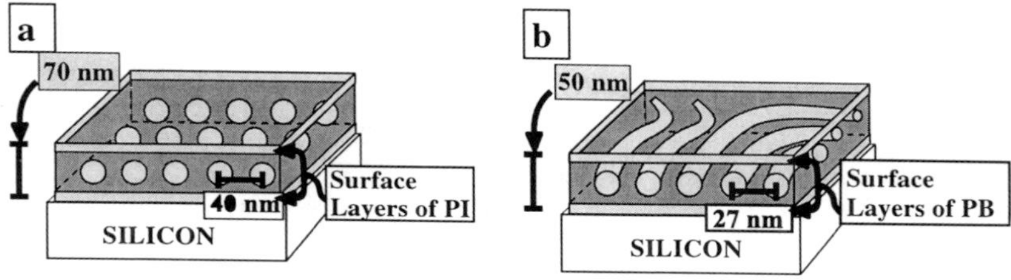

Figure 2: (a) A monolayer of PI spheres in a PS matrix, with accompanying free and confined surface PI wetting layers. (drawn for SI 68/12) (b) A monolayer of PB cylinders in a PS matrix, with accompanying free and confined surface PB wetting layers. (drawn for SB 36/11)

The microdomain template as shown above (Figure 2) is not an effective mask for RIE because the PS, PI, and PB blocks etch approximately at the same rate under either CF_4 or CF_4/O_2 RIE, which we find to be the most effective etching process for pattern transfer. Therefore, further modification of the microdomains is necessary to make a useful mask. By allowing the monolayer to act as a template and taking advantage of the different chemical properties of the component blocks, we found that the template could function as either a *positive* or *negative* resist for pattern transfer.

Sample Preparation

Asymmetric PS-PB and PS-PI diblock copolymers were synthesized (designated SB 36/11 and SI 68/12 respectively, with the molecular weights of the blocks in kilograms per mole) by standard high-vacuum anionic techniques(*25*). In bulk, SB 36/11 microphase separates into a cylindrical morphology and produces hexagonally ordered PB cylinders in a PS matrix. SI 68/12 adopts a spherical morphology and produces PI spheres in a PS matrix with body-centered-cubic order, as shown in Figure 1b. Thin polymer films were produced by spin-coating solutions of polymer dissolved in toluene onto various substrates, and the film thickness was controlled by varying the spinning speed and polymer concentration. The films were annealed in vacuum between 130°C and 170°C, a temperature above the glass transition temperatures of both blocks, for 24 hours to obtain well-ordered morphologies.

Pattern Transfer as a Positive Resist

For pattern transfer, thin films of spherical or cylindrical microdomain monolayers were directly spin coated on silicon wafers. Previous work on imaging the microdomain pattern on silicon wafers with a combination of low-voltage, high resolution SEM and a RIE allowed us to determine the optimum film thickness deposited by spin coating to form a monolayer of spheres or cylinders(*21*). To make a monolayer of spheres with SI 68/12, a 70 nm thick film was required, and for a monolayer of cylinders using SB 36/11, a 50 nm film was required (see Figure 2). For silicon nitride patterning, ~60 nm of silicon nitride was deposited on silicon wafers at 250°C by plasma enhanced chemical vapor deposition (PECVD), followed by spin coating polymer films. To pattern germanium, we first evaporated germanium on silicon wafers, then deposited a ~10 nm isolation layer of PECVD silicon nitride, and subsequently spin coated polymer solutions. We found it necessary to protect the germanium during the ozonation (discussed in the following paragraph) process with silicon nitride to prevent the formation of germanium oxides which damage the sample. During pattern transfer, the microdomain pattern was etched through the silicon nitride and into the germanium underneath.

By selectively degrading and removing the PI or PB microdomains, the template functioned as a positive resist on the substrate, making a mask of microdomain voids. The positive resist was created by placing the coated wafers in an aqueous environment through which ozone was bubbled for four minutes(*26*). The ozone cleaved the carbon-carbon double bonds of the diene component and the degradation fragments dispersed in the water(*27,28*). The PS was crosslinked (as evidenced by insolubility in various solvents) by the ozone though not topographically altered. The removal of the diene component after ozonation to produce microdomain voids was confirmed by examination of films placed on thin silicon nitride windows by transmission electron microscopy.

In order to transfer the submicron features from our polymer template to the underlying material, low-pressure, low-power anisotropic CF_4 based RIE was used to minimize lateral etching(see Figure 3)(*29*). The regions in the polymer film with microdomain voids have an effective thickness less than the total film thickness and were more quickly etched through with CF_4 RIE, transferring the pattern to the substrate underneath. After patterning the substrate, any remaining polymer film was stripped away with O_2 RIE.

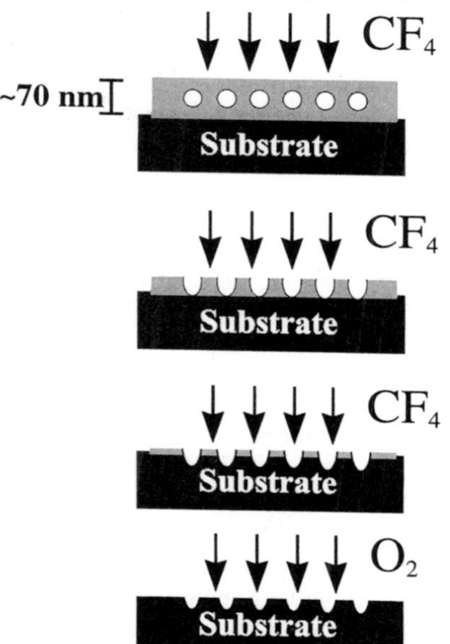

Figure 3: CF_4 RIE is used to transfer the pattern of a monolayer of spherical voids (created by ozonation) into the substrate, followed by an O_2 strip to remove any remaining polymer residue.

To demonstrate that this process scales to large areas and is compatible with standard wafer sizes, we uniformly patterned a three inch silicon wafer with approximately three trillion holes. Shown in Figure 4 is a scanning electron microscope (SEM) image of a representative region of the patterned wafer. The image shows a hexagonal lattice of holes, where the contrast is produced by the topography of the etched silicon. The holes appear darker than the non-etched honeycomb-like matrix. The holes are approximately 20 nm wide and spaced 40 nm apart. At each step of fabrication - pattern formation, PI degradation with ozone, and CF_4 etching - the process was accomplished in parallel, allowing us to quickly pattern macroscopic areas. As a comparison, serial techniques such as probe lithography would have been enormously time-consuming. At a rate of one millisecond per hole, production of this pattern would have required approximately 100 years.

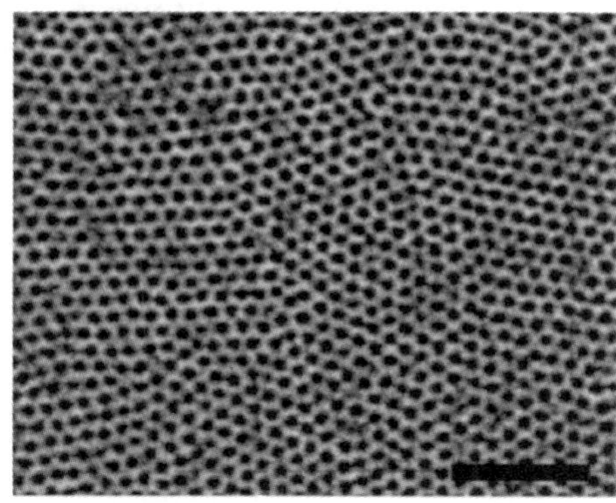

Figure 4: A silicon wafer patterned with a hexagonal lattice of 20 nm diameter holes at a density of 7×10^{10} holes/cm^2. Bar = 400 nm. (SEM parameters: 1 kV, SE + BSE signal)

Shown in Figure 5 are two SEM images of a germanium substrate that has been patterned with this technique. The plan view (5(a)) shows a hexagonal lattice of holes similar to the patterned silicon wafer in Figure 4. The cross-sectional view (5(b)) was obtained by cracking the wafer and examining the side-profile. The lattice of holes can be seen on the top surface and the 20 nm deep holes can be seen in the cleaved face. Since the diameter of the holes is also 20 nm, the aspect ratio is approximately 1.

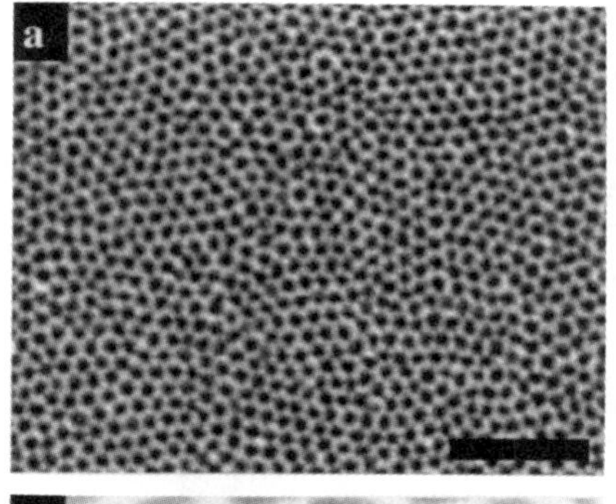

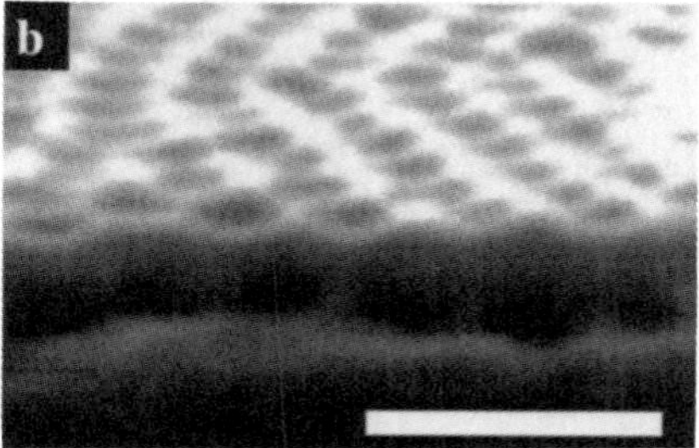

Figure 5: (a) SEM image of a lattice of holes patterned in germanium, fabricated with a polymer template which functioned as a positive resist. Bar = 200 nm (b) Cross-sectional view of patterned germanium on a substrate that has been cracked and viewed end-on. The depth of the holes is approximately 20 nm. Bar = 100 nm. (SEM parameters: 1 kV, SE + BSE signal) (5(b) Reproduced with permission from reference *20*)

Pattern Transfer as a Negative Resist

By staining the spherical microdomains with vapors of osmium tetroxide (OsO_4), the template functioned as a negative resist, making a mask of etch-resistant microdomains. With an RIE gas combination of 10:1 CF_4:O_2, we found that the PS etched twice as fast as the osmium-stained PB or PI microdomains, enabling the pattern to be transferred into the substrate (Figure 6). Previous electron beam exposure to the polymer mask, such as with a transmission electron microscope (TEM), proved necessary for optimal pattern transfer. Upon pattern transfer via CF_4 RIE, regions of the polymer film that were not previously exposed to a TEM electron beam showed little recognizable pattern in the silicon nitride underneath. This could be useful for patterning selected regions of a wafer by directed exposure of an electron beam. Starting with a mask of OsO_4-stained spherical microdomains (SI 68/12) which have been spin coated onto a silicon nitride membrane(*30*), we fabricated an ordered lattice of silicon nitride posts (Figure 7(a)) with an average spacing of 40 nm. Similarly, with a mask of OsO_4 stained cylindrical microdomains (SB 36/11), we fabricated ~15 nm deep troughs which are shown with a spacing of approximately 27 nm (Figure 7(b)).

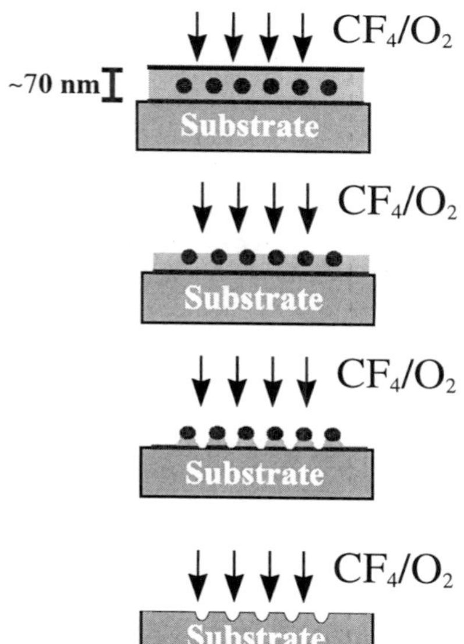

Figure 6: 10:1 CF_4:O_2 RIE is used to transfer the pattern of a monolayer of OsO_4 stained spheres into the substrate, followed by an O_2 strip to remove any remaining polymer residue.

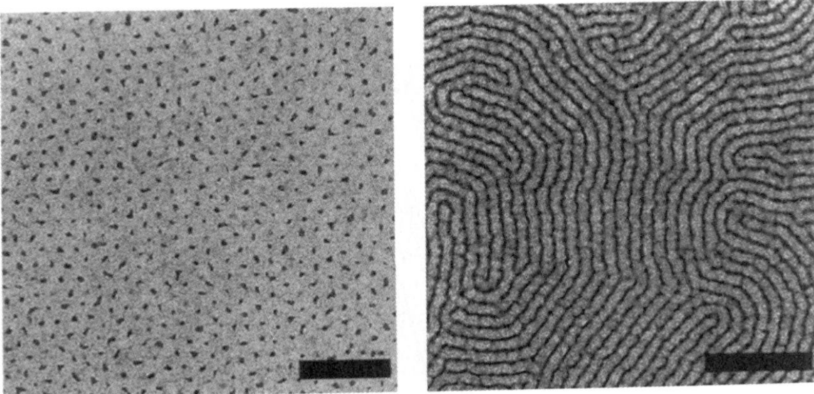

Figure 7: (a) TEM image of patterned silicon nitride posts (darker), or "dots," fabricated with an OsO_4-stained polymer template which functioned as a negative resist. (b) TEM image of patterned silicon nitride troughs that have been fabricated with an OsO_4-stained cylindrical polymer template. Bar=200 nm (TEM energy: 200 keV)

Future Work

Though our research has focused on PS-PI and PS-PB block copolymer thin films to form useful masks for patterning, their poor etch resistance under CF_4 RIE limits the aspect ratio of fabricated features to no greater than 1. Future work will include the exploration of other block copolymer systems, such as those embedded with etch resistant metal clusters(*31*) or silicon-containing block copolymers(*32*). Etch resistant polymers or microdomains would enable the fabrication of features with larger aspect ratios which could be advantageous for filtration devices and memory storage.

Acknowledgements

We would like to thank the staff of the Cornell Nanofabrication Facility, without whom this project could not have succeeded. In addition we would also like to thank Michael J. Valenti of the Advanced Technology Center for Photonics and Optoelectronic Materials (ATC/POEM) of Princeton University for his assistance in sample characterization and useful discussions. We also thank Joseph Horvath for assistance in germanium evaporation. This work was supported by the National Science Foundation through the Princeton Center for Complex Materials (DMR-9400362).

Literature Cited

[1] Sze, S. M. *VLSI Technology*; McGraw Hill: New York, NY; 1988 pp 142-154.
[2] Gabor, A. H.; Ober, C. K. In *Microelectronics Technology: Polymers for Advanced Imaging and Packaging*; Reichmanis, E.; Ober, C.K.;MacDonald, S.A.; Iwayanagi, T.; Nishikubo, T., Eds.; ACS Symposium Series 614, **1995**, pp 283-298.
[3] Chang, C. Y.; Sze, S. M. *ULSI Technology*; McGraw Hill: New York, NY; **1996,** pp xxi-xxiii.
[4] Chou, S. Y.; Krauss, P. R.; Renstrom, P. J. *Science* **1996**, *272*, pp 85.
[5] Deckman, H. W.; Dunsmuir, J. H. *App. Phys. Lett.* **1982**, *41*, pp 377.
[6] Park, M.; Harrison, C.; Chaikin, P. M.; Register, R. A.; Adamson, D. H. *Science* **1997**, *276*, pp 1401.
[7] Hofstadter, D. *Phys. Rev. B* **1976,** *14*, pp 2239.
[8] Thouless, D. J.; Kohmoto, M.; Nightingale, M. P.; den Nijs, M. *Phys. Rev. Lett.* **1982,** *49*, pp 405.
[9] Weiss, D.; Roukes, M. L.; Menschig, A.; Grambow, P.; von Klitzing, K.; Weimann, G. *Phys. Rev. Lett.* **1991,** *66*, pp 2790.
[10] Smith, C. G.; Chen, W.; Pepper, M.; Ahmed, H.; Hasko, D.; Ritchie, D. A.; Frost, J. E. F.; Jones, G. A. C. *J. Vac. Sci. Tech. B.* **1992**, *10*, pp 2904.
[11] Takahara, J.; Nomura, A.; Gamo, K.; Takaoka, S.; Murase, K.; Ahmed, H. *Jpn. J. Appl. Phys.* **1995**, *34*, pp 4325.
[12] Kang, W.; Stormer, H. L.; Pfeiffer, L. N.; Baldwin, K. W.; West, K. W. *Phys. Rev. Lett.* **1993**, *71*, pp 3850.
[13] Volkmuth, W. D.; Austin, R .H. *Nature* **1992**, *358*, pp 600.
[14] Bates, F. S.; Fredrickson, G. H. *Ann. Rev. Phys. Chem.* **1990**, *41*, pp 525.
[15] Henkee, C. S.; Thomas, E.L; Fetters, L. J. *J.Mater. Sci.* **1988**, *23*, pp 1685.
[16] Anastasiadis, S. H.; Russell, T. P.; Satija, S. K.; Majkrzak, C. F. *Phys. Rev. Lett.* **1989**, *62*, pp 1852.
[17] Mansky, P.; Russell, T. P. *Macromolecules* **1995**, *28*, pp 8092.
[18] Mansky, P.; Chaikin, P. M.; Thomas, E. L. *J. Mater. Sci.* **1995**, *30*, pp 1987.
[19] Park, M.; Harrison, C.; Chaikin, P. M.; Register, R. A.; Adamson, D. H. *Symposium Proceedings v BB: Morphological Control in Multiphase Polymer Mixtures, Materials Research Society, 1996 Fall Meeting., pp 179*
[20] Harrison, C.; Park, M.; Chaikin, P. M.; Register, R. A.; Adamson, D. H. *J. Vac. Sci. Tech. B.*, in press.

[21] Harrison, C.; Park, M.; Chaikin, P. M.; Register, R. A.; Adamson, D. H.; Yao, N. *Polymer*, in press.
[22] PB at the free surface layer has previously been observed. See, for example, Hashimoto, T.; Hasegawa, H. *Polymer* **1992**, *33*, pp 475.
[23] Ahagon, A.; Gent, A. N. *J. Polym. Sci., Polym. Phys. Ed.* **1975**, *13*, pp 1285.
[24] Jones, R. A. L.; Norton, L. J.; Shull, K. R.; Kramer, E. J.; Felcher, G. P.; Karim, A.; Fetters, L. J. *Macromolecules* **1992**, *25*, pp 2539.
[25] Morton, M.; Fetters, L. J. *Rubber Chem. Technol.* **1975**, *48*, pp 359.
[26] Mansky, P.; Harrison, C. K.; Chaikin, P. M.; Register, R. A.; Yao, N. *Appl. Phys. Lett.* **1996**, *68*, pp 2586.
[27] Lee, J. S.; Hirao, A.; Nakahama, S. *Macromolecules* **1989**, *22*, pp 2602.
[28] Smith, D. R.; Meier, D. J. *Polymer* **1992**, *33*, pp 3777.
[29] We used a customized Applied Materials RIE at the Cornell Nanofabrication Facility. The parameters used were 10 SCCM of CF_4, 2 millitorr, and 0.08 watts/cm^2.
[30] Silicon nitride membranes have been used previously in the study of block copolymers. See Morkved, T. L.; Lu, M.; Urbas, A. M.; Ehrichs, E. E.; Jaeger, H. M.; Mansky, P.; Russell, T. P. *Science* **1996**, *273*, pp 931
[31] Spatz, J. P.; Roescher, A.; Moller, M. *Adv. Mater.* **1996**, *8*, pp 337.
[32] Chu, J. H.; Rangarajan, P.; Adams, J. L.; Register, R. A. *Polymer* **1995**, *36*, pp 1569.

Chapter 2

Inorganic Nanostructures on Surfaces Using Micellar Diblock Copolymer Templates

Joachim P. Spatz[1], Thomas Herzog[2], Stefan Mössmer[1], Paul Ziemann[2], and Martin Möller[1,3]

[1]Organische Chemie III/Makromolekulare Chemie and [2]Abt. Festkörperphysik, Universität Ulm, D-89081 Ulm, Germany

We present a general method for generating rather regular arrays of nanometer sized noble metal and metal oxide clusters on flat substrates by the use of a polymer template. In a first step micelles from polystrene-block-poly(2-vinylpyridine) and polystyrene-block-poly (ethylene oxide) in toluene are used as nanocompartments which are loaded by a defined amount of $HAuCl_4$ or $LiAuCl_4$. In a subsequent step the Au(III) ions can be reduced in such a way that exactly one gold particle is formed in each micelle. When a flat substrate is dipped into a dilute solution of the Au or Au(III) loaded micelles, a monomicellar film is deposited in which the gold is arranged in a mesoscopic near to hexagonal 2-D lattice. Treatment by an oxygen plasma allows the removal of the polymer completely, and irrespective of whether the gold was reduced first, naked clusters of the elementary metal are deposited on the substrate with uniform sizes between 1 and 15nm and within tunable distances of 10-200nm. In a further step such regular micellar films are used as masks for nanolithography by argon sputtering. This way the micellar pattern can be used to generate a surface relief in an underlying layered semiconducting substrate. The inorganic / organic block copolymer film can be tailored from a negative to a positive resist.

Periodic and aperiodic microstructures prepared by lithography form the basis for a vast number of applications such as electronic circuits, sensors, actuators, etc. *(1,2)*. As the dimensions of the structures are pushed to ever smaller limits, size dependent quantum effects are encountered opening new fields of interesting physics and

[3]Corresponding author.

nanotechnology like quantum dot photoluminescence, resonant tunnelling devices and single electron transistors *(1-3)*. However, for structures smaller than 100nm, the application of conventional lithography techniques becomes increasingly difficult and there is an obvious need for alternative approaches that are based on a size controlled growth of inorganic structures *(3)* and on molecular concepts from organic and macromolecular chemistry *(4-12)* .

The point-like or linear pattern of alternating chemical composition observed in the diblock copolymer films possess an intriguing potential for the application as lithographic masks. An A-B diblock copolymer consists of two chemically different macromolecular chain segments. If these do not mix, microdomains are formed whose size and mesomorphic order depends largely on the lengths of the blocks and the molecular weight distribution *(7-10)*. Typical periodicities are in the range of 10-200nm. In thin films, the domain structures are affected by the interfacial energies and the geometrical constraints introduced by the flat interfaces *(13-16)*. Depending on the chemical and physical interaction which can be undergone by the monomer units, the microdomains can bind other components, e.g., selective swelling by a solvent or complexation or σ-binding of a transition metal compound *(18-21)*. The latter method has been employed extensively in order to achieve a strong contrast in transmission electron microscopy images *(22)*.

Only recently, Chaikin et al. demonstrated the imprint of a diblock copolymer microdomain pattern into an underlying silicon nitride substrate. Complementary techniques based on reactive ion etching enabled the preparation of 20nm wide holes with a periodicity of 40nm as well as a corresponding inverse structure, i.e., interconnected channels of 30nm width separated by 15nm wide elevations *(23,24)*.
Here we report an approach where uniform noble metal or metal oxide particles with a diameter between 1-15nm can be arranged on flat substrates within regular distances of a few nm to 150nm. Besides flatness, the only requirement for the substrate is that it must be resistant to oxygen. Thus, a vast number of substrates can be employed, e.g. gold, glass, sapphire, $SrTiO_3$, mica or also materials with surfaces which can be passivated by an oxygen layer like silicon, GaAs, InP or Si_3N_4.

The approach is based on the self-assembly and formation of mono micellar films of a diblock copolymer which serve as a template for the nanoparticle formation. Diblock copolymer micelles are loaded with a suitable transition metal salt and deposited as a mono micellar film on the substrate. In a subsequent step the diblock copolymer is removed by means of an oxygen plasma, and leaving behind regularly arranged uniform nanoparticles of the noble metal or a metal oxide. Besides the deposition of nanoscopic, inorganic clusters on surfaces, the self-assembled micellar diblock copolymer films are used in combination with etching techniques for the preparation of periodic point pattern in an inorganic substrate, i.e., GaAs, InGaAs and InP ultimately allowing aspect ratios larger than one.

Such structures are highly relevant regarding nanometer scaled electrooptical devices, like light emitting quantum dots, nano diodes and electrical contacts on the nanometer scale for single molecular contacts.

Inorganic Modification and Film Formation of Diblock Copolymer Micelles

When Poly(styrene-b-2-vinylpyridine), (PS-b-P2VP), or polystyrene-b-poly(ethylene oxide), (PS-b-PEO), diblock copolymers differing in the molecular weight get dissolved in toluene the diblock copolymers associate to micelles at rather low concentration *(18-21)*. Toluene dissolves preferentially the PS block while the P2VP or PEO is almost insoluble. Such solutions were treated with solid $HAuCl_4$ or $LiAuCl_4$. As the pyridine units are protonated or the PEO blocks complex the Lithium cations, $AuCl_4^-$ ions are bound as counterions to the polar core of the micelles. The aurate loaded micelles are formed in equilibrium and the amount of gold per micelle varies only in narrow limits *(18,19)*. Typically up to one or half an equivalent of $AuCl_4^-$ per 2VP or EO unit, respectively, can be taken up by the micellar solution.

In the next step a suitable flat substrate is dipped into the solution and drawn back at a controlled rate. Depending on the concentration of the solution and the rate by which the substrate is pulled out of the solution, micellar films of different thickness can be prepared. After evaporation of the solvent these films do not necessarily represent an equilibrium structure. In the case the core/corona volume ratio is larger than 0.15-0.2, vitrification and the extremely long relaxation times of diblock ionomer micelles ensure that the micellar structure is preserved *(25,26)*. Long range van der Waals interactions, capillary forces acting between the micelles upon evaporation of the solvent in the adherent film, and the intrinsic stability of a relatively thick mono micellar layer favours the formation of a closed film of densely packed micelles *(27)*. This way, mono micellar continuous films extending over an area of up to $3x3cm^2$ were obtained by pulling a mica or glass plate or also a carbon coated Cu-grid with a rate of 10mm/min out of a micellar solution of 5mg polymer per milliliter toluene (Scheme 1). Smaller pulling rates resulted in incomplete coverage of the substrates and the formation of small islands. The film structure was studied by transmission electron microscopy (Philips, EM 400T at 80keV) and by scanning force microscopy (Digital Instruments, Nanoscope III, contact mode).

Figure 1a shows a schematic drawing of a mono micellar film formed on a flat substrate together with a corresponding electron micrograph of such a film prepared on a carbon coated copper grid. In this case, PS-b-PEO diblock copolymer was used with a PS block consisting of about 540 monomer units and a PEO block with about 220 monomer units, PS(540)-b-PEO(220). The micellar solution in toluene had been treated with 0.3 equivalents of $LiAuCl_4$ per EO-unit. When the film was bombarded by the electrons during TEM, the aurate ions were reduced and small gold particles formed inside the core of each micelle. These are depicted as small dark spots in Figure 1a and are surrounded by the "unstained" shell of polystyrene.

Figure 1b shows a film of a PS-b-P2VP diblock copolymer with two equally long blocks consisting of about 300 monomer units each, PS(300)-b-P2VP(300). In this case, the micellar solution in toluene had been treated with 0.5 equivalents of $HAuCl_4$ per 2VP unit. In addition, the micellar solution was treated with anhydr. N_2H_4 in order to reduce the gold already in solution. The details of the reduction procedure have been described before *(18,19)*. Reduction in solution yielded one single gold particle in each micelle. A monomicellar film which was obtained from this solution is

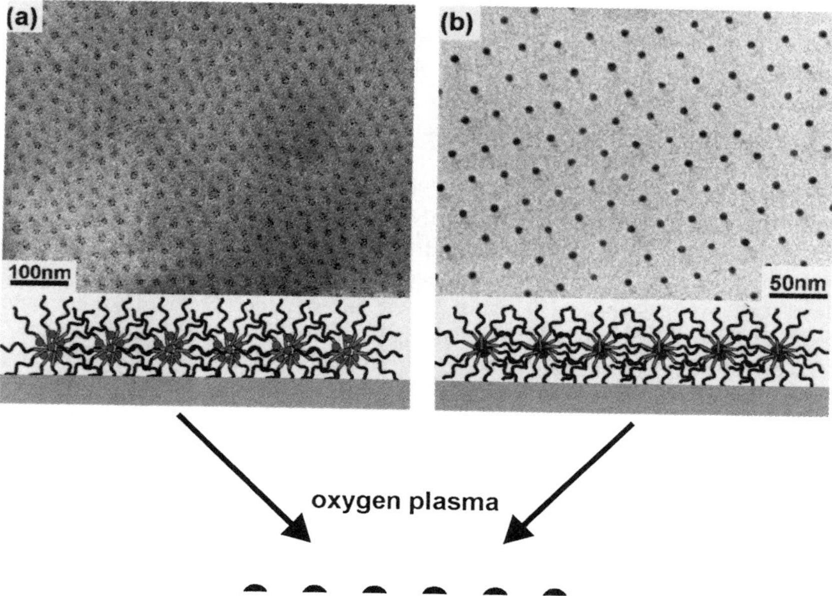

Figure 1. TEM micrographs of regular mono micellar films (a) loaded with (a) $LiAuCl_4$, and (b) where the aurate ions were converted to a single Au cluster per micelle prior to film casting. The scheme illustrates that in both cases a single gold particle remains at the place of the micelle.

shown in Figure 1b and demonstrates the regular arrangement of the gold particles with an interparticle distance of 30nm. The size distribution of the particles is remarkably small with a mean diameter of 6nm.

Naked Metal Nanodot Pattern by Applying Oxygen Plasma

Both types of films have been treated by an oxygen plasma resulting in the following two effects: (i) complete removal of the polymer and (ii) conversion of the aurate to a single gold particle in the case (a) of Figure 1. The reduction can be attributed to intermediate oxidation products of the polymer such as CO and free electrons which emerge due to ionization of the plasma. Thus, irrespective of whether the original polymer film already contained single gold particles or still contained the aurate, the plasma treatment resulted in the deposition of small gold crystallites whose size and inter-particle distance was controlled by the structure imposed by the diblock copolymer. This is schematically shown in Figure 1 and experimentally demonstrated in Figure 2. The scanning force micrograph image of Figure 2a depicts a mono micellar film of a PS(1700)-b-P[2VP$(HAuCl_4)_{0.5}$(450)] diblock copolymer on glass loaded with $HAuCl_4$ ($L = HAuCl_4 / 2VP = 0.5$). The topographic SF-micrograph displays the outer contour of the polymer micelles with the elevations shown as lighter areas. The average lateral periodicity of 80nm and the height corrugations of the diblock copolymer micelles is demonstrated by the height profile along a horizontal line running over several micelles as indicated in Figure 2a. Variations in the maximum height origin predominantly from a mismatch of the horizontal line with the location of the micelles. Figure 2b shows the SFM topography of the same sample after it has been treated with an oxygen plasma running for 20 min at 200 Watt. The white spots represent naked Au particles 8nm in height and about 10nm in lateral dimensions (corrected for the tip radius *(28)*). The height profile of a horizontal line running over several naked Au clusters is indicated in the SF-image. The lateral periodicity of 80nm is unchanged and again the height corrugations shown in the profile mainly comes from the mismatch of the horizontal line with the location of the Au clusters. XPS experiments did not give the slightest indication of residual polymer. Exactly the same pattern was formed when a mono micellar film of PS(1700)-b-P[2VP$(HAuCl_4)_{0.5}$(450)] was deposited on glass directly treated by an oxygen plasma without prior formation of gold crystallites by reduction with anhydr. N_2H_4 in solution. In all cases the gold particles were strongly bound to the glass and could not be removed either when the surface was repeatedly scanned with higher tip forces or by washing with different solvents or rubbing with a soft tissue. Also the temperature stability appears to be exceptional. By the same procedure, Au particles of 4nm in diameter have been deposited onto a $SrTiO_3$ wafer. When the sample was heated for 20min. to 800°C in an Argon/Oxygen atmosphere, the SFM images demonstrated that the point pattern remained the same. Also in this case no indication of cluster coagulation was observed. Furthermore, these results prove that there is no polymer left between the Au islands and the substrate.

Scanning force micrographs in Figure 3 demonstrate the possibility to vary the particle periodicity and the particle size by employing diblock copolymers of different molecular weight. Figure 3a depicts a sample prepared from a toluene solution of PS(800)-b-P[2VP$(HAuCl_4)_{0.5}$(860)], Figure 3b a sample from PS(325)-b-P[2VP$(HAuCl_4)_{0.5}$(75)] and Figure 3c a sample from PS(1700)-b-

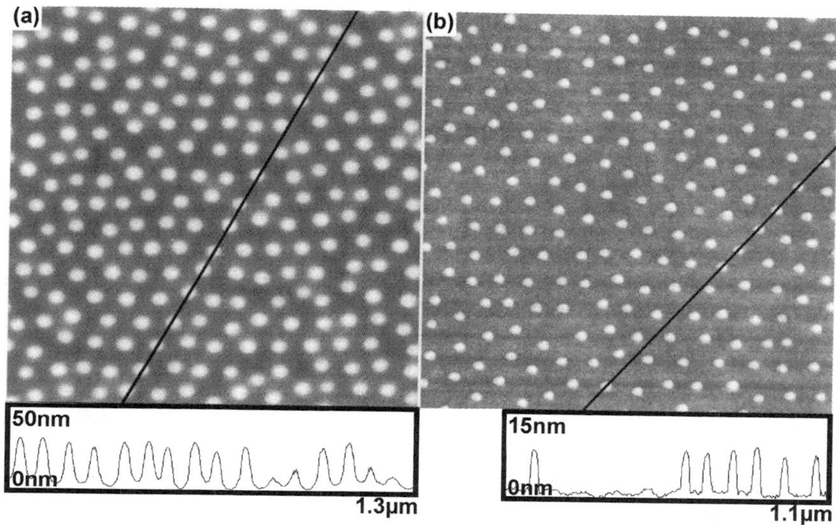

Figure 2. (a) SFM topography image of a mono micellar film with a lateral periodicity of 80nm. (b) Same sample as in (a) but after oxygen plasma treatment: The image depicts naked Au particles on the glass substrate. The horizontal width of the images is 1.1μm.

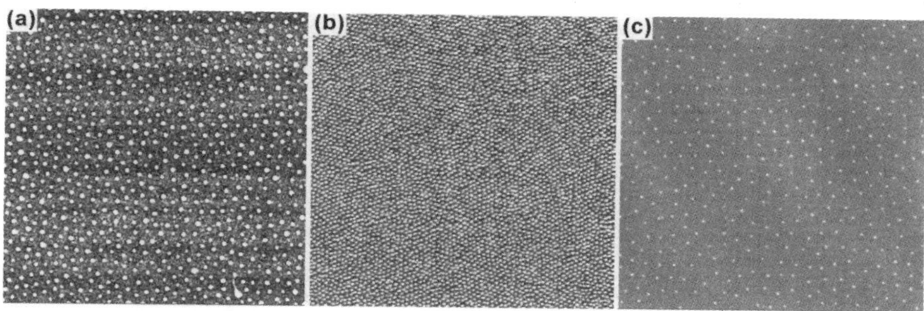

Figure 3. SFM topography images of naked Au cluster on mica obtained after the oxygen plasma treatment. The different interparticle distances are obtained by varying the lengths and loading of the polymer blocks as described in the text. The width of each image corresponds to 3μm.

$P[2VP(HAuCl_4)_{0.1}(450)]$. All samples were prepared on a freshly cleaved piece of mica, and the polymer matrix had been totally removed by oxygen plasma treatment. The respective periodicities were 80nm, 30nm and 140nm. The gold particles exhibit a uniform height of 12nm, 2nm and 1nm, respectively. In this images, the apparent diameter of the particles is exaggerated due to the finite radius of the tip apex which was determined to be approximately 15nm *(28)*.

The approach described here represents a remarkably simple procedure for the preparation of nanometer sized dots in a rather regular pattern. Without further effort we have prepared ordered films of $3x3cm^2$. In principle there is no limitation in size. Besides the diblock copolymers which we employed, also other amphiphilic diblock copolymers form reverse micelles in a non-polar solvent and can be used to bind a large variety of metal compounds like H_2PtCl_6, $Pd(Ac)_2$, $TiCl_4$, $FeCl_3$, etc.. Chemical transformation of the inorganic compound can be performed before deposition of the micellar film or upon removal/oxidation of the film. While noble metals are deposited in the elementary state, less noble compounds can only be deposited in their oxidic form, i.e., TiO_2, Fe_2O_3, InO_x.

Inorganic-Polymer Micellar Hybrid Systems - A Tool for Nanolithography

As discussed briefly in the introduction, their is a need for novel and easy procedures to structure semiconductor materials in the range below 100nm. Because of their regular variation in composition and thickness, the mono micellar films described above can be used as a mask for lithographic etching of the underlying substrate and to create a surface relief which corresponds to the point pattern of the film. Because the formation of mono micellar films does not require a substrate with special properties besides flatness and wettability by the solvent, the films can be applied to practically all semiconductor materials. This includes wafers which have been overgrown by nanometer thick layers of different composition and different electronic properties. The loading of micelles by an inorganic compound can provide particularly large etching contrasts, e.g., argon ions sputter TiO_2 or SiO_2 at rates 10 times lower than organic films and 26 times lower than GaAs *(29)*. Such a high etching contrast is essential for the preparation of structures with an aspect ratio of one and larger one as needed for quantum dot and quantum wire devices.

In the following we demonstrate how the lateral structure of a micellar PS-b-$P2VP(HAuCl_4)_{0.4}$ diblock copolymer film as the one shown in Figure 1 is transferred to an underlying substrate and how the etching rates and the etched surface structures can be altered by variations of the inorganic compound. Here we employed Argon sputtering as a means to erode the surface. Other possibilities are offered by reactive ion etching with CF_4 or Cl_2 plasma *(29)*. Because the etching contrast does not only depend on local variations in the polymer film composition but also on local variations in film thickness, it is necessary to come back to the surface topography of the mono micellar films, either filled with an inorganic component or not.

The topographical SFM image in Figure 4a depicts a mono micellar film of a PS(1700)-b-P2VP(450) which was deposited on glass. The diblock copolymer was not treated with $HAuCl_4$. The overall layer thickness was about 20nm. The height profile along the horizontal line indicated in the SFM micrograph demonstrates the surface roughness with non periodic corrugations of 3-4nm. The profile does not obviously

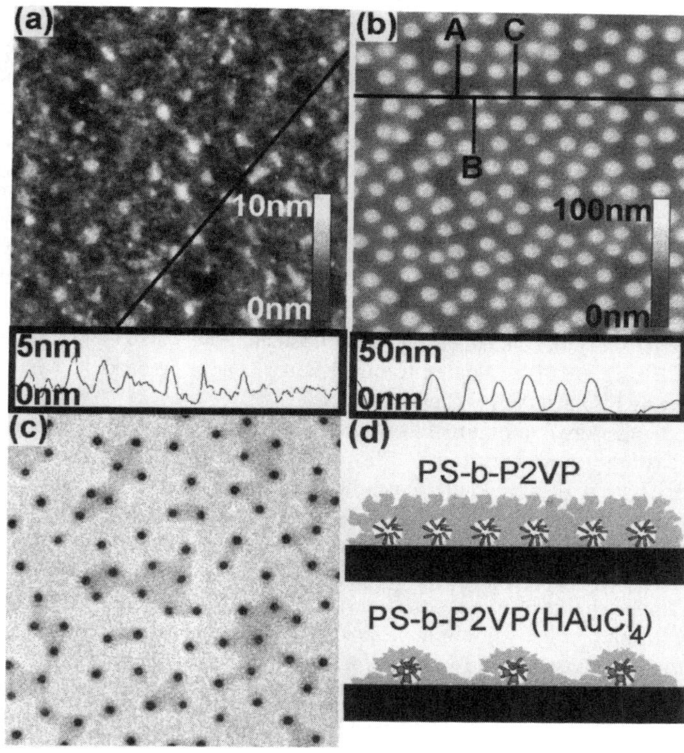

Figure 4. SFM micrographs (1x1μm^2) of (a) a PS-b-P2VP film and (b) a PS(1700)-b- P[2VP($HAuCl_4$)$_{0.4}$(450)] film cast from a micellar solution. A, B, and C mark structural variations as described in the text. The TEM micrograph (c) shows the same micellar film as in (b). (d) The scheme depicts the structural difference of the film from PS-b-P2VP and PS(1700)-b-P[2VP($HAuCl_4$)$_{0.4}$(450)] respectively.

correspond to the topography expected for densely packed spherical micelles. Annealing at 150°C above the glass temperature of both polymer blocks caused further flattening of the film.

In contrast, well developed periodic corrugations were observed for a film loaded with $HAuCl_4$, i.e., PS(1700)-b-P[2VP$(HAuCl_4)_{0.4}$(450)] on glass. The height profile along the horizontal line drawn in the SFM micrograph displays a maximum height of 30nm (point B) above the glass surface (point A), a lateral periodicity of 80nm, and a height of the polymer film between two adjacent micelles of 9nm (point C). Obviously, neutralization of the vinylpyridine core by $HAuCl_4$ has a significant influence on the film structure. While the micelles of the pure diblock copolymer can reorganize in such a way that a rather flat film with a minimum surface energy is formed, the associated structure of the micelles formed in solution is largely fixed when they are loaded by the gold acid *(30,31)*.

In order to check the film structure by transmission electron microscopy, a carbon coated copper grid was dipped into the PS(1700)-b-P[2VP$(HAuCl_4)_{0.4}$(450)] diblock copolymer solution according to the same procedure. Figure 4c depicts the transmission electron micrograph of the resulting mono micellar film. Induced by the electron irradiation the Au^{3+} ions were instantaneously reduced to elemental Au and the core of the micelles got stained darkly by ultrasmall gold clusters *(18,19,30)*. The diameter of the core area is evaluated to be Ø=25nm. The grey disks around the dark spots in Figure 4c displays the PS shell.

The scheme drawn in Figure 4d summarizes the structure of the films formed by solvent casting from PS(1700)-b-P2VP(450) and PS(1700)-b-P[2VP$(HAuCl_4)_{0.4}$(450)]. In the latter case the P2VP chains in the core volume cannot rearrange because of the ionic interactions *(30,31)*. Still, the system minimizes the surface energy by reducing the surface curvature which is done by stretching PS chains laterally. In the case of the PS(1700)-b-P2VP(450) micelles, structural relaxation is not hindered by internal ionic interactions and energy consuming curved surfaces are prevented by forming rather smooth micellar film surfaces.

Figure 5 demonstrates how the gold modified micellar films can be employed for nanolithography. We employed a film of PS(1700)-b-P[2VP$(HAuCl_4)_{0.4}$(450)] micelles and micelles of the same diblock copolymer containing each a gold crystal of 12nm diameter. For this purposes the Au^{3+} was reduced to Au with N_2H_4 prior to film casting.

A GaAs, an InP substrate, or a layered quantum well substrate (GaAs/InGaAs/GaAs) grown by Metal Organic Molecular Beam Epitaxy *(29)* was dipped into a toluene solution containing 5mg/mL of the diblock copolymer and pulled out at a rate of 5mm/min. By this procedure the wafer was covered by a discontinuous mono micellar film. The samples were exposed to a 1.1 keV Ar^+ beam (12μA/cm^2) for 15min in order to erode the organic-inorganic micellar layer.

Figure 5a depicts a SFM micrograph of a GaAs surface which had been covered by a non-continuous PS(1700)-b-P[2VP$(HAuCl_4)_{0.4}$(450)] micellar film. The sputtering resulted in complete removal of the coating and locally varying etching of the GaAs substrate. Islands of 25nm in height and of 20nm in diameter are observed. Height variations in the surface of GaAs can be correlated to the structure of the originally applied inorganic/polymer hybrid coating: (i) Substrate surface areas which had not been covered by the polymer were most deeply etched (most dark areas marked by A in Figures 4b, 5a, and 5b); (ii) substrate areas which had been covered by a 9nm thick layer of PS were etched to an intermediate depth (areas marked by C in Figure 4b, 5a, and 5b), and (iii) the elevating islands of 25nm in diameter were those areas which were covered by the core volume of the micelles consisting of P2VP/$HAuCl_4$ plus a 5nm thick layer of PS (marked by B in Figure 4b and 5a).

A different result occurred, however, when the GaAs surface was coated with each micelle containing one elementary gold particle of Ø=12nm. Such a `resist´ corresponds to the film shown in Figure 1b in which case the Au^{3+} was reduced to Au prior to film formation. Instead of formation of an elevation at the places where the substrate was shielded by the micellar core, holes of 10nm in depth and 10nm in diameter were obtained after sputtering (marked by B´ in Figure 5b).

Figures 5a and 5b demonstrate that the structure of the diblock copolymer micelles can readily transferred into the underlying substrate. In addition, the lithographic contrast can be reversed by altering the inorganic component. The quantitative height information from scanning force microscopy allows to determine the corresponding Ar^+ etching rates E for the different micellar domains in relation to the Ar^+-sputtering rate of GaAs, i.e., E_{GaAs}=30Å/min (1.1 keV, 12 μA/cm^2) *(29)*. Based on this value and the thickness variations of the micellar film determined from SFM and TEM analysis in Figure 4, the following etching rates were calculated with the surface profiles shown in Figure 5: E_{PS}=15 Å/min, $E_{P2VP/HAuCl4}$=36 Å/min, and $E_{P2VP/Au}$=360 Å/min (Ar^+ ion beam of 1.1 keV energy and 12 μA/cm^2 current density, normal to the substrate). No differences in the etching rates were observed between PS and P2VP layers.

The rates summarize quantitatively that the $HAuCl_4$ modified core is eroded about a factor of 2 faster than a pure PS or P2VP layer of the same thickness. This can be attributed to the fact that the sputter probability is proportional to Z^2 where Z is the atomic weight. The ten times higher etching rate found at areas covered by a micellar core containing an elemental gold particle is remarkable. In this case we have to consider that the formation of a solid Ø=12nm Au cluster in each single micelle decreases the interaction between the polymer and the gold. Furthermore, it is apparent that the nanocrystals are sputtered faster than bulk gold. The sputter rate of bulk gold taken from literature is about 6 times slower than the value we observed for the P2VP/Au core *(29)*. Because of the large number of Au atoms located on the surface of the nanoparticles their stability is considerably reduced. This is also stated by the decrease of the melt temperature of nanoparticulated gold in comparison to bulk gold *(32)*. Summarizing the results described above, the Scheme describes the formation of holes and islands. If the micelles contain an elementary gold particle holes are formed in the underlying GaAs substrate as the core volume now etches significantly faster than the 9nm PS layer between adjacent core volumes.

In the case a substrate is covered by a continuous micellar film (pulling rate of 10mm/min out of a 5mg/mL polymer weight concentrated solution), argon etching results in the formation of arrays of islands or of holes with a periodicity of 80nm as demonstrated by the three dimensional SFM micrographs in Figure 6. Using this approach GaAs/InGaAs/GaAs quantum well layers have been structured with islands as demonstrated in Figure 6a and schematically drawn in Figure 6c. First measurements

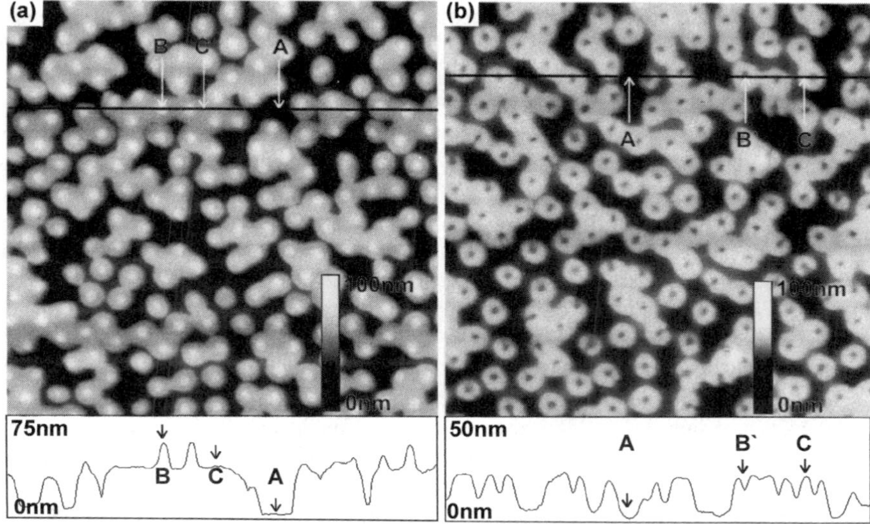

Figure 5. SFM micrographs ($1.2x1.2\mu m^2$) of a pure GaAs surface structured by (a) islands of 25nm in height and 20nm in diameter and (b) holes of 10nm in depth and 10nm in diameter. The height profiles refer to the lines drawn in the micrographs. A, B, B`, and C mark surfaces areas corresponding to those marked in Figure 4. The inversion of the structure was obtained as the Au^{3+} was reduced to a single gold particle of Ø=12nm in each micelle.

Scheme

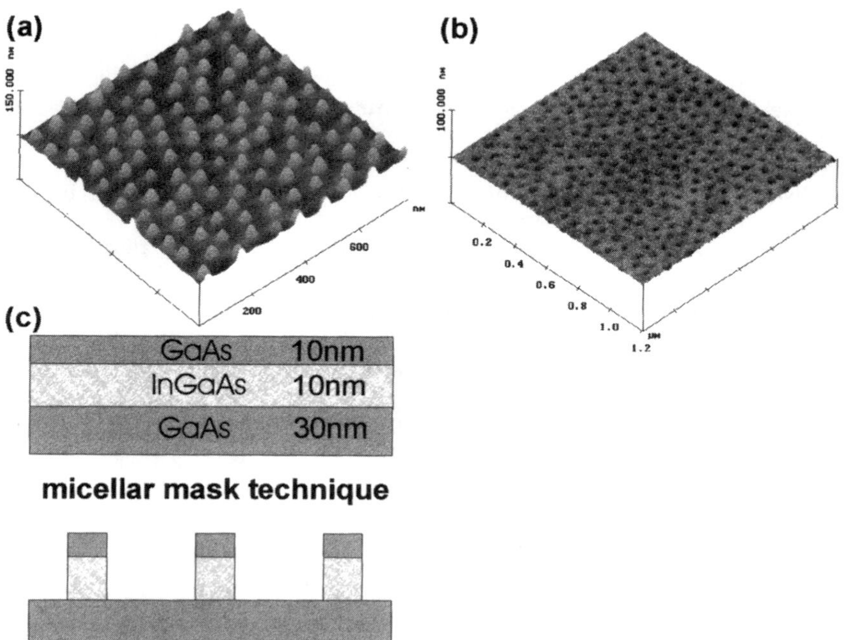

Figure 6. Three dimensional SFM images of arrays of (a) islands of 25nm height, 20nm diameter ($800x800nm^2x150nm$) and (b) holes of 10nm depth, 10nm diameter separated by ca. 80nm ($1.2x1.2\mu m^2x100nm$). (c) Schematic drawing of the fabrication of light luminescent quantum dots of a GaAs/InGaAs/GaAs wafer as shown in (a).

revealed photo luminescence of light from such quantum dot arrays. These investigations are subject of ongoing work.

So far, the enormous potential of the diblock copolymer approach has been demonstrated which is based on three particular contributions: (i) self-organization of block copolymers with periodicities down to a few ten nanometers, (ii) easy application of structurally well controlled thin films over large areas of various substrates, and (iii) the highly selective etching contrast which can be achieved by the incorporation of suitable inorganic components. Most importantly, the latter will allow to prepare nanostructures in semiconductors with an aspect ratio not yet conceivable by other parallel processing methods.

Acknowledgement

The authors are greatful to the financial support by the DFG (SFB 239), the BMBD (Supramolekulare Chemie), and the Fonds der Chemischen Industrie, Frankfurt.

Literature cited

(*1*) Goldhaber-Gordon, D.; Montemerlo, M.S.; Love, J.Ch.; Opiteck, G.J.; Ellenbogen, J.C. *Proc. Of the IEEE* **1997**, *85*, 521

(*2*) Collins, P.G.; Zettl, A.; Bando, H.; Thess, A.; Smalley, R.E. *Science* **1997**, *278*, 100

(*3*) Fendler, J.H; Meldrum, F.C. *Adv. Mater.* **1995**, *7*, 607

(*4*) Pearson, D.H.; Tonucci, R.J. *Adv. Mater.* **1996**, *8*, 1031

(*5*) Elghanian, R.; Storhoff, J.J.; Mucic, R.C.; Letsinger, R.L.; Mirkin, C.A. *Science* **1997**, *277*, 1078

(*6*) Chung, J.C.; Gross, D.J.; Thomas, J.L.; Tirrell, D.A.; Opsahl-Ong, L.R. *Macromolecules* **1996**, *29*, 4636

(*7*) Bates, F.S.; Frederickson, G.H. *Ann. Rev. Phys. Chem.* **1990**, *41*, 525

(*8*) Milner, S.T. *Science* **1991**, *251*, 905

(*9*) Leibler, L. *Macromolecules* **1980**, *25*, 4967

(*10*) Yu, S.M.; Conticello, V.P.; Zhang, G.; Kayser, C.; Fournier, M.J.; Mason, T.L.; Tirrell, D.A. *Nature* **1997**, *389*, 167

(*11*) Lehn, J.M. *Supramolecular Chemistry*; Ed. 1; VCH Verlagsgesellschaft: Weinheim, 1995

(*12*) Kim, E.; Xia, Y.; Zhao, X.M.; Whitesides, G.M. *Adv. Mater.* **1997**, *9*, 651

(*13*) Kellogg, G.J.; Walton, D.G.; Mayes, A.M.; Lambooy, P.; Russell, T.P.; Gallagher, P.D.; Satija, S.K. *Phys. Rev. Lett.* **1996**, *76*, 2503

(*14*) Morkved, T.L.; Lu, M.; Urbas, A.M.; Ehrichs, E.E.; Jaeger, H.; Mansky, P.; Russell, T.P. *Science* **1996**, *273*, 931

(*15*) Mansky, P.; Russell, T.P.; Hawker, C.J.; Mays, J.; Cook, D.C.; Satija, S.K. *Phys. Rev. Lett.* **1997**, *79*, 237

(*16*) Mansky, P.; Liu, Y.; Huang, E.; Russell, T.P.; Hawker, C. *Science* **1997**, *275*, 1458

(*18*) Spatz, J.P.; Mößmer, S.; Möller, M. *Chem. Eur. J.* **1996**, *2*, 1552

(*19*) Spatz, J.P.; Roescher, A.; Möller, M. *Adv. Mater.* **1996**, *8*, 337

(*20*) Cheong Chan, Y.,Ng.; Schrock, R.R.; Cohen, R.E. *J. Am. Chem. Soc.* **1992**, *114*, 7295
(*21*) Moffit, M.; Eisenberg, A. *Chem. Mater.* **1995**, *7*, 1178
(*22*) Schwark, D.W.; Vezie, D.L.; Reffner, J.R.; Thomas, E.L.; Annis, B.K. *J. Mater. Sci. Lett.* **1992**, *11*, 352
(*23*) Mansky, P.; Harrison, C.K.; Chaikin, P.M.; Register, R.A.; Yao, N. *Appl. Phys. Lett.* **1996**, *68*, 2586
(*24*) Park, M.; Harrison, C.K.; Chaikin, P.M.; Register, R.A.; Adamson, D.H. *Science* **1997**, *276*, 1401
(*25*) Zhang, L.; Eisenberg, A. *Science* **1995**, *268*, 1728
(*26*) Spatz, J.P.; Mößmer, S.; Möller, M. *Angew. Chemie Int. Ed. Engl.* **1996**, *35*, 1510
(*27*) Kralchevsky, L. *Langmuir* **1994**, *10*, 23
(*28*) Sheiko, S.S.; Möller, M.; Reuvekamp, E.M.C.M.; Zandbergen, H.W. *Phys. Rev. B* **1993**, *48*, 5675
(*29*) Brodie, I., Muray, J.J. *The Physics of Micro/Nano-Fabrication*; Plenum Press: New York, N.Y., 1992
(*30*) Spatz, J.P.; Roescher, A.; Sheiko, S.S.; Krausch, G.; Möller, M. *Adv. Mater.* **1995**, *7*, 731
(*31*) Spatz, J.P.; Sheiko, S.S.; Möller, M. *Macromolecules* **1996**, *29*, 3220
(*32*) Lai, S.L.; Guo, I.Y.; Petrova, V.; Remeneth, G.; Allen, L.H. *Phys. Rev. Lett.* **1996**, *77*, 99

Chapter 3

Synthesis of Stereoregular Polymers as Precursors to Highly Conducting Carbon for Use in Applications in Micro- and Nanolithography

C. B. Gorman, R. W. Vest, J. L. Snover, T. L. Utz, and S. A. Serron

Department of Chemistry, North Carolina State University, Box 8204, Raleigh, NC 27695–8204

Organic conductors have received virtually no attention in lithographic processes and may be uniquely suited in these applications. In particular, it should be possible to design polymers that can be converted from electrical insulators into electrical conductors in one, dry step. This process is anticipated to save time and minimize the use of organic solvents in processing. Moreover, very small ($\leq$ 100 nm) features of organic conductors can be covalently linked thoughout and should not suffer from migration processes (particularly electromigration) found in metallic structures with small size features. Up to the present, it does not appear that polymers capable of facile direct conversion from an insulating phase to a conducting phase are available. Potential routes to lithographically defined, graphitic carbon from precursor polymers based on stereoregular poly(acrylonitrile) and poly(cyanoacetylene) are illustrated. As stereoregular versions of these polymers are not available, routes for their preparation are presented and discussed.

There are a number of fundamental challenges in improving lithographic methods. The process is time consuming, requiring a number of steps just to fabricate electrical interconnects between circuit elements. Omission of some coating and developing steps will lower the time required to fabricate circuitry and reduce the waste generated by the process. At smaller length scales, a number of process issues arise, most notably the need for new methods to write information rapidly and with good contrast at sub-micron sizes. In addition, materials issues may limit the reliability of such nano-circuitry. Metals used in device fabrication, such as copper, migrate, particularly under the application of an electric field (electromigration) (*1,2*).

For metals such as copper, this process inevitably results in the loss of electrical continuity for wires of < 100 nm width.

Organic conductors have a real potential to overcome some of the processing problems as well as some of the materials problems mentioned above. It is envisioned that they may be uniquely suited for certain nanolithography applications. Graphitic organic polymers can display high electrical conductivities (> 1000 $\Omega^{-1}cm^{-1}$). To date, several routes have been explored for their fabrication in bulk structures such as films and fibers (*3,4*). Comparatively little is known about how to fashion structures on small length scales. Moreover, most routes to intrinsically conductive organics involve high temperature (500 - 900 °C) carbonization processes, and few routes based on processes such as UV or electron beam lithography have been described.

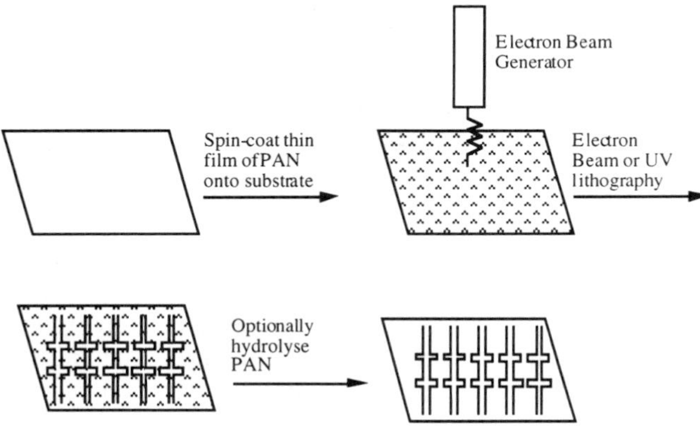

Figure 1. Schematic for the fabrication of a conductive organic circuit pattern. PAN refers to any in a series of nitrile-based polymers discussed below. Unconverted PAN may be an acceptable insulating phase. Alternatively, brief acid or base hydrolysis might convert it into the insulating poly(amide).

The use of a precursor polymer offers several potential advantages:

• Precursor polymers may be applicable in direct writing applications (*5*). That is, they will be convertable, either by UV, X-ray or electron beam radiation from an insulating phase to a conducting phase in one step. Reduction in processing steps is of interest in any lithographic process.

• The processing could be performed without organic solvents. Although organic vapors will not be completely excluded, these conditions should ameliorate disposal costs and concerns.

• An ideal process could be designed so that the conversion from an insulating to a conducting polymer proceeds efficiently under irradiation but does not proceed in the absence of a threshold flux. These considerations will permit the simultaneous fabrication of highly electrically conductive structures with large contrast between insulating and conducting regions.

• Graphitic polymers would be linked by strong bonds between the atoms. Unlike in metals, where metal-metal bonding is weak, these covalent structures should not be susceptible to electromigration or other processes that could change the shape of a nanometer scale fabricated structure under the influence of an electrical or magnetic field.

Here, we describe potential routes to new precursor polymers for nanolithography applications. Specifically, two known polymers, poly(acrylonitrile) and poly(cyanoacetylene), have some promise in this area. It is argued, however, that their current lack of applicability as acceptable graphitic precursors stems from a lack of stereoregularity along the polymer backbone. Correct alignment of reactive groups in a precursor polymer in the solid state is critical for graphitization. Incorrect alignment of reactive groups can completely prevent a desired solid-state reaction. As the stereoregularity of the polymer chain will determine this alignment, this parameter is the focus of the work described below.

Poly(acrylonitrile) as a Graphitic Precursor Polymer

Polyacrylonitrile (PAN, Figure 2, top left) can be subjected to thermal treatment to produce a conjugated, conducting pyrolysed structure (*6-9*). An idealized depiction of the chemical changes that result upon heating PAN is shown in Figure 2. Particularly the second step of this process in which the rings formed in the first step are aromatized to form the completely unsaturated structure does not occur without a number of side reactions, particularly in the presence of oxygen (*7*). Conjugated ladder structures such as pyrolysed PAN have nevertheless commanded a great deal of attention because of their rigid-rod nature and their electronic, electrochemical and nonlinear optical properties. Thermal conversion in bulk has resulted in fibers, sheets and foams (*3*) with modest electrical conductivities (*4*). Maximum conductivites, however, are reached after very high temperature conversion required to carbonize the structure.

Because of its potential to be converted from an insulator to a conductor, PAN and its derivatives are an attractive starting point for use in direct writing schemes. Reports in the literature, however, unambiguously illustrate that the scheme shown in Figure 2 is extremely idealized and shows only part of what actually occurs upon heating the polymer. It is unclear that a conjugated ladder structure must be extensively crosslinked (graphitized) in order to become highly electrically conducting. Rather, defects produced during such high temperature operations may serve to limit conductivity, requiring higher temperatures to further carbonize the structure. Conjugated, acene-type polymers have potential to act as intrinsic electrical conductors if their structure is regular enough (*10-12*). At minimum, they should be easily oxidized extrinsic conductors using photooxidants or treatment with chemical oxidants.

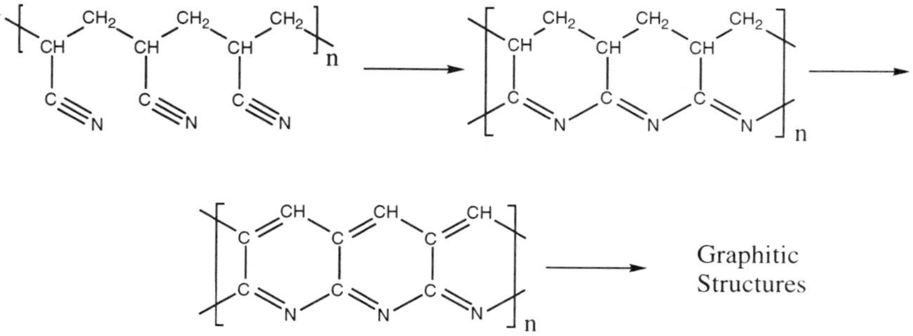

Graphitic Structures

Figure 2. Model for the conversion of PAN to a conjugated ladder polymer structure.

PAN can, in principle, exist with its cyano (e.g. -C≡N) groups in a variety of configurations with respect to each other. For example, one can consider an atactic chain (Figure 3c) in which the groups have a random stereo-orientation with respect to each other and compare to isotactic (Figure 3a) or syndiotactic (Figure 3b) structures in which the groups are on the same side of the polymer chain or alternate regularly between the two sides of the chain, respectively. Increasing the stereospecificity of these patterns can dramatically change the three-dimensional geometry of the chain and the bulk morphology of the polymer. These differences can often lead to large changes in glass-transition temperatures, crystallization behavior, and phase behavior when the polymer is blended or when the structure is employed in a block copolymer. For example, as the isotactic dyad fraction in poly(methyl methacrylate) is increased from 0.18 to 1.00, its glass transition temperature changes from 126 °C to 48 °C (*13*).

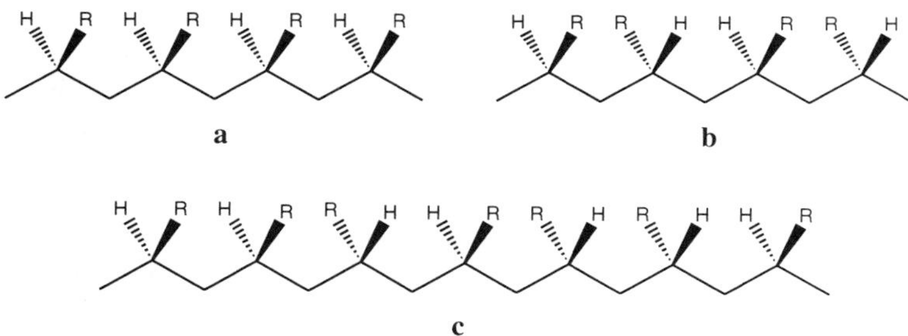

Figure 3. Representations of the all head-to-tail polymerization of a vinyl monomer. (a) isotactic sequence, (b) syndiotactic sequence (c) atactic sequence.

An X-ray study of the structure of PAN fibers (*14*) indicated the following sequence structures (Figure 4). These structures suggest that, in isotactic sequences, the nitrile groups are not in a position to undergo the desired conversion. The syndiotactic chain is, however, poised to convert to the ladder structure.

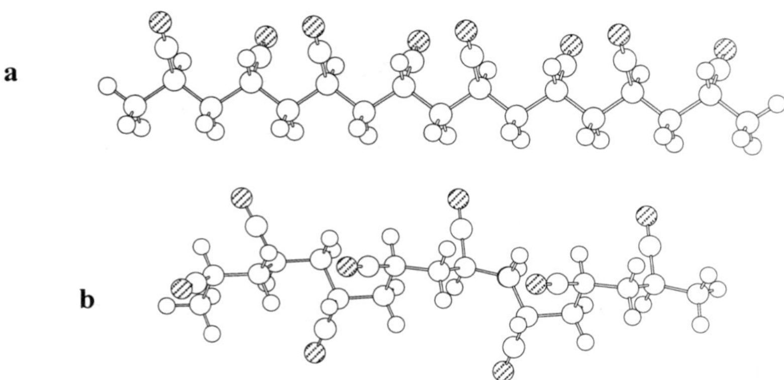

Figure 4. Models of the chain conformation in PAN including (a) syndiotactic and (b) isotactic sequences (*14*).

Thus, it is possible that relatively high temperatures (ca. 200-300 °C) are required to effect the first transformation shown in Figure 2 are mostly necessary to obtain the correct chain conformation in the solid state. It is hypothesized that highly syndiotactic PAN can be converted to a conjugated ladder structure (result of the first step in Figure 2) under milder conditions than commercially available, atactic PAN. These milder conditions will be those of lower temperature and under UV irradiation conditions employed in conventional lithography. Partial photoconversion of PAN, probably to the -(-C=N-)_n conjugated system (result of Step 1, Figure 2) has been illustrated but not apparently exploited in lithography (*15*).

Few stereospecific polymerizations of acrylonitrile are known (*16*) and none give a predominantely syndiotactic structure. Our first attempt to produce stereoregular PAN involved the use of low temperature, stereoselective anionic polymerizations that were based on procedures known to give reasonable yields of PAN. A number of attempts were made to adapt these for the synthesis of syndiotactic PAN. A typical result of several attempts is shown in Table I below for the precedented butyl lithium/sparteine system (*17-19*). As indicated, although syndio-rich poly(methyl methacrylate) (e.g. a high percentage of rr dyads as evidenced by ^1H or ^{13}C NMR) can be produced under these conditions, polymerization of acrylonitrile under these conditions affords essentially atactic PAN (e.g. a 1:2:1 collection of rr/rm(mr)/mm dyads) (*20*). We hypothesize that the biggest impediment in this synthesis is that the small size of the nitrile group (or, alternatively, of acrylonitrile itself), provides an insufficient stereoelectronic bias in the two diasteromeric transition states for acrylonitrile addition to the propagating chain. Addition of bulky Lewis acid auxiliaries to the acrylonitrile polymerization reaction also did not influence the tacticity of the resulting polymer.

Table I. Results of Anionic Polymerization with Chiral-Auxiliary

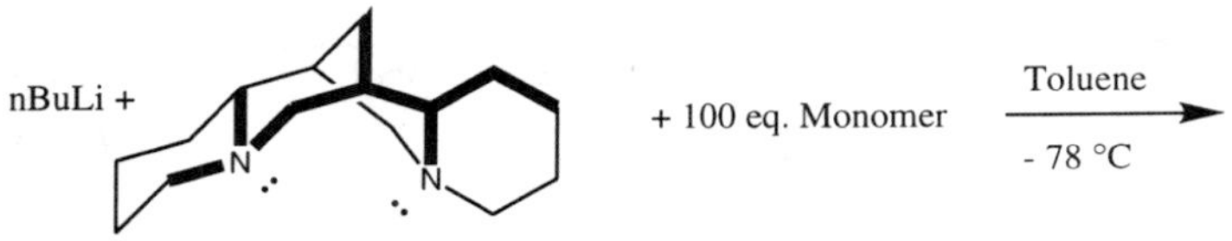

nBuLi + [chiral diamine] + 100 eq. Monomer —Toluene, - 78 °C→

Monomer	% Yield of Polymer	%rr	Tacticity %rm or mr	%mm
Methyl Methacrylate	99 %	0.66	0.20	0.13
Acrylonitrile	79 %	0.26	0.49	0.25

An alternative to a direct synthesis of a stereoregular nitrile polymer is the conversion of pendant groups on a preformed, stereoregular polymer to nitrile groups. This approach has the challenge that, as any reaction on a polymer (*21*), transformations must occur in high yield to prepare a modified polymer without a large proportion of unreacted groups. We are in the process of implementing a scheme for preparation of tactic poly(methacrylonitrile) polymers from tactic poly(trimethylsilyl methacrylate). This route offers the advantage that routes to highly syndiotactic and highly isotactic poly(trimethylsilyl methacrylate) are known (*22*). The route for the conversion of the silyl ester to the nitrile is shown in Scheme I.

Scheme I

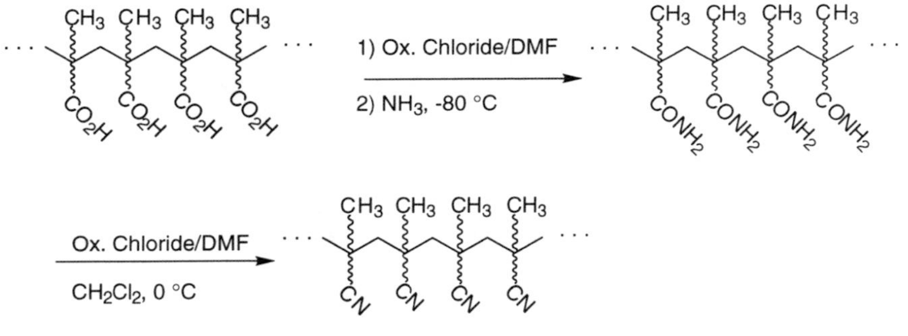

Polymerization of trimethylsilyl methacrylate to alternatively form the isotactic, syndiotactic and atactic polymers has been described previously (*22*). The molecular weight and tacticity of these polymers could be determined after acidic workup of the polymerization reaction followed by isolation of the polymer and treatment with diazomethane to form poly(methyl methacrylate). The molecular weight and tacticity of these polymers are given in Table II. The values obtained represent dramatic differences in tacticity as a function of preparation method and thus provide a reasonable basis for the testing the effect of polymer microstructure on its subsequent behavior.

Table II. Tacticity and Molecular Weights for Poly(methyl methacrylate) from Poly(trimethylsilyl methacrylate)

Conditions[a]	Tacticity produced	Mn & Mw[b]	PDI[b]	% rr[c]	% rm or mr[c]	% mm[c]
t-BuLi / MeAl(OBHT)$_2$[d], -78 °C	Syndiotactic	6,200 11,000	1.78	0.81	0.10	0.09
t-BuLi, -78 °C	Isotactic	40,700 51,800	1.28	0.02	0.07	0.92
AIBN, 60 °C	Atactic	16,500 33,300	2.02	0.59	0.32	0.09

[a]Conditions for preparation of poly(trimethylsilyl methacrylate). All reactions were run in toluene for 24 hours using a monomer:catalyst ratio of 50:1. [b]As determined by gel permeation chromatography in tetrahydrofuran versus poly(methyl methacrylate) standards. PDI = Mw/Mn. [c]As determined by ^{1}H NMR spectroscopy. [d]OBHT = -O-(2,6-di-t-butyl)-4-methylbenzene (butylated hydroxy toluene)

Poly(methacrylic acid) was obtained as the result of hydrolysis of poly(trimethylsilyl methacrylate) during isolation of this polymer. This polymer was subsequently converted to poly(methacroyl chloride) and then, to poly(methacrylamide) as follows. First, poly(methacrylic acid) was suspended in tetrahydrofuran and treated at 0 °C with an excess (3-fold equivalents based on the number of carboxylic acid groups) of oxalyl chloride with a catalytic amount of dimethyl formamide present. This intermediate acid chloride polymer was precipitated by addition of pentane, briefly dried under vacuum and dissolved in acetone. The resulting solution was added dropwise to a solution of liquid ammonia at -80 °C (*23*). After 3 hours of reaction at this temperature, the solution was allowed to warm to room temperature with stirring overnight. Poly(methacrylamide) was isolated by adding a small amount of water to the reaction followed by addition to tetrahydrofuran at 0 °C. In the conversion of the amide, it was found that the temperature was critical in acheiving the desired results. Figure 5 illustrates the result of this reaction under two conditions as monitored by infrared spectroscopy. Polymer formed when this reaction was run as described above (Figure 5b) displayed a peak at 1665 ± 2 cm^{-1} indicative for formation of the amide. Few to no carboxylic acid groups remained on the polymer chain after this transformation as evidenced by a lack of a carbonyl stretch due to this group at ca. 1700 cm^{-1}. The same infrared spectrum was obtained on commercially available, atactic sample of poly(methacrylamide). Polymer obtained by running this reaction at higher temperature (greater than room temperature, Figure 5c) displayed a peak at 1696 ± 2 cm^{-1} indicating a different side group on the polymer. This group has been attributed to the intramolecular formation of an imide based on comparison with the published infrared spectra of small molecule imides. Thus, reactions were run at this low temperature. Syndiotactic, isotactic and atactic poly(methacrylic acid) were converted to poly(methacrylamide) in this fashion.

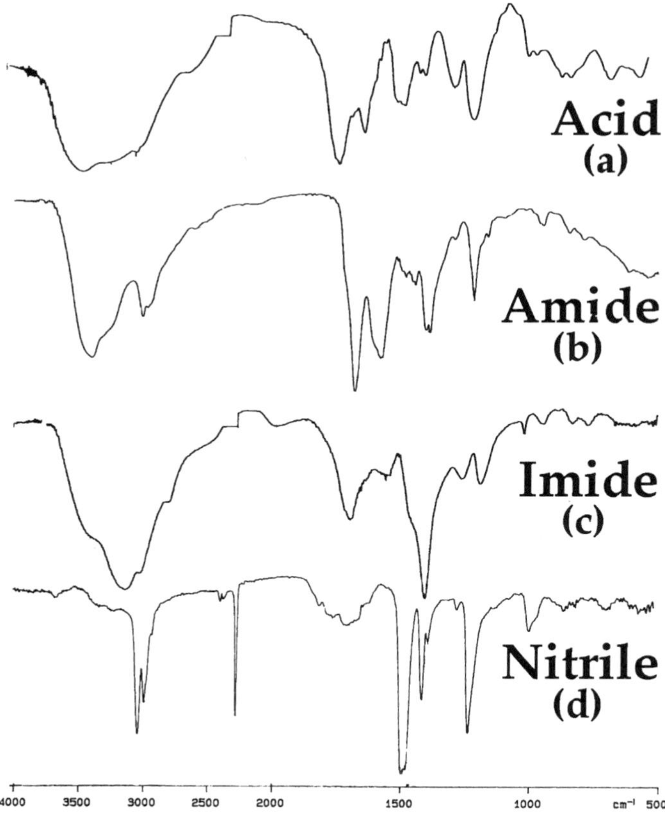

Figure 5. Infrared spectra of (a) poly(methacrylic acid) (b) poly(methacrylamide) (c) poly(methacrylamide) containing imide groups and (d) poly(methacrylonitrile) prepared as described in the text.

Poly(methacrylamide) could subsequently be dehydrated to poly(methacrylonitrile) using a large excess of oxalyl chloride and dimethyl formamide in methylene chloride at 0°C, warming to room temperature overnight as indicated in Scheme I. To date, this reaction has only been followed by infrared spectroscopy on the atactic material. Appearance of a peak at 2238 cm^{-1} is indicative of formation of nitrile groups in this polymer.

Currently, the viability of this route to form poly(methacrylonitrile) has been illustrated. NMR spectroscopy is in progress to verify that the starting tacticity of the polymer chain has been maintained during this synthesis. As none of these reactions should involve a mechanism in which the pseudo-asymmetric centers along the polymer backbone are influenced, it is anticipated that this illustration will be possible.

Poly(cyanoacetylene) as a Graphitic Precursor Polymer

Polymers of cyanoacetylene have been reported (*24-33*). To date, however, little information is available about the molecular weight of these materials or their microstructure. Figure 6 indicates four possible microstructures that are possible for poly(cyanoacetylene). Of these four structures, only one, the trans-transoid structure, is poised to undergo a facile cyclization reaction in the solid state such as that suggested in step one of Figure 2 for poly(acrylonitrile). The remaining structures may not even be planar and thus highly conjugated due to steric interactions between neighboring groups. The cis-cisoid structure offers a highly non-planar, helical structure.

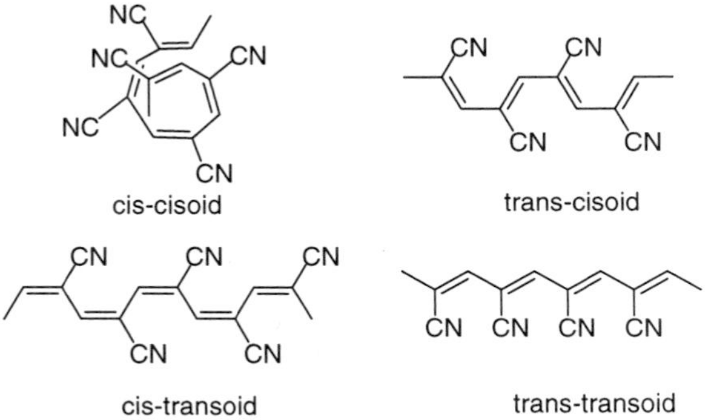

cis-cisoid

trans-cisoid

cis-transoid

trans-transoid

Figure 6. Four microstructures possible for poly(cyanoacetylene) assuming a strictly head-to-tail polymerization

Recently several reports have indicated the possibility of employing late transition metals such as rhodium (*34-36*), palladium, and nickel (*37*) for polymerization of phenylacetylene derivatives. Such catalysts had not previously been employed in the polymerization of cyanoacetylene. Hypothesizing that these catalysts might afford polymers with different microstructures than those already reported, a series of late transition metal catalysts were examined as initiators for cyanoacetylene polymerization.

Indeed, several of these catalysts were found to be active in this polymerization. The resulting polymers are compared to poly(cyanoacetylene) prepared using triethylamine as initiator. Triethylamine had previously been reported to produce poly(cyanoacetylene) (*32*). The resulting polymers were of various colors (a significance highlighted below) and appeared to be soluble in polar aprotic solvents such as dimethyl sulfoxide and dimethyl formamide. Yields and molecular weights of these polymers (as determined by gel permeation chromatography versus polystyrene standards) are reported in Table III. It is striking that the "traditional" initiator, triethylamine forms low molecular weight polymer — in fact, this is an oligomeric material as GPC was able to partially resolve the different oligomers in this sample. The nickel and palladium catalysts ((norbornadiene)RhCl dimer was also tried but gave very little polymer) produced much higher molecular weight materials. Neither the oxidation state nor the ligand environment of the metal did not dramatically influence the results. Both M(0) and M(II) catalysts were employed with similar results (M = Ni, Pd). The nickel-derived catalysts, however, produced much higher molecular weight materials than did the palladium-derived catalysts.

Table III, Percent Yield and Molecular Weights of Poly(cyanoacetylene) Prepared using Several Catalysts

Catalyst[a]	% Yield[d]	Mn & Mw[f]	PDI[f]
$(PPh_3)_4Ni$	23	25,100 72,300	2.87
$(PPh_3)_2NiCl_2$	35	18,800 50,600	2.69
$(dppe)NiCl_2$ [b]	30	11,500 21,900	1.91
$Pd(CH_3CN)_4(BF_4)_2$	46	8,200 21,300	2.59
$Pd_2(dba)_3$ [c]	21	7,900 16,700	2.12
Et_3N	40[e]	2,500[g] 14,000	5.72

[a]Conditions — all reactions performed in dimethyl sulfoxide at room temperature for 24 hours at a monomer:catalyst ratio of 100:1. [b]dppe = bis-diphenylphosphino ethane [c]dba = dibenzylidene acetone. [d]Yield after precipitation into an excess of ethanol, centrifugation and drying under dynamic vacuum at room temperature overnight [e]A maximum value as complete removal of the solvent from this low molecular weight material was not possible. [f]As determined by gel permeation chromatography in dimethyl formamide/0.05 M LiBr versus poly(styrene) standards. PDI = Mw/Mn. [g]Integration over several partially resolved oligomeric peaks limits the validity of this value.

The difference between the nickel and palladium catalyzed cyanoacetylene polymerizations is also reflected in the UV/VIS spectra of each material. Figure 7 shows UV/VIS spectra of three polymers prepared with nickel, palladium and triethylamine catalysts, respectively. Those polymers initiated by triethylamine and by palladium(II) were brown in solution and have relatively high energy optical absorptions, indicating a low effective conjugation for these polymers. In contrast, the polymer initiated by nickel(II) (those initiated by nickel(0) catalysts were similar) was purple to blue in color, indicating a much higher effective conjugation for these polymers. It is speculated that, in addition to being of much higher molecular weight, the nickel catalysed polymerizations resulted in a different microstructure for these polymers as compared to those initiated by palladium or triethylamine.

This difference in microstructure is reflected by the dramatic red shift due to increased planarity and thus a higher effective conjugation for this former class of materials.

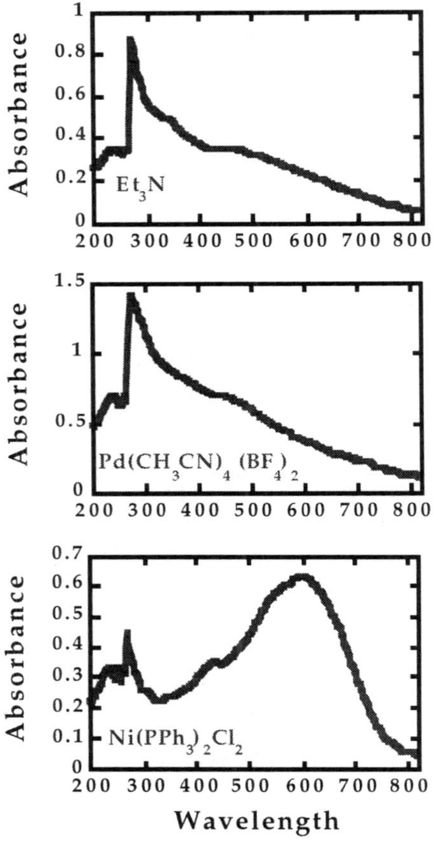

Figure 7. UV/VIS spectra of poly(cyanoacetylene) prepared using Et_3N, $Pd(CH_3CN)_4(BF_4)_2$, and $Ni(PPh_3)_2Cl_2$ as catalysts.

Infrared spectroscopy also indicates some of the features of the poly(cyanoacetylene) polymers produced. All polymers have a characteristic cyano stretch at ca. 2200 cm^{-1}. This feature indicates that the cyano groups in the monomer are largely preserved in the polymer — propagation occurs through the acetylene rather than the cyano units in the monomer. The peak at ca 1500 cm^{-1} indicate C=C stretching in the polyconjugated backbone of the polymer. In line with the respective molecular weights and effective conjugation lengths indicated by GPC and UV/VIS above, the triethylamine-catalyzed material has the smallest, highest energy C=C stretch and the nickel-catalyzed material has the largest, lowest energy C=C stretch. Also present are peaks at ca. 3400 - 3450 cm^{-1} that are possibly due to acetylinic C-H stretching due to residual monomer in the sample and/or perhaps olefinic C-H stretching due to end-groups in the polymer. This former interpretation appeared unlikely as no residual acetylinic proton signal was observed in the ^1H NMR spectra of any of these polymers. If it is due to end groups, the relative size of this peak corresponds to the trend in molecular weight as the polymers prepared from these three catalysts are compared.

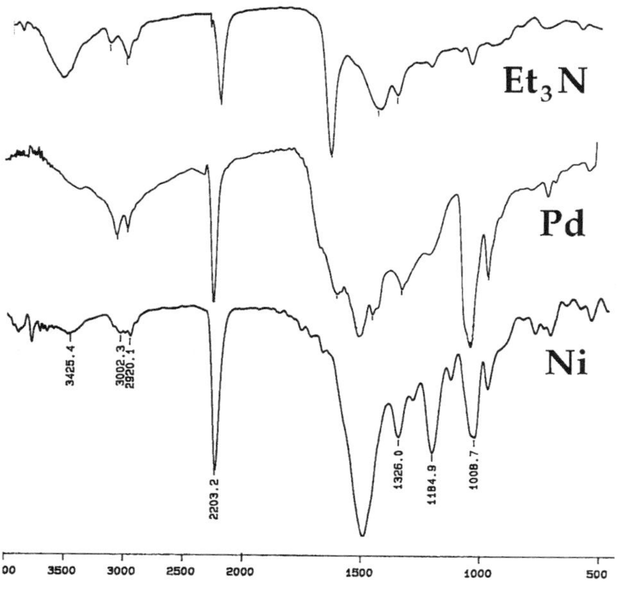

Figure 8. Infrared spectra of poly(cyanoacetylene) prepared using Et_3N, $Pd(CH_3CN)_4(BF_4)_2$, and $(PPh_3)_4Ni$ as catalysts.

Our current work in this area is focusing on further characterization of these materials. Solution ^1H NMR has, to date, been of little value in characterizing these polymers. Solid-state NMR, solution ^{13}C NMR and thermal analysis of these polymers is currently in progress.

Acknowledgments

Acknowledgment is made to the North Carolina State University (Startup Funds), the Office of Naval Research (Young Investigator Grant) the Kenan Institute (with funds provided by Hoechst-Celanese) and the donors of The Petroleum Research Fund, administered by the ACS, for support of this research.

Literature Cited

(1) Thompson, C. V.; Lloyd, J. R. *MRS Bull* **1993**, *18*, 19-24.
(2) Tao, J.; Cheung, N. W.; Hu, C. *IEEE Electron Dev. Lett.* **1993**, *14*, 249.
(3) Renschler, C. L.; Sylwester, A. P. *Mat. Sci. Forum* **1989**, *52/53*, 301-322.
(4) Jobst, K.; Sawtschenko, L.; Schwarzenberg, M.; Wuckel, L. *Synth. Met.* **1991**, *41-43*, 959-962.
(5) Guay, D.; Tourillon, G.; Viel, P.; Lecayon, G. *J. Phys. Chem.* **1992**, *96*, 5917-5921.
(6) Chung, T.-C.; Schlesinger, Y.; Etemad, S.; Macdiarmid, A. G.; Heeger, A. J. *J. Polym. Sci. Polym. Phys. Ed.* **1984**, *22*, 1239-1246.
(7) Clarke, A. J.; Bailey, J. E. *Nature* **1973**, *243*, 146-150.
(8) Coleman, M. M.; Petcavich, R. J. *J. Polym. Sci. Polym. Phys. Ed.* **1978**, *16*, 821-832.
(9) Enzel, P.; Bein, T. *Chem. Mater.* **1992**, *4*, 819-824.
(10) Brédas, J. L.; Street, G. B. *Acc. Chem. Res.* **1985**, *18*, 309-315.
(11) Brédas, J. L. *J. Phys. Chem.* **1985**, *82*, 3808-3811.
(12) Brédas, J. L.; Heeger, A. J.; Wudl, F. *J. Chem. Phys.* **1986**, *85*, 4673-4678.
(13) Lenz, R. W. in *Preparation and Properties of Stereoregular Polymers*Lenz, R. W. and Ciardelli, F.; Eds.; D. Reidel Publishing Company: Boston, 1979, pp 163-184.
(14) Liu, X. D.; Ruland, W. *Macromolecules* **1993**, *26*, 3030-3036.
(15) Sergides, C. A.; Chughtai, A. R.; Smith, D. M.; Schissel, P. *Macromolecules* **1986**, *19*, 1448-1453.
(16) Polymerization of acrylonitrile in a urea canal structure using γ-ray initiation results in predominantely isotactic PAN: Minagawa, M.; Yamada, H.; Yamaguchi, K.; Yoshii, F. *Macromolecules* **1992**, *25*, 503-510.
(17) Oishi, T.; Yamasaki, H.; Fujimoto, M. *Polym. J.* **1991**, *23*, 795-804.
(18) Basu, A.; Beak, P. *J. Am. Chem. Soc.* **1996**, *118*, 1575-1576.
(19) Beak, P.; Kerrick, S. T.; Wu, S.; Chu, J. *J. Am. Chem. Soc.* **1994**, *116*, 3231-3239.
(20) Kamide, K.; Yamazaki, H.; Okajima, K.; Hikichi, K. *Polym. J.* **1986**, *18*, 277-280.
(21) Platé, N. A. *Pure & Appl. Chem.* **1976**, *46*, 49-59.
(22) Kitayama, T.; He, S.; Hironaka, Y.; Iijima, T.; Hatada, K. *Polym. J.* **1995**, *27*, 314-318.
(23) Schulz, R. C.; Elzer, P.; Kern, W. *Makromol. Chem.* **1960**, *42*, 189-196.
(24) Byrd, N. R.; Kleist, F. D.; Rembaum, A. *J. Macromol. Sci. (Chem.)* **1967**, *A1*, 627-633.
(25) Chien, J. C. W.; Carlini, C. *J. Polym. Sci. Polym. Chem. Ed.* **1985**, *23*, 1383-1393.
(26) Chien, J. C.; Carlini, C. *Makromol. Chem., Rapid Commun.* **1984**, *5*, 597-602.
(27) Deits, W.; Cukor, P.; Rubner, M.; Jopson, H. *Ind. Eng. Chem. Prod. Res. Dev.* **1981**, *20*, 696-704.
(28) Fouletier, M.; Armand, M.; Audier, M. *Mol. Cryst. Liq. Cryst.* **1985**, *121*, 333-336.

(29) Ho, T. H.; Katz, T. J. *J. Mol. Catal.* **1985**, *28*, 359-367.
(30) Hsieh, B. T.; Cheng, H. M.; Jiang, G. J. *Synth. Met.* **1990**, *37*, 13-21.
(31) Matsumura, K.; Tsukamoto, J.; Takahasi, A. *J. Mater. Sci. Lett.* **1985**, *4*, 509-512.
(32) Wallach, J.; Manassen, J. *J. Polym. Sci. A-1* **1969**, *7*, 1983-1996.
(33) Manassen, J.; Wallach, J. *Macromolecules* **1965**, *87*, 2671-2677.
(34) Kishimoto, Y.; Miyatake, T.; Ikariya, T.; Noyori, R. *Macromolecules* **1996**, *29*, 5054-5055.
(35) Tabata, M.; Sadahiro, Y.; Nozaki, Y.; Inaba, Y.; Yokota, K. *Macromolecules* **1996**, *29*, 6673-6675.
(36) Tang, B. Z.; Poon, W. H.; Leung, S. M.; Leung, W. H.; Peng, H. *Macromolecules* **1997**, *30*, 2209-2212.
(37) Sen, A.; Lal, T.-W. *Organometallics* **1982**, *1*, 415-417.

Chapter 4

Metallization on Poly(tetrafluoroethylene) Substrate by Excimer-Laser-Induced Surface Reaction and Chemical Plating

Hiroyuki Niino and Akira Yabe

National Institute of Materials and Chemical Research (NIMC), Higashi 1-1, Tsukuba, Ibaraki 305-8565, Japan

Surface of poly(tetrafluoroethylene) [PTFE] film was modified chemically by an ArF excimer laser-induced reaction in a hydrazine gas atmosphere. The polymer surface defluorinated with the photolyzed hydrazine showed hydrophilicity high enough to be metallized by chemical plating. The mechanism for the modification was investigated by XPS, FTIR, and SIMS analyses. A patterned nickel plating having a sufficient adhesion strength was carried out by the area-selective laser-treatment. This processing provides a fully-additive method for fabrication of printed wiring boards with PTFE substrates.

Surface modification and metallization of poly(tetrafluoroethylene) (PTFE) has attracted a considerable attention from viewpoints of fundamental science and applied technology. PTFE has a low surface free energy while it shows excellent thermal and chemical stability. Therefore, the chemical and physical inertness of PTFE makes metallization an extremely difficult process. At present a chemical treatment using a sodium naphthalenide solution (*1-4*), a radio frequency plasma process (*5-8*), and electron/ion beam irradiation (*9-11*) have been employed for the modification of PTFE surface.

On the other hand, as another new technique by laser material processing, there is an increasing interest in surface modification of materials (*12-14*). In recent years several investigations on surface modification of fluorocarbon polymers by laser processing have been reported (*15-24*). In a series of these investigations we found that the surface of PTFE was modified into a hydrophilic state by an excimer laser-induced chemical reaction (*17-22*). Our processing consists of the irradiation with an ArF excimer laser under a hydrazine atmosphere, and the mechanism is explained by the defluorination from PTFE surface and the incorporation of amino group and hydrogen atoms into PTFE chain. The laser-treated surface showed remarkably favorable property for metallization by electroless plating (*18, 19*). Then, this processing provides a fully-additive method for fabrication of printed wiring boards with PTFE substrates, which is free from a drastic treatment with dangerous chemicals and a patterning by using photolithography (*25-27*), as shown in Figure 1. We wish to report the surface modification of PTFE film by our processing, and to show its application into a new fabrication method of printed wiring boards by using chemical plating for laser-treated PTFE films.

Surface Modification upon ArF Excimer Laser Irradiation

In a series of our investigations on the surface reaction of fluorocarbon polymers, we found an effective hydrophilic modification for PTFE surface. We have studied the reactivity on the films with hydrazine (N_2H_4), ammonia (NH_3), and hydrogen peroxide (H_2O_2) which are excited upon irradiation with excimer lasers. The most effective modification to a hydrophilic surface was achieved with the ArF laser irradiation of the films in a hydrazine gas atmosphere.

The radiation apparatus used for the experiments was an ArF excimer laser (λ = 193 nm) with the pulse duration of 14 ns (*17*). The PTFE film was 30 mm in diameter and 2 mm thick. Anhydrous hydrazine (Tokyo Kasei Kogyo Co. Ltd.) in a small glass tube (0.5 ml) was used after degassing with liquid nitrogen to prevent bumping. A reaction chamber having a quartz window was filled with hydrazine vapor under its saturation vapor pressure (1.2×10^3 Pa) at the room temperature (Figure 2). Since the distance between the sample and the window was 1 mm, more than 90% of incident laser beam could reach onto the sample surface.

Characterization of Laser-Treated Surface. When the irradiation of ArF laser to a PTFE film was conducted at fluences of 27 $mJ{\bullet}cm^{-2}{\bullet}pulse^{-1}$ with 10 Hz in a hydrazine vapor atmosphere, the PTFE surface became hydrophilic. Water contact angle (CA) was measured as the easiest but sensitive method for characterization of our laser processing. Compared with CA of 130° on PTFE before the irradiation, CA was changed to 25° upon irradiation of 3000 pulses at 27 $mJ{\bullet}cm^{-2}{\bullet}pulse^{-1}$ (Table I).

Within the fluences of 1 - 27 $mJ{\bullet}cm^{-2}{\bullet}pulse^{-1}$, the contact angle decreased with an increase of pulses, and it is interesting that the change in the contact angle showed a remarkable dependency on the fluence (*19*). Figure 3 shows the CA on PTFE after the laser treatment with 1 and 27 $mJ{\bullet}cm^{-2}{\bullet}pulse^{-1}$. Although the same dose ratio was applied onto PTFE surface in the hydrazine, the irradiation of 27 $mJ{\bullet}cm^{-2}{\bullet}pulse^{-1}$ was effective for the hydrophilic modification. These results suggested that reactive species to modify PTFE surface were produced effectively upon the laser irradiation with a higher fluence. In addition, the irradiation at 100 $mJ{\bullet}cm^{-2}{\bullet}pulse^{-1}$ deviated from this dependence, indicating that it was too strong for the chemical modification by laser-induced surface reaction. The modified layer on the PTFE film would be etched by irradiation with the higher fluence. These results agree with the results from X-ray photoelectron spectroscopy (XPS) as mentioned later.

XPS Analysis of PTFE Surface. Table I shows the atomic ratios of fluorine, nitrogen, oxygen, and carbon determined by XPS with monochromatic AlKα radiation before and after the laser irradiation of PTFE at different fluences and pulses. In the case of 27 $mJ{\bullet}cm^{-2}{\bullet}pulse^{-1}$, it is very clear that the ratio of fluorine decreases while that of nitrogen increases with an increase in laser pulses. An exceptional result for nitrogen in the case of 3000 pulses will be explained by the photodecomposition of N-containing group. A small portion of oxygen came from the reaction between ambient oxygen and reactive sites on the treated PTFE surface because of our *ex-situ* XPS measurement.

In a fine XPS measurement of irradiated film, the binding energy of the C_{1s} shifted to a lower energy (Figure 4) because an electron donating group of hydrogen and N-containing group was substituted for the electron withdrawing group of

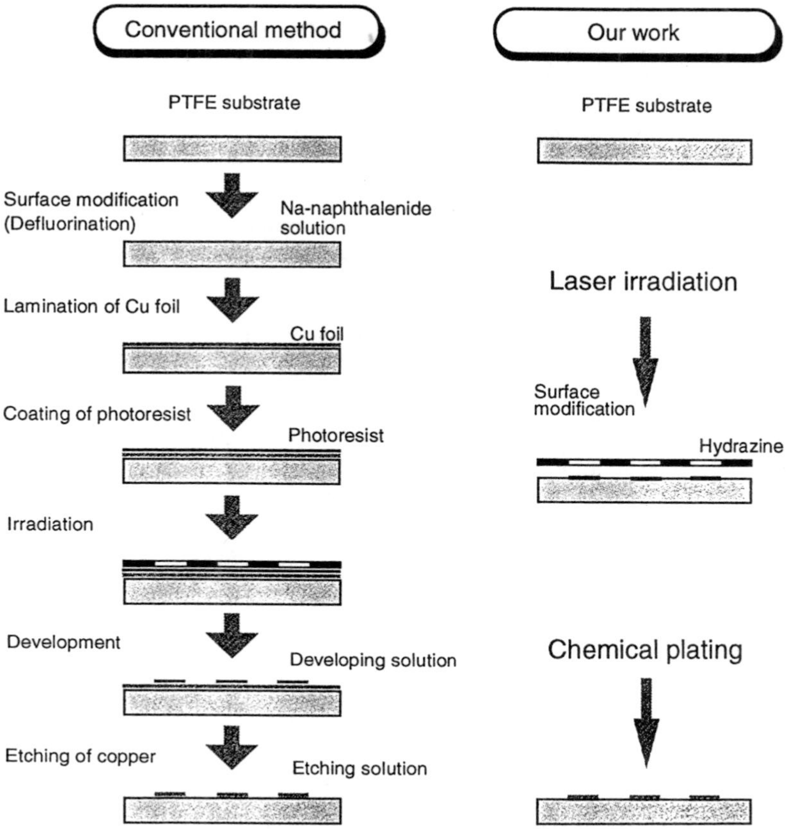

Conventional method

PTFE substrate

Surface modification (Defluorination)

Na-naphthalenide solution

Lamination of Cu foil

Cu foil

Coating of photoresist

Photoresist

Irradiation

Development

Developing solution

Etching of copper

Etching solution

Our work

PTFE substrate

Laser irradiation

Surface modification

Hydrazine

Chemical plating

Figure 1. Fabrication process of printed wiring boards using PTFE substrates.

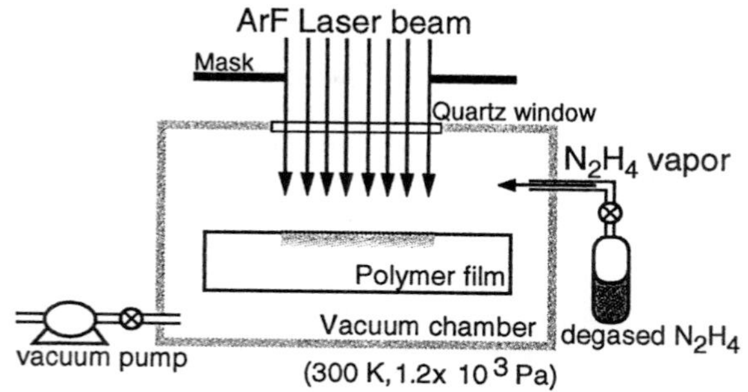

Figure 2. Reaction chamber for PTFE surface modification.

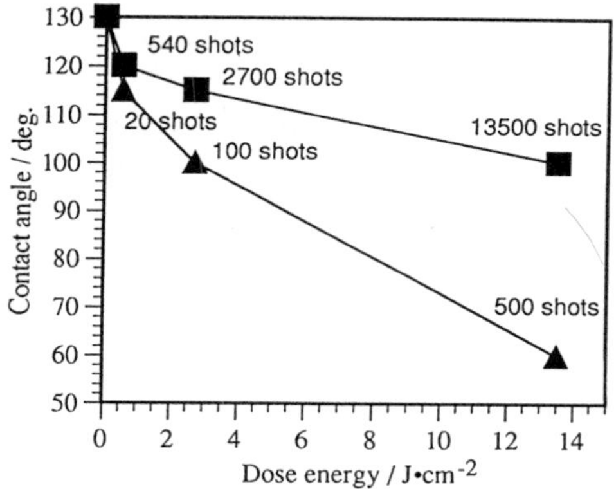

Figure 3. Water contact angle on PTFE film; dependence on radiation dose energy of laser irradiation at 1 $mJ \cdot cm^{-2} \cdot pulse^{-1}$ (■) and 27 $mJ \cdot cm^{-2} \cdot pulse^{-1}$ (▲).

Table I Atomic ratio of carbon(C) : fluorine(F) : nitrogen(N) : oxygen(O) and contact angle of water on the PTFE surface after ArF laser treatment in hydrazine vapor (Adapted from ref. 19).

Fluence [$mJ \cdot cm^{-2} \cdot pulse^{-1}$]	Shot number [shot]	Atomic ratio: Carbon	Fluorine	Nitrogen	Oxygen	Contact angle [deg]
Before irradiation		1.0	2.1	—— a	—— a	130
1	540	1.0	1.8	—— a	—— a	120
	2700	1.0	1.4	0.017	0.023	115
	13500	1.0	1.2	0.031	0.040	100
5	100	1.0	1.8	0.026	0.0053	110
	3000	1.0	0.29	0.18	0.044	60
27	20	1.0	1.7	0.11	—— a	115
	100	1.0	0.99	0.19	0.033	100
	500	1.0	0.34	0.27	0.033	60
	1000	1.0	0.21	0.28	0.029	30
	3000	1.0	0.016	0.19	0.033	25
100	1000	1.0	2.0	0.054	—— a	120

[a] Not detected

fluorine. Hydrogen substitution was confirmed by secondary ion mass spectrometry (SIMS) measurement. We also estimated the formation ratio of unsaturated bonds on the surface by immersion of the irradiated film into a CCl_4 solution of bromine. New peaks ascribed to Br_{3d} and Br_{3p} appeared in the XPS spectrum of the film, indicating addition of bromine to an unsaturated carbon bond. Since the atomic ratio of Br : C on the surface of PTFE film after the irradiation of 1000 pulses was 0.16 : 1.0, the substitution by hydrogen atom was presumed to be about 75%.

A change with the passage of time for the chemically modified surface by our laser processing was studied by XPS (*18*). The atomic ratios of F/C and N/C were almost the same on standing after the laser treatment. However, the ratio of O/C increased after hundreds' days because the sample was stored in a desiccator filled with dry air and some oxidative reactions might proceed on the active sites remaining on the surface.

FTIR-ATR Analysis of PTFE Surface. Fourier transform infrared spectroscopy with attenuated total reflection (FTIR-ATR; prism: KRS-5, detection angle: 45°) did not show a significant difference before and after the laser treatment, because the analytical depth of the ATR method is too deep to detect the modified layer on the film. Our FTIR-ATR measurement provided sub-microns of the analytical depth in a conventional IR region. On the other hand, by a chemical derivatization method (*29*) using a reaction of the irradiated PTFE surface with trifluoroacetic anhydride $[(CF_3CO)_2O]$ gas, a new absorption peak due to a carbonyl group was observed at 1700 cm^{-1} (*18*), indicating the formation of an amide moiety by a reaction between an amino group and trifluoroacetic anhydride (*eq.* 1). Therefore it was confirmed that the N-containing group shown in the XPS spectra was an amino group.

$$R_1\text{-}NH_2 + (CF_3CO)_2O \longrightarrow R_1\text{-}NHCOCF_3 + CF_3COOH$$

(R_1: Polymer chain) (*eq.* 1)

SIMS Analysis of PTFE Surface. Static SIMS spectra of PTFE for positive ions produced with an ion source of Xe^+ at 4 keV are shown in Figure 5. While the fragments due to CF^+ [mass number: 31], CF_3^+ [30], and C_nF_m [n, m= 1-5] were observed in the spectrum before laser irradiation (Figure 5(a)), a spectrum characteristic of poly(hydrocarbons) like polyethylene was given from the sample after laser treatment, indicating that the modified surface retained a linear chain in the structure (Figure 5(b)) (*28*). Moreover, N-containing ions such as NH_4^+ [18], $CH_2NH_2^+$ [30], and $C_3H_7NH_3^+$ [60] were observed in Figure 5(b). This is another evidence of amino groups on the laser treated PTFE surface. On the other hand, negative SIMS spectra showed F^- [19] and F_2^- [38] ions for a fresh PTFE film, CH^- [13], C_2H^- [25] and CN^- [26] ions were detected from the treated PTFE film (Figure 6).

Figure 7 shows SIMS spectra of PTFE treated by alkali etching with a commercially available 1:1 complex of sodium and naphthalene (Tetra-Etch). The PTFE surface, where the atomic ratio of C : F : O was determined to be 1.0 : 0.0 : 0.12 by XPS, was extensively defluorinated by the alkali attack (*10*). The positive SIMS spectrum having dominant peaks of Na^+ [23], $C_2H_3^+$ [27], $C_2H_5^+$ [27], $C_3H_3^+$ [39], and $C_3H_5^+$ [41], was distinguished from that of the laser-treated PTFE. In addition, the negative SIMS spectrum shows C_n^- [12•n], C_nH^- [12•n+1], O^- [16], and OH^- [17]. These results indicated that the chemically etched surface consisted of a graphite-like structure partially having hydrogen and hydrogen oxide. Although the PTFE surface defluorinated by the alkali etching possessed hydrophilicity (CA: 70°), its surface structure was completely different from that of the laser treatments.

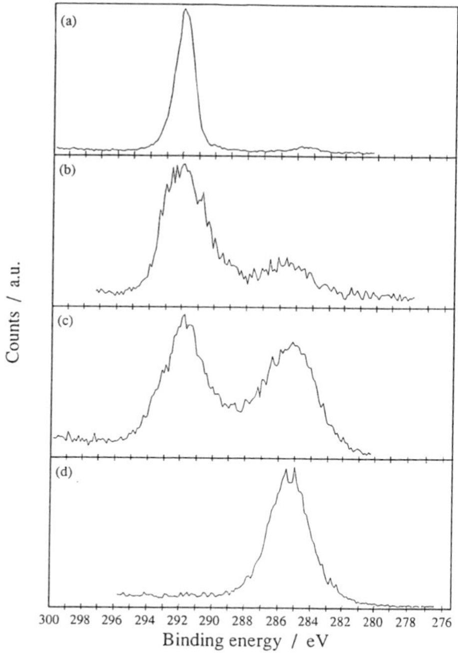

Figure 4. XPS spectra of C_{1s} on PTFE film after laser-treatment; (a) 0 shot, (b) 20 shots, (c) 100 shots, (d) 3000 shots.

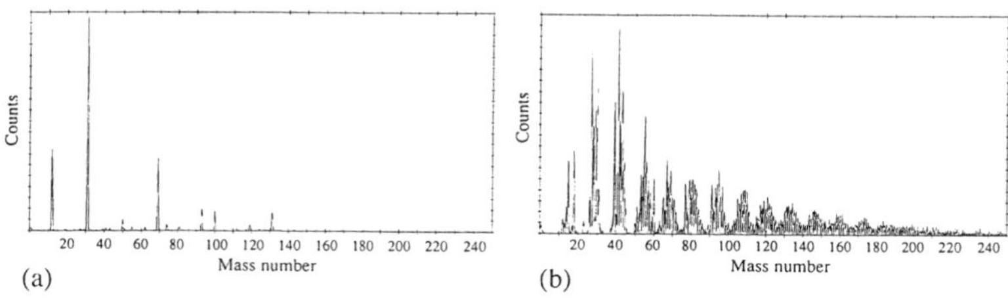

Figure 5. Static SIMS spectra of PTFE film for positive ions, (a) fresh surface, (b) laser-treated surface (ArF laser 27 $mJ{\bullet}cm^{-2}{\bullet}pulse^{-1}$, 3000 pulses).

From these surface analyses, our modification provided a linear hydrocarbon chain like a polyethylene structure having an amino group as a hydrophilic site.

It is worth noting that mass signals of SIMS spectra are strongly dependent on surface chemical structures (*30*). In our measurements, native PTFE surface gave higher signals compared with modified surfaces. The scale factor of our figures is Fig. 5(a) : Fig. 5(b) : Fig. 7(a) = 1.0 : $2.6x10^{-2}$: $9.2x10^{-3}$ for positive ions and Fig. 6(a) : Fig. 6(b) : Fig. 7(b) = 1.0 : $5.5x10^{-2}$: $1.7x10^{-1}$ for negative ions. On the surface of a partially modified PTFE (*e.g.*, 500 shots' irradiation at 27 mJ•cm^{-2} •pulse^{-1}), original signals of PTFE were dominantly observed in a SIMS spectrum.

Mechanism for Chemical Surface Modification. Since PTFE films having no significant absorption at 193 nm were not excited upon the irradiation with an ArF laser, hydrazine is only photodissociated by the irradiation. Photolysis of hydrazine upon the ArF irradiation has been investigated in detail by Hawkins (*31*) and Lindberg (*32*). The irradiation of hydrazine in a gas phase produces reactive species such as $•N_2H_3$, $•NH_2$, and •H radicals and :NH nitrene in high efficiency. By assuming that these species attack the PTFE surface, we studied a preliminary reaction path for these reactive species and perfluorocarbons, that is, which species can react with perfluoroalkyl molecules.

Semi-Empirical Calculation for a Preliminary Reaction of the Modification. In order to investigate the reaction mechanism, semi-empirical calculation was performed by using MOPAC ver. 6.10 in CAChe Worksystem (Sony-Tektronix) (*33*). We used perfluoroalkyl compounds such as CF_4 and C_5F_{12} instead of PTFE in order to reduce calculation time (*eq.* 2). The saddle point and enthalpy of formation in model reactions was obtained using PM3 as Hamiltonian (*19*). Table II shows formation enthalpy (ΔH) and activation energy (ΔEa). Fluorine-abstraction by hydrogen radical is an exothermic reaction (ΔH: -23 kcal•mol^{-1}) to form a stable product, hydrogen fluoride (HF). This abstraction has a low activation energy, ΔEa= -8.7 kcal•mol^{-1}, suggesting that hydrogen radical can readily abstract fluorine atoms from the PTFE surface. On the other hand, the amino radical can not abstract fluorine atoms because of large ΔEa and an endothermic reaction having a large ΔH. The other species of the N-containing intermediates were similar to $•NH_2$. It may be concluded that F-abstraction by hydrogen radical plays an important role in a primary stage of the reaction. However, the concentration of •N2H3, •NH2, and also the radicals produced on the surface could be high enough to induce the saturation of the surface modification reaction by intense laser irradiation. This may be related to the deviation on the results upon irradiation at 100 mJ•cm^{-2}•pulse^{-1}.

$$C_2F_5\text{-}CF_2\text{-}C_2F_5 + •R_2 \longrightarrow C_2F_5\text{-}CF\text{-}C_2F_5 + R_2\text{-}F$$

(R_2: intermediate) (*eq.* 2)

Model Reaction in Gas Phase. To confirm the results from the above calculation, the irradiation of the gas mixture of C_6F_{14} and hydrazine with the laser was examined. After the photolysis of the mixture, the formation of HF was detected by a chemical analysis (*34*). It is presumed that the hydrogen radical generated by the photolysis of hydrazine abstracts the fluorine atom on the surface of a PTFE film to form HF, and that reactive species such as N-containing $•NH_2$ and $•N_2H_3$ can be substituted in the C-chain after F-abstraction.

Primary Process for the Modification. On the basis of the above results, it should be noted that the chemical reaction proceeded actually on the surface of the PTFE film. The irradiation was performed under the condition that more than 90 % of incident laser reached on the surface. After the laser treatment, we confirmed the

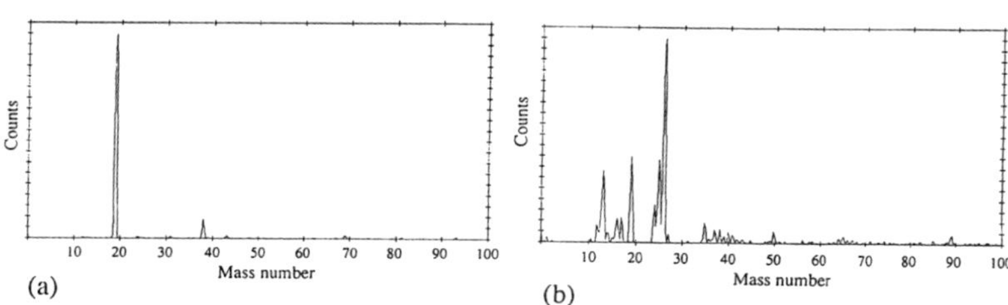

Figure 6. Static SIMS spectra of PTFE film for negative ions, (a) fresh surface, (b) laser-treated PTFE surface (ArF laser 27 mJ•cm^{-2}•pulse^{-1}, 3000 pulses).

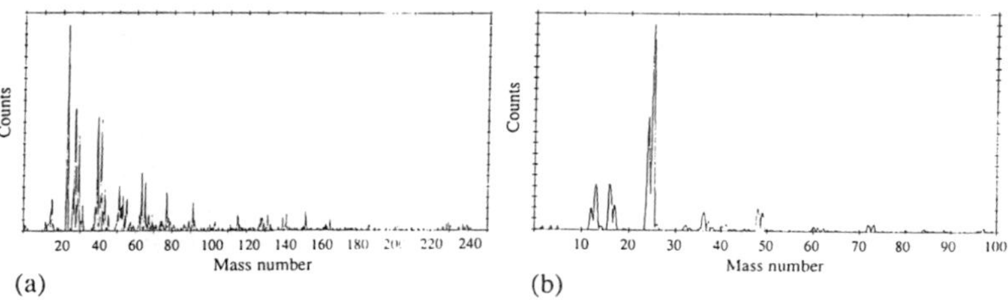

Figure 7. Static SIMS spectra of PTFE film after alkali etching, (a) positive ion spectrum, (b) negative ion spectrum.

Table II Calculated formation enthalpy (ΔH) and activation energy (ΔEa) in a fluorine-abstraction of fluorocarbons using MOPAC (PM3) [unit: kcal•mol^{-1}] (Adapted from ref. 19)..

	CF_4		C_5F_{12}	
intermediate	ΔEa	ΔH	ΔEa	ΔH
•H	8.69	-23.28	6.11	-36.23
•NH_2	67.87	51.74	58.73	38.80
•N_2H_3	78.58	51.74	67.70	55.08
:NH (triplet)	66.14	44.24	63.04	31.29

presence of a deposited layer comprised of HF-N_2H_4 salt by chemical analysis. On the surface the salt comprised of hydrazine in the condensed phase would be photolyzed effectively. Therefore it is concluded that the effective reaction for PTFE film upon the irradiation occurs mainly through the photolysis of the HF-N_2H_4 salt adsorbed on the surface, not through the reactive species generated in the vapor phase, as shown in Figure 8. This scheme is supported by the fact that the chemical composition of PTFE surface showed no change by an ArF laser irradiation in a vacuum and in the air (*35*).

Chemical Plating of Laser-Treated Films

By the laser treatment, the surface of PTFE films could be made hydrophilic and the surface energy increased in comparison with that of a native PTFE film. As the results, we intended to extend the hydrophilic modification technique into metallization on the PTFE film by using a chemical (electroless) plating technique. If the activators for the plating will deposit only on the hydrophilic region, area-selective metallization will be performed on the surface of PTFE films modified upon the laser-irradiation through a mask by a projection method.

Nickel Chemical Plating of Films. The procedure of nickel metallization consists of four steps as shown in Figure 9. First, the hydrophilic PTFE films were dipped into 0.5 M aqueous H_2SO_4 for 30 s as a pre-dipping, followed by rinsing with de-ionized water at room temperature. Second, the dipped film was immersed into an aqueous solution of palladium(II) sulfate (KAT-450, C.Uyemura Co.) for 2 min at 35 °C in order to produce a catalyst for electroless plating. Third, as a post-dipping, the substrates were dipped into 0.5 M aqueous H_2SO_4 and de-ionized water at room temperature again. Finally, the catalyzed films were immersed in an electroless nickel plating bath (Nimuden-SX, C.Uyemura Co.) for 3-30 min at 60 °C.

Figure 10 (a) shows the PTFE surface metallized by chemical plating after the ArF laser irradiation of 3000 pulses at 27 $mJ \cdot cm^{-2} \cdot pulse^{-1}$. The nickel layer on the PTFE surface showed metallic luster and had an adhesion force strong enough not to be peeled off in an adherence test using a conventional tape test (tape: Nitto Denko Co. No. 401). From a pull-stud test, the adhesive strength was determined to be 15 MPa for nickel plating. This adhesive strength is also supported by the adhesive strength (max. 14 MPa) of PTFE to a carbon steel rod through a cyanoacrylate adhesive (*20*). It is known in the literature that the tensile strength of PTFE itself is 14-34 MPa (*36*). The strength of around 15 MPa given from our test is close to that of PTFE itself. It means that the modified layer given by our laser processing had a strong intimacy with PTFE itself and metal layer.

With use of a resolution pattern mask, we examined preliminarily the resolution of a metallized pattern given by our process. Figure 10 (b) shows a picture of the metallized PTFE analyzed with a scanning electron microscope. The line and space of 100 µm was produced as the highest resolution. Both the laser irradiation and plating processes should be related to the factors determining resolution. In the present work only the single lens was used for projection of mask. By further precise treatments using a superior optical set-up and an optimal control of plating, we can expect a resolution higher than 100 µm, the order of 10 µm as the best.

Copper Chemical Plating of Films. We employed four steps for the copper metallization. The first step is pre-dipping in 0.5 M aqueous H_2SO_4 for 1 min at room temperature. The second is an activation with Sn^{2+}-Pd^{2+} ion for 4 min at room temperature (AT-105, PED-104 (promoter)). The third is an acceleration to produce Pd(0) colloid for 2 min at room temperature (AL-106). The fourth is an electroless copper plating for 45 min at 60 °C (ELC-SP).

1. Photolysis of hydrazine: generation of H radical

$$N_2H_4 \longrightarrow N_2H_3 + \cdot H$$

2. Attack of F on PTFE surface by H radical

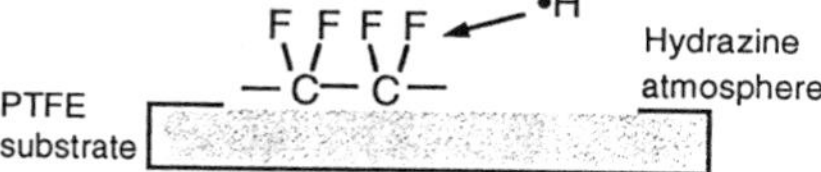

3. Abstraction of F and formation of HF

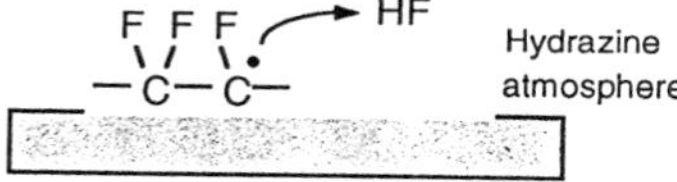

4. Formation of HF-N_2H_4 salt

Hydrazine atmosphere

$HF\text{-}N_2H_4$

5. Photolysis of HF-N_2H_4 salt: formation of reactive species

($\cdot H$, $\cdot NH_2$, $:NH$, $\cdot N_2H_3$)

$HF\text{-}N_2H_4$ $h\nu$

6. Substitution by H and NH_2

($\cdot H$, $\cdot NH_2$, $:NH$, $\cdot N_2H_3$)

F F H H F H H NH_2

Figure 8. Schematic mechanism on the PTFE surface induced by laser photolysis of hydrazine.

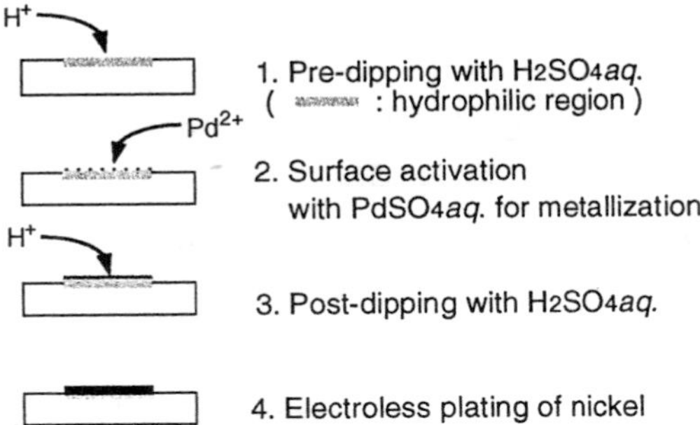

Figure 9. Schematic diagram on selective-area electroless plating of nickel metal on the laser-treated PTFE film (adapted from ref. 19).

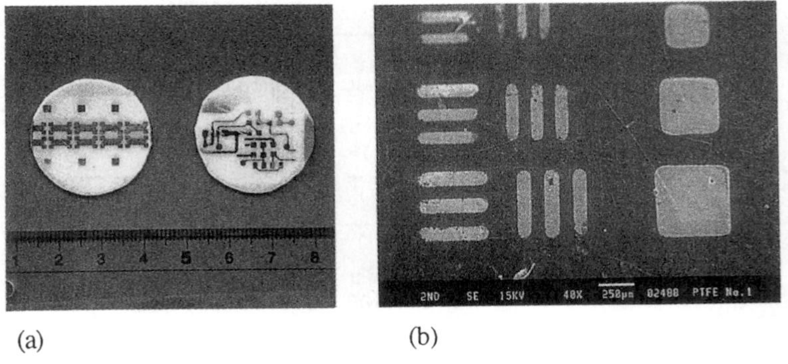

(a) (b)

Figure 10. Nickel metallization on the laser-treated PTFE film, (a) optical picture, (b) SEM picture (adapted from ref. 19).

The time required for the deposition of copper metal on PTFE film after immersion into the plating bath was dependent on the number of laser shots employed in the surface modification; when the film was irradiated with 3000, 1000, and 500 shots, it required 3, 6, and 12 s, respectively. In this process, differing from the case of nickel, the copper metal was plated on the entire PTFE surface, indicating that the activators were deposited on the entire surface by the promoter to improve the adhesion of plating metal and substrate. However, the well-defined image of copper metal appeared on the film after the adherence test with the tape, due to a clear difference between irradiated and un-irradiated regions in adhesion. The adhesive strength between the copper metal and PTFE film was estimated to be 7.8 MPa by the pull-stud test.

Conclusion

PTFE is one of the most interesting and challenging materials in industry because of its outstanding properties and difficult treatments. Previously many researches have been devoted to the surface modification and structuring of PTFE films. Our processing consists of highly effective laser photolysis of hydrazine on the PTFE surface, as the results of the abstraction of fluorine atoms by hydrogen radicals and the formation of N_2H_4-HF salt to be photolyzed effectively for the surface reaction. Because the defluorination from PTFE and the incorporation of amino groups into PTFE chain proceeded under very mild conditions upon ArF laser irradiation, the degree of surface modification gradually changed from the upper layer to the inner layer of PTFE.

On the other hand, our method constitutes only one step of laser irradiation for both surface modification and selective patterning. Moreover, the chemical modification with photolyzed hydrazine shows more reliable and stronger adhesion with a metal layer, compared with drastic chemical etching with a sodium solution. In conclusion, our processing provides a simple effective processing for fabrication of printed wiring boards with PTFE substrates.

Acknowledgment

The authors acknowledge many useful discussions and experiments on chemical plating provided by S. Matsumura and T. Murao of C. Uyemura Co., and thank Nitto Denko Co. for supplying PTFE films. We also thank H. Okano and K. Inui of Kubota Corporation for the discussion of adhesion experiments.

References

1. Rye, R. R. *J. Polym. Sci., Lett. Ed.* **1988**, *26*, 2133.
2. Rye, R. R.; Marrtinez, R. J. *J. Appl. Polym. Sci.* **1989**, *37,* 2529.
3. Soukup, L.; *Int. Polym. Sci. Technol.*, **1978**, *5(9)*, T/19.
4. Ha, K.; McClain, S.; Suib, S. L.; Garton, A. *J. Adhesion*, **1991**, *33*, 169.
5. Hall, J. R.; Westerdahl, C. A. L.; Bodnar, M. J.; Levi, D. W. *J. Appl. Polym. Sci.*, **1972**, *16,* 1465.
6. Morra, M.; Occhiello, E.; Garbassi, F. *Langmuir*, **1989**, *5*, 872.
7. Vargo, T. G.; Gardella, J. A.; Meyer, A. E.; Baier, R. E. *J. Polym. Sci. A: Polym Chem.* **1991**, *29*, 555.
8. Hook, D. J.; Varga, T. G.; Gardella, J. A.; Litwiler, K. S.; Bright, F. V. *Langmuir*, **1991**, *7*, 142.
9. Vargo, T. G.; Thompson, P. M.; Gerenser, L. J.; Valentin, R. F.; Aebischer, P.; Hook, D. J.; Gardella, J. A. *Langmuir*, **1992**, *8*, 130.
10. Rye, R. R. *Langmuir*, **1990**, *6*, 338.
11. Lappan, U.; Lunkwitz, K. *Zeit. Phys. Chem.*, **1995**, *191*, 209.

12. Bäuerle, D. *Laser Processing and Chemistry 2nd Ed.;* Springer Series in Material Sciences vol. 1; Springer: Berlin, 1996; 472-478.
13. Rabek, J. F.; Photodegradation of Polymers; Springer: Berlin, 1996; 146-160.
14. Yabe, A.; Niino, H. *Laser Ablation of Electronic Materials*; Fogarassy, E.; Lazare, S., Eds.; European Materials Research Society Monographs, vol. 4; Elsevier Science Publishers B.V.: Amsterdam, 1992; 199-212.
15. Okoshi, M.; Murahara, M.; Toyoda, K. *J. Mater. Res.* **1992**, *7*, 1912.
16. Murahara, M.; Toyoda, K. *J. Adhesion Sci. Technol.* **1995**, *9*, 1601.
17. Niino, H.; Yabe, A. *Appl. Phys. Lett.* **1993**, *63*, 3527.
18. Niino, H.; Murao, T.; Matsumura, S.; Yabe, A. *Mol. Cryst. Liq. Cryst.* **1995**, *267*, 365.
19. Niino, H.; Yabe, A. *Appl. Surf. Sci.* **1996**, *96-98*, 365.
20. Niino, H.; Okano, H.; Inui, K.; Yabe, A. *Appl. Surf. Sci.* **1997**, *109-110*, 259.
21. Heitz, J.; Niino, H.; Yabe, A. *Appl. Phys. Lett.* **1996**, *68*, 2648.
22. Heitz, J.; Niino, H.; Yabe, A. *Jpn. J. Appl. Phys.* **1996**, *35*, 4110.
23. Ichinose, N.; Maruo, M.; Kawanishi, S.; Izumi, Y.; Yamamoto, T. *Chem. Lett.* **1995**, 943.
24. Ichinose, N.; Kawnishi, S. *Macromolecules* **1996**, *29*, 4155.
25. Rye, R. R.; Howard, A. J.; Ricco, A. J. *Thin Solid Films*, **1995**, *262*, 73.
26. Perry, W. L.; Chi, K. M.; Kodas, T.; Hampden-Smith, M.; Rye, R. *Appl. Surf. Sci.* **1993**, *69*, 94.
27. Chang, C.-A.; Baglin, J. E. E.; Schrott, A. G.; Lin, K. C. *Appl. Phys. Lett.*, **1987**, *51*, 103.
28. T. A. Hohlt *ed.*, " *Static SIMS Handbook of Polymer Analysis* "; Perkin Elmer Co., Minnesota, 1991.
29. Chilkoti, A.; Ratner, B. D. *Surface Characterization of Advanced Polymers,* Sabbatini, L.; Zambonin, P. G., Eds.; VCH Verlagsgesellschaft mbH: Weinheim Germany, 1993; 221-256.
30. Reed, N. M.; Vickerman, J. C. *Surfac Characterization of Advanced Polymers,* Sabbatini, L.; Zambonin, P. G., Eds.; VCH Verlagsgesellschaft mbH: Weinheim Germany, 1993; 83-162.
31. Hawkins, W. G.; Houston, P. L. *J. Phys. Chem.* **1982**, *59*, 704.
32. Lindberg, P.; Raybone, D.; Salthouse, J. A.; Watkinson, T. M.; Whitehead, J. C. *Mol. Phys.* **1987,** *62*, 1297.
33. Stewart, J. J. P. *QCPE Bull.* **1989**, *9*, 10.
34. The salt of HF-N_2H_4 was determined by several analyses; (i) its water solution showed acidity, (ii) The FTIR spectrum was similar with that of N_2H_4, (iii) F^- ion in the water solution was detected by an indicator (Alfusom®: Wada, H.; Mori, H.; Nagasawa, G. *Anal. Chim. Acta* **1985**, *172*, 297.)
35. Girardeaux, C.; Idrissi, Y.; Pireaux, J. J.; Caudano, R.; *Appl. Surf. Sci.* **1996**, *96-98*, 586.
36. Agranoff, J. Ed. " *Modern Plastics Encyclopedia*, 1983-1984 ", McGraw-Hill, 1983; 470-501.

Chapter 5

An Inorganic Approach to Photolithography: The Photolithographic Deposition of Dielectric Metal Oxide Films

Ross H. Hill and Sharon L. Blair

Department of Chemistry, Simon Fraser University, Burnaby, British Columbia, Canada, V5A 1S6

The photochemistry of thin films of inorganic complexes has been investigated as a means to produce thin films of inorganic materials. Thin films of the precursor complexes are cast by spin coating. The ultraviolet exposure of these complexes leads to the production of metal oxides. Films of the complexes, Mn(II) 2-ethylhexanoate, Zr(IV) 2-ethylhexanoate and $Ti(acac)_2(i\text{-}prop)_2$ can all be constructed. The photolysis of these films yields metal oxide films of the following compositions, MnO, ZrO_2 and TiO_2. By using precursor films constructed of mixtures of precursors more complex metal oxide films may be produced. Using this methodology films of $PbZr_{0.5}Ti_{0.5}O_3$, $BaTiO_3$ and $YMnO_3$ were all prepared. The heating of the sample of $PbZr_{0.5}Ti_{0.5}O_3$ resulted in the formation of the ferroelectric phase of this material.

Typically lithographic processes require two steps. The first is the formation of a mask using a polymer while the second is the deposition of the material of interest. Our work has concentrated on the direct photochemical deposition of inorganic materials. This process has the advantage that it is inherently photolithographic in nature. In this paper we outline the process for the deposition of a range of ferroelectric materials.

Ferroelectric materials have numerous microelectronics applications, including capacitors (*1,2*), nonvolatile memory devices (*2-4*), electrooptic devices (*1,4*) and many others (*5*). The ferroelectrics described in this paper are $Pb(Zr_xTi_{1-x})O_3$ (PZT), $BaTiO_3$, and $YMnO_3$. Here we investigate the use of a photochemical method for the direct deposition of these complex materials.

The deposition method we will utilize involves the initial deposition of a precursor film by spin casting followed by the irradiation of this film to yield the target materials (*6,7*). In order for the process to work we need to identify precursors which both form optical quality films and undergo the required photochemistry. We previously found that metalcarboxylates (*8-9*) and metal acetylacetonates (*10*) were useful in this process. The photolysis of thin films of these precursors led to the loss of the carboxylate and acac ligands. If the photolysis was conducted in an air atmosphere, impurity-free metal oxide films were obtained. From these results, it was postulated that a variety of metal compounds containing these ligands would be useful precursors for the deposition of pure metal oxide films. If the metal complexes were stable towards each other then this method could be used to deposit mixed metal oxide films such as ferroelectric materials. This is the premise for the study described here.

Results

Spectroscopic Data for the Complexes. FTIR spectroscopic data for amorphous films of $M(O_2CC_7H_{15})_x$ ($O_2CC_7H_{15}$ = $O_2CCH(C_2H_5)C_4H_9$), M(X)= Zr(IV), Pb(II), Mn(II), Ba(II) and Y(III) are reported in Table I. The frequency difference between the asymmetric and the symmetric CO_2 stretches, ($\Delta\nu = \nu a(CO_2) - \nu s(CO_2)$) is used to distinguish the types of bonding of the carboxylate with the metal. Several types of bonding types are available for this type of complex, shown in Figure 1. In general the $\Delta\nu$ values decrease in the order unidentate > ionic > bridging > bidentate bonding of the carboxylate ligand with metals (*11*). For metal acetate complexes, $\Delta\nu$ for ionic acetates is 164 cm^{-1}. Unidentate acetates have $\Delta\nu$ values in the range 200-300 cm^{-1}. Bidentate acetates have $\Delta\nu$ values between 40 and 80 cm^{-1}, and bridging acetates have $\Delta\nu$ values of approximately 140-170cm^{-1}.

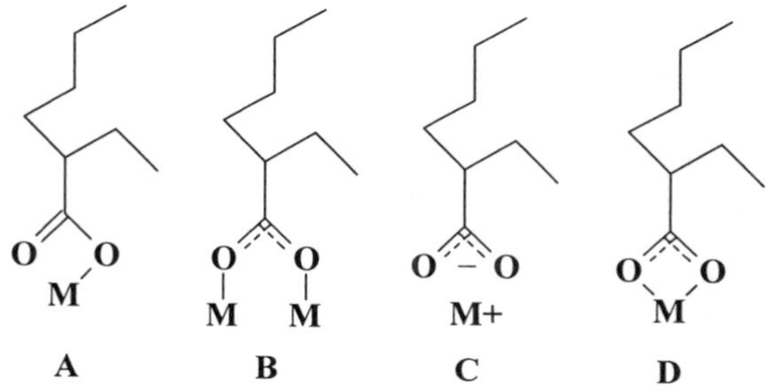

Figure 1. Bonding modes for the 2-ethylhexanoate ligand illustrating A) unidentate, B) bridging, C) ionic and D) bidentate bonding.

Table I. FTIR spectroscopic data for symmetric and antisymmetric carboxylate stretches for complexes of the formula $M(O_2CC_7H_{15})_x$.

M	Zr(IV)	Pb(II)	Mn(II)	Ba(II)	Y(III)	Assignment
	1701		1684	1687		$\nu a(CO_2)$ -unidentate
	1578	1518	1589	1543	1539	$\nu a(CO_2)$ -bridging
	1551	1518	1537sh	1543	1539	$\nu a(CO_2)$ -bidentate
	1462	1458	1462	1460	1450	$\nu s(CO_2)$ -bidentate
	1425	1414	1412	1416	1430	$\nu s(CO_2)$ -bridging
	1321	1319	1316	1290		$\nu s(CO_2)$ -unidentate

The νa(CO_2) and νs(CO_2) values used to calculate Δν for the complexes in this chapter were obtained from the data in Table I and the Δν values are shown in Table II. From this table it is evident that there are a number of different types of bonding of the carboxylate to each of the metals (refer to Figure 1). All complexes exhibit covalent bonding, with one or more of the three following types: unidentate, bridging bidentate, and bidentate. No evidence was observed for ionic bonding, where we would expect a Δν value in the range between the Δν values for unidentate and bridging types of approximately 247 cm^{-1}.

Table II. Bonding type and characteristic frequency difference values, Δν= νa(CO_2)-νs(CO_2) from the FTIR spectra of metal complexes. The data is reported in Table I.

	Zr(IV)	Pb(II)	Mn(II)	Ba(II)	Y(III)
νa-νs (cm^{-1}) -unidentate	380		368	397	
νa-νs (cm^{-1}) -bridging	153	104	177	127	109
νa-νs (cm^{-1}) -bidentate	89	60	75	81	89

The both lead (II) and yttrium (III) 2-ethylhexanoates showed a splitting of the symmetric stretching band which we attribute to the existence of both bridging and chelating carboxylates with a νa-νs of 104 and 109 cm^{-1} for the bridging and 60 and 89 cm^{-1} for the chelating. In the FTIR of the lead complex a peak consistent with the symmetric stretch of a unidentate carboxylate was observed at 1319 cm^{-1} but no corresponding antisymmetric stretch was observed leading to the supposition that that transition is due to another vibration. Based on the frequency difference values in Table II, films of all of the other metal 2-ethylhexanoate complexes consist of the unidentate, bidentate and bridging bidentate metal-carboxylate bonds. This indicates that most of the complexes exist in dimeric or polymeric form in amorphous films. The films containing a mixture of these materials likely have carboxylate groups which bridge two different metals.

The FTIR spectroscopic data for $Ti(CH_3COCHCOCH_3)_2(OCH(CH_3)_2)_2$ (defined as $Ti(acac)_2(i\text{-}prop)_2$) was also collected. The FTIR spectrum of the $Ti(acac)_2(i\text{-}prop)_2$ was assigned by analogy with the assignments of various metal acetylacetonate and metal alkoxide complexes (*11*). Absorptions at 1013 and 993 cm^{-1} are assigned to the ν(CO) bands of the isopropoxide ligands. An absorption band at 930 cm^{-1} is assigned to ν(CC) + ν(CO) of the isopropoxide. The remaining absorptions are assigned to the acac ligand. A band at 1589 cm^{-1} is assigned to ν(CC) coupled with ν(CO) and a band at 1524 cm^{-1} is assigned to ν(CO) coupled with ν(CC). At lower energy near 1380 cm^{-1} are bands associated with the methyl substituent. The acac ligand is bidentate, and the isopropoxide ligands are cis oriented (*12-13*). The spectrum was consistent with only monomeric $Ti(acac)_2(i\text{-}prop)_2$ in the film.

Photochemistry of Manganese (II) 2-Ethylhexanoate films. An amorphous film of manganese (II) 2-ethylhexanoate was prepared by spin coating a solution of the complex in CH_2Cl_2 onto a silicon substrate. The photolysis of the complex was monitored by FTIR spectroscopy. Figure 2 shows the photoinduced decay of the FTIR bands at 1684, 1589 and 1537 cm^{-1} associated with the $\nu a(CO_2)$ and 1462, 1412 and 1316 cm^{-1} associated with the $\nu s(CO_2)$ stretches. Concurrently, a broad band centered at 1550 cm^{-1} is observed to grow in intensity and then decay following further photolysis. This indicates that irradiation of manganese (II) 2-ethylhexanoate yields a thermally stable, photosensitive product, which loses all ligands upon prolonged photolysis. The photoreaction observed is consistent with equations 1 and 2 for the idealized case of a monomeric complex. The film was analyzed by Auger spectroscopy to determine the non-volatile products of photolysis.

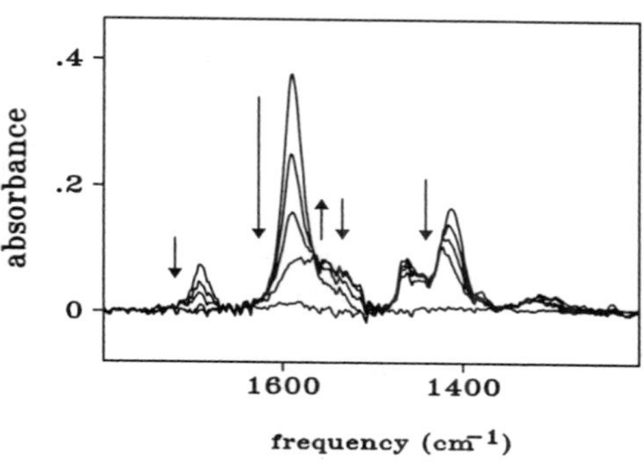

Figure 2. Spectroscopic changes associated with photolysis of Manganese (II) 2-ethylhexanoate for 0, 15, 30, 60 and 1440 min.

$$Mn(O_2CC_7H_{15})_2 \xrightarrow{h\nu} Mn(O_2CC_7H_{15}) + CO_2 + \bullet CH(C_2H_5)C_4H_9 \qquad (1)$$

$$Mn(O_2CC_7H_{15}) \xrightarrow{h\nu} Mn + CO_2 + \bullet CH(C_2H_5)C_4H_9 \qquad (2)$$

The surface of the film, cleaned by sputtering the surface with Ar^+ ions for 5 seconds, contained 45.5 (±9)% manganese and 54.5 (±11)% oxygen. No signal due to carbon was observed. This confirms that the manganese formed upon ligand loss combines with oxygen. The composition of the film is within error of both MnO and Mn_2O_3, stable oxides of manganese (*14*). The results of Auger analysis are summarized in Table III.

Photochemistry of Zirconium (IV) 2-Ethylhexanoate Films. Photolysis of a film of Zr(IV) 2-ethylhexanoate resulted in the loss of all FTIR absorptions associated with the complex. No new absorptions were observed throughout the experiment. However, the rate of decay of the absorption bands at 1578 and 1551 cm^{-1} decayed at a slower rate than the bands at 1462 and 1425 cm^{-1}. This indicates that a thermally stable primary photoproduct has absorptions coincident with the starting material. Loss of all IR absorptions was observed on prolonged photolysis. These results are consistent with equations 3 and 4 for the idealized case of a monomeric molecule.

$$Zr(O_2CC_7H_{15})_4 \xrightarrow{h\nu} Zr(O_2CC_7H_{15})_3 + CO_2 + \bullet CH(C_2H_5)C_4H_9 \qquad (3)$$

$$Zr(O_2CC_7H_{15})_3 \xrightarrow{h\nu} Zr + 4CO_2 + 4\bullet CH(C_2H_5)C_4H_9 \qquad (4)$$

Table III. Auger analysis of films resultant from photolysis of metal 2-ethylhexanoates (M= Zr, Pb, Mn) and $Ti(acac)_2(i\text{-}prop)_2$ with 254 nm light.

Precursor	sputter time(sec)	% M	% O	% C	product (calcd)
$Mn(O_2CC_7H_{15})_2$	0	35.8 (±7)	47.7 (±10)	16.5 (±3)	Mn_3O_4
	5	45.5 (±9)	54.5 (±11)	--	Mn_5O_6
	10	45.5 (±9)	54.5 (±11)	--	Mn_5O_6
$Zr(O_2CC_7H_{15})_4$	0	21.2 (±4)	67.8 (±14)	11.0 (±2)	Zr_5O_{16}
	5	27.4 (±5)	72.6 (±15)	--	Zr_3O_8
	10	26.5 (±5)	73.5 (±15)	--	Zr_4O_{11}
	20	28.5 (±6)	71.5 (±14)	--	Zr_2O_5
$Ti(acac)_2(i\text{-}prop)_2$	0	23.7 (±5)	76.3 (±15)	--	Ti_5O_{16}
	10	28.8 (±6)	71.2 (±14)	--	Ti_2O_3
	15	29.9 (±6)	70.1 (±14)	--	Ti_3O_7

The results of Auger analysis of a film produced from photolysis of zirconium (IV) 2-ethylhexanoate in air are summarized in Table III. The surface of the film contained 11(±2) % carbon, originating presumably from the atmosphere. After 20 sec of sputtering the surface contained 28.5(±6) % Zr and 71.5(±14) % O. The composition of the film was within error of ZrO_2, a known stable form of zirconium oxide (*14*).

Photochemistry of Titanium Bis-acetylacetonate Di-isopropoxide Films. The photolysis of an amorphous film of $Ti(acac)_2(i\text{-}prop)_2$ was monitored by FTIR spectroscopy. Loss of all IR absorptions was observed on photolysis. No evidence consistent with the formation of an intermediate was observed throughout the reaction. The photoreaction is shown in equation 5.

$$Ti(CH_3COCHCOCH_3)_2(OCH(CH_3)_2)_2 \xrightarrow{h\nu} Ti + 2\bullet CH_3COCHCOCH_3 + \bullet OCH(CH_3)_2 \quad (5)$$

The film resultant from photolysis was analyzed by Auger spectroscopy. Following a 10 sec sputter, the film contained 28.8(±6)% Ti and 71.2(±14)% O. No signal due to carbon was observed throughout the film. The composition of this film was TiO_2, within error. The results of Auger analysis are summarized in Table III.

Preparation of Various Mixed Metal Oxides. In order to deposit the mixed metal oxide, $Pb(Zr_{0.5}Ti_{0.5})O_3$ a 2:1:1 stoichiometric solution of the lead, zirconium and titanium precursors was prepared. An amorphous film of these precursors was prepared by spin coating on silicon.

Figure 3 illustrates the loss of intensity of all absorptions due to the ligands in the FTIR spectra of a) $Pb(O_2CC_7H_{15})_2$, b) $Zr(O_2CC_7H_{15})_4$, c) $Ti(acac)_2(i\text{-}prop)_2$ and d) 2: 1: 1 Pb: Zr: Ti precursor mixture films following photolysis. It is apparent that when the precursors are mixed, the same spectral changes occur on photolysis as that of the separate precursors (Fig. 3 a), b), c)). This is based on the observation that all bands due to the starting materials decrease to zero intensity, with subsequent production of a broad band around 1400 cm^{-1} coincident with the product of the lead precursor. No evidence for a unique intermediate is observed for the process.

It should be noted that the efficiency of decomposition of the multicomponent precursor solutions appears to be different from that of single component systems. In particular, the efficiency of decomposition of the precursor mixtures used to prepare $PbZrO_3$ and $Pb(Zr_{0.5}Ti_{0.5})O_3$, is greater than the efficiency of decomposition of the Zr precursor alone. For example, the half-life for photolysis of the lead, zirconium and titanium precursors was 38.5, 73.3 and 47.5 min., respectively. The half-life for the combined precursors, for the deposition of $Pb(Zr_{0.5}Ti_{0.5})O_3$, was 44.3 min. under the same conditions. Possible reasons for this behavior will be presented in the discussion.

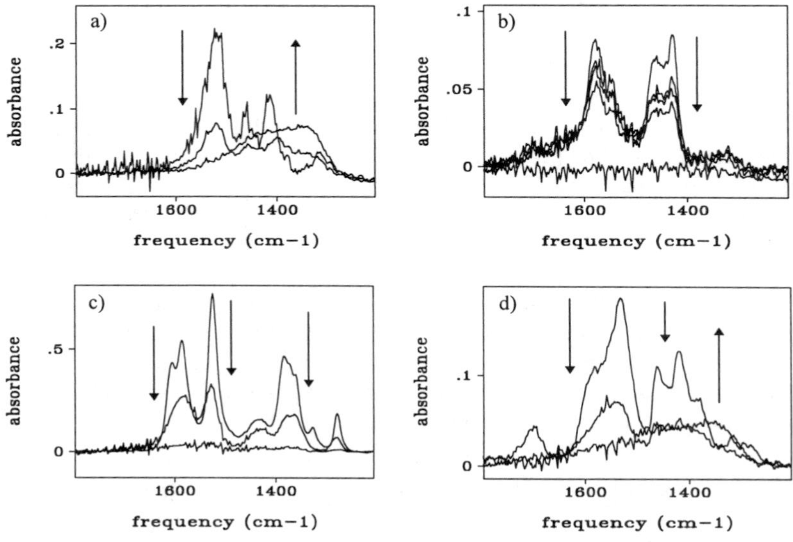

Figure 3. Spectroscopic changes associated with photolysis with 254 nm light of films of the a) Pb precursor for 0, 1 and 20 hr; b) Zr precursor for 0 min., 10 min., 190 min., 1 day, 1 week; c) Ti precursor for 0, 1, 20 hr d) 2:1:1 Pb:Zr:Ti precursors for 0 min., 20 min., 20 hr.

Following the photolysis the surface of the film was analyzed by Auger electron spectroscopy. No carbon was found in the film indicating that all ligands were ejected from the film during photolysis. The composition of the film calculated from the Auger spectrum (*15,16*) was $Pb_{1.1}(Zr_{0.48}Ti_{0.52})O_{2.9}$. This composition was within the range expected from the initial precursor mixture. The results are reported in Table IV.

Table IV. Auger analysis[a] of films resultant from photolysis of films composed of metal 2-ethylhexanoates (M= Zr(IV), Mn(II), Pb(II), Ba(II), Y(III)) and Ti(acac)$_2$(i-prop)$_2$ with 254 nm light and the dose to print.

Precursor M,M',M'' (ratio)	sputter time(sec)	%M	%M'	%M''	%O	%C	composition (calcd)	dose (J/cm^{-2})
Ba, Ti	0	24	14		52	10	$Ba_5Ti_3O_{11}$	7.8
(1:1)	10	23	18		59	--	$BaTiO_3$	
Y, Mn	0	7	17		46	30	$YMn_{2.4}O_{6.6}$	2.6
(1:1)	20	23	22		45	10	$YMnO_2$	
Pb, Zr	0	18	14		35	33	$Pb_{1.3}ZrO_{2.5}$	1.3
(1:1)	20	23	20		57	--	$Pb_{1.2}ZrO_{2.9}$	
Pb, Ti	0	17		10	56	17	$Pb_{1.7}TiO_{5.6}$	2.6
(1:1)	20	14		20	66	--	$Pb_{0.7}TiO_{3.4}$	
Pb, Zr, Ti (2:1:1)	20	23	10	10	57	--	$Pb_{1.1}Zr_{0.5}Ti_{0.5}O_{2.9}$	2.6
Pb, Zr, Ti	0	26	6.3	6.7	55	6	$Pb_{2.0}Zr_{.48}Ti_{.52}O_2$	
(2:1:1)[b]	20	13	6.1	6.7	74	--	$Pb_{1.0}Zr_{.48}Ti_{.52}O_{5.8}$	

[a] error estimated at approximately 20 %, the sensitivity factor of 0.32 for Ba was obtained from a standard BaO film, the sensitivity factor for Pb was obtained from reference 16 and all others from reference 15.
[b]annealed for 30 min. at 650°C in an air atmosphere.

Thermal annealing experiments were performed on the photoproduced amorphous film. To determine whether the perovskite $Pb(Zr_{0.5}Ti_{0.5})O_3$ phase could be formed, a film on Pt(111) after photolysis was annealed according to literature procedures, at 650°C for 30 minutes (2). The film was then analyzed by Auger spectroscopy. On the surface of this film, some carbon contamination was observed. Sputter cleaning the surface with argon ions revealed a carbon-free surface with a composition of $Pb_{1.0}(Zr_{0.48}Ti_{0.52})O_{5.8}$. The surface was lead rich, consistent with the diffusion of lead to the surface during the annealing process (*17*). This film was analyzed by grazing incidence X-ray powder diffraction.

The polycrystalline film consists of the tetragonal phase of PZT with peaks arising at approximately $2\theta = 22°, 31°, 39°, 45°$, and $55°$. These peaks correspond to the (100), (110), (111), (200), and (211) orientations, respectively. A broad peak centered around $2\theta = 30°$ was also evident. This may correspond to the metastable pyrochlore phase. The pyrochlore phase is lead deficient, such that Pb, labeled x, in $Pb_x(Zr_{0.48}Ti_{0.52})O_3$, is less than one, consistent with the loss of lead oxide during the anneal at 650°C. Lead oxide is known to evaporate from films during the thermal annealing conditions described here (*17,18*).

The tetragonal form has a unit cell that c >(a = b). The lattice parameters were calculated from the 2θ values. The 2θ values at 43.55 and 44.9, corresponding to the (002) and (200) (hkl) orientations respectively, were used in the calculation. The value for c was calculated as 4.152, and a= 4.033. The c/a ratio is 1.030, within error of the literature value for $Pb(Zr_{0.52}Ti_{0.48})O_3$ of 1.027 (*19,20*).

The other mixed metal oxides were prepared in the same way. Stoichiometric solutions of $Ba(O_2CC_7H_{15})_2$ and $Ti(acac)_2(i\text{-}prop)_2$ were used to spin coat the amorphous precursor film for the deposition of $BaTiO_3$. The stoichiometric solutions used for the preparation of $PbTiO_3$ and $PbZrO_3$ were made with 1:1 mole ratios of Pb:Ti and Pb:Zr respectively. The preparation of $YMnO_3$ was done from films cast from a stoichiometric mixture of $Y(O_2CC_7H_{15})_3$ and $Mn(O_2CC_7H_{15})_2$. In each case the surface resultant from photolysis was analyzed by Auger spectroscopy and the results are summarized in Table IV.

Lithography. Photolithography experiments were conducted with amorphous films of the compounds to determine minimum dose requirements to leave an insoluble pattern on the surface. A series of amorphous films of manganese (II) 2-ethylhexanoate on silicon was made by spin coating. Each film was then partially covered and the sample was irradiated with 254 nm light. The samples were exposed to doses of 0.044, 0.26, 1.3, 2.6, 7.9, and 32 J/cm^2. The samples were then developed by dipping in ethanol. Both the exposed and the unexposed portions of the film dissolved for doses of less than 2.6 J/cm^2. For films exposed to 2.6 J/cm^2 or more only the unexposed regions dissolved. This indicates that films of this complex required 2.6 J/cm^2 dose of 254 nm light to leave a pattern on the surface. It should be noted that this dosage is an upper limit for the minimum dose required. The films used for this experiment were thin films such that the light absorption by the film was minimal. Under these conditions the dose required was independent of film thicknes.

Dose experiments were conducted for each complex. The titanium precursor required a dose of 7.9 J/cm^2 and the zirconium precursor a 32 J/cm^2 dose. The doses for the mixed systems were less than required for the pure precursors. The required dose for the 2:1:1 Pb: Zr: Ti and the 1:1 Pb: Ti mixture was the same (2.6 J/cm^2) as that for the titanium precursor alone and 60 times slower than the lead precursor alone. However, the required dose for patterning the 1:1 Pb: Zr film was 1.3 J/cm^2, approximately 26 times lower a dose than when zirconium is patterned alone. This presumably is a result of radical chemistry but may also reflect solubility effects.

In order to determine the compatibility of our process with current lithography techniques, an attempt was made to pattern PZT on silicon. For this experiment a thin film of a stoichiometric mixture of the 2:1:1 ratio of precursors as earlier was deposited on a Si(100) substrate. The film was irradiated through a mask for 90 min. and, after rinsing the unexposed starting material with ethanol, 2 μm wide lines are visible (see Figure 4).

Film Morphology. A brief study of the film quality available from this method was made. The uniformity of the precursor films cast by spin coating was measured by ellipsometry. Over a 4" silicon wafer coated with $Zr(O_2CC_7H_{15})_4$ using a spin-speed of 2000 rpm the thickness was 3668 and the RMS variation was 38 Å. By using faster spin speed or lower concentration solutions thinner films were easily cast.

The surface of a film of amorphous ZrO_2 produced by the room temperature photolysis of a $Zr(O_2CC_7H_{15})_4$ film was imaged by atomic force microscopy. The

RMS roughness of the film was 5.7 Å and 90% of the surface height fell within a 20 Å range over a square micron area.

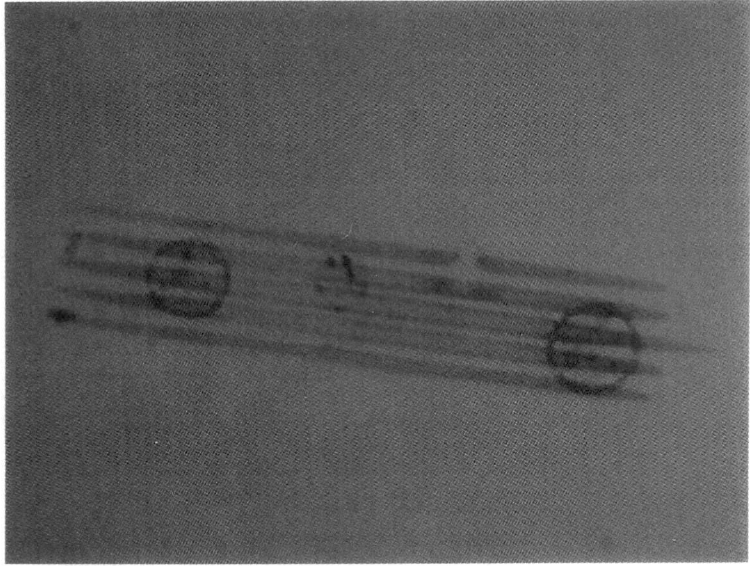

Figure 4. Optical micrograph of lithographed amorphous PZT. The scale is given internally by the 50μm length of the four shorter lines.

Discussion

All of the complexes studied here could be cast as optical quality films by spin coating. The films appeared amorphous optically and no evidence for microcystallinity was observed. The films are probably not simple complexes and in many cases evidence for multiple bonding modes of the complexes were obtained. These may be due in part to polymeric structures such as shown in Figure 5.

Amorphous films of metal 2-ethylhexanoate complexes underwent primary photochemical reactions according to equation 6. The photon results in a ligand to metal charge transfer transition. This results in the weakening of the metal oxygen bonds. As a result the 2-ethylhexanoate ligand is ejected as a radical and decarboxylates producing the products CO_2 and a heptyl radical. The radical can either combine with another heptyl radical, forming $C_{14}H_{30}$, or abstract a hydrogen from another radical forming 2-heptene or 3-heptene and heptane (equation 7). These products have been observed previously on photolysis of thin films of copper (II) 2-ethylhexanoate (*8*) and uranyl (VI) 2-ethylhexanoate (*10*).

$$M(O_2CC_7H_{15})_x \xrightarrow{h\nu} \bullet M(O_2CC_7H_{15})_{x-1} + CO_2 + \bullet CH(C_2H_5)C_4H_9 \quad (6)$$

$$M(X) = Zr(IV), Pb(II), Mn(II)$$

$$6 \bullet CH(C_2H_5)C_4H_9 \longrightarrow C_{14}H_{30} + C_2H_5CHCHC_3H_7 + CH_3CHCHC_4H_9 + 2C_7H_{16} \quad (7)$$

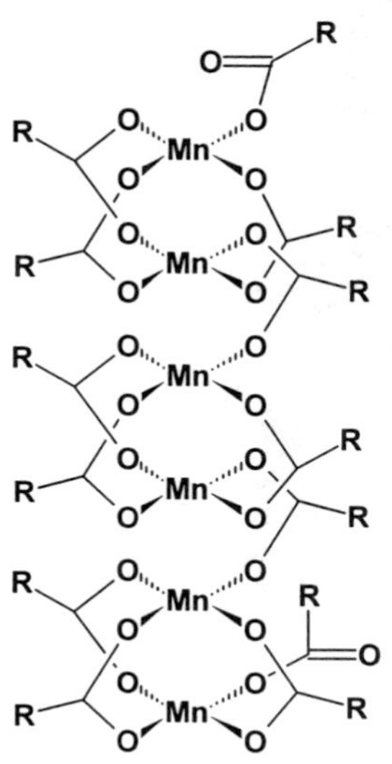

Figure 5. Representation of the polymeric structure of the film.

Subsequent reactions of the metal 2-ethylhexanoate fragments were either thermal or photochemical depending on the metal. For metals which have stable oxidation states other than the initial state such as Mn(II) and Zr(IV) thermally stable intermediates are formed of Mn(I) and Zr(III). In contrast no intermediates were observed for Pb(II) or Cr(III) studied previously (*9*). Alternatively the intermediate may be a result of polymeric form of the material. Upon further photolysis, the polymeric form breaks down and decomposes (equation 8).

Photolysis of $Ti(acac)_2(i\text{-}prop)_2$ resulted in loss of all ligands as indicated by the loss of all FTIR bands associated with the complex. Presumably, the primary photoreaction is loss of an acetylacetonate ligand (acac) as a radical (equation 9). This product is thermally unstable, and rapidly decomposes to yield titanium (equation 10). The ejected acac and isopropoxyl radicals may react, resulting in the abstraction of a hydrogen from an isopropoxyl radical, yielding acacH and acetone (equation 11). Hydrogen abstraction by an ejected acac radical has been observed previously, following photolysis of an amorphous film of uranyl acetylacetonate hydrate (*10*).

$$(RCO_2)x[\bullet M(O_2CR)y]n(O_2CR)x \xrightarrow{h\nu} nM + (ny+2x)CO_2 + (ny+2x)\bullet R \quad (8)$$

$$M(X) = Mn(II), Zr(IV);\ R = CH(C_2H_5)C_4H_9$$

$$Ti(CH_3COCHCOCH_3)_2(OCH(CH_3)_2)_2 \xrightarrow{h\nu} \bullet Ti(CH_3COCHCOCH_3)(OCH(CH_3)_2)_2 + \bullet CH_3COCHCOCH_3 \quad (9)$$

$$\bullet Ti(CH_3COCHCOCH_3)(CH(CH_3)_2)_2 \xrightarrow{\Delta} Ti + \bullet CH_3COCHCOCH_3 + \bullet OCH(CH_3)_2 \quad (10)$$

$$\bullet CH_3COCHCOCH_3 + \bullet OCH(CH_3)_2 \longrightarrow CH_3COCH_2COCH_3 + OC(CH_3)_2 \quad (11)$$

Photolithography experiments demonstrated that direct patterning of metal oxides from amorphous films of inorganic precursors can be achieved. Minimum dose measurements indicate that the selective photolysis of lead(II) 2-ethylhexanoate films require 44 mJ/cm^2 (*9*). This is comparable to dosage requirements for photoresists. A typical dose for patterning of a diazonaphthaquinone (DNQ)/novolak resist system is approximately 55 mJ/cm^2 (*21*). It should be noted that the patterns made during dose experiments are a result of solubility change. It does not necessarily infer reaction completion. The most important factor for us to minimize is the dose because the equipment used in industry for photolithography has a very high capital cost. The resultant patterns can later be converted to pure material following lithography by either nonselective irradiation or heat, depending on the application.

The precursors presented here, with the exception of lead, require 2.6 to 32 J/cm^2, significantly more than current photoresists. It should be noted that, although these are the required doses given the conditions of the experiment, some conditions may be altered to reduce the required dose. One obvious improvement could be made by optimization of the developing solvent. In our protocol we use a solvent in which the precursor is very soluble and the product is insoluble. By utilizing a solvent which the precursor is less soluble in a differential solubility may be observed at a smaller dose.

Monitoring of both the photochemistry and the dose requirements indicated that the mixed systems such as PZT may react more efficiently than at least some of the individual components react in pure films. This may be due to the effect of the mixture on the structure of the films or due to a radical chain component to the reaction. If a radical chain component is present then radicals photogenerated by a precursor such as lead which reacts efficiently may serve to initiate the reaction of less photosensitive systems such as the zirconium precursor. This result is consistent with the higher dose requirement for the lead in the presence of the other components. These results indicate that a mixture of precursors may improve the efficiency of photodecomposition. Further enhancements may be available by using additives which are photoradical generators, however, these may also alter the purity of the products obtained.

Photolithography of a 2: 1: 1 Pb: Zr: Ti precursor mixture resulted in the formation of 2 x 50 μm lines. The quality of the lithography indicates that the precursor film is sufficiently mobile to heal any defects which form during the photolysis. This also results in an extremely flat deposited surface. AFM of the surface failed to reveal any microstructure.

Annealing of the amorphous material results in the formation of the ferroelectric form of PZT. This demonstrates that the amorphous films which are deposited may be annealed to produce the active crystalline materials. Using rapid thermal annealing it should be possible to generate oriented materials by this process (*2*).

Conclusions

In this study , it has been demonstrated that elementally pure amorphous metal oxide thin films of Zr, Ti, and Mn may be prepared photochemically from thin film precursors at room temperature by our method. By using these and other compounds which are compatible were able to prepare amorphous mixed metal oxides. In the case of thin films of $Pb(Zr_xTi_{1-x})O_3$ thermal annealing studies indicated that the polycrystalline tetragonal phase of $Pb(Zr_{0.48}Ti_{0.52})O_3$ could be produced from the amorphous films.

Optical lithography of 2 x 50 μm lines of PZT demonstrated the compatibility of this process with current lithography technology. The drawback of this approach is the high dose requirement for the systems under study. The photochemical studies indicate an increase in reaction efficiency in the presence of some species. This result indicates that it may be possible to significantly improve the photosensitivity of these films. The developing chemistry of these films is also in its infancy and optimization in this area may also result in an further lowering of the doses required for printing these materials.

Experimental details

Instruments and Materials. The silicon wafers were obtained from Wafernet Incorporated. Si (100) surfaces were used in these studies and the wafers were p-type silicon with tolerances and specifications as per SEMI Standard M1.1.STD.5 cut to the approximate dimensions 10 x 12 mm as needed. The substrates used for PZT deposition were Si(111) wafers coated with 1000 Å of titanium and 2000 Å of Pt(111). These wafers were prepared and donated by Northern Telecom Canada.

The Fourier transform infrared (FTIR) spectra were obtained with 4 cm^{-1} resolution using a Bomem Michelson 120 FTIR spectrophotometer. FTIR spectra of the thin films on silicon were obtained in air with Si (100) used as a reference.

The photolysis beam was the output from a UVP Inc. model UVG-54 short wave UV-254 nm lamp. All photolysis experiments were performed in air. Light intensity measurements were made using an International Light IL 1350 radiometer.

Auger spectra were obtained using a PHI double pass CMA at 0.85 eV resolution at the Surface Physics Laboratory, Department of Physics, Simon Fraser University. The chromium plated quartz lithography mask was kindly donated by IBM Corp. The light source used for lithography experiments was an Osram 150 W high pressure Xe lamp. Grazing incidence X-ray diffraction patterns were obtained using a Siemens D5000 X-ray diffractometer. The X-ray source was a Cu Kα (1.54 Å) beam.

Preparation of Amorphous Precursor Films. A solution of $Mn(O_2CC_7H_{15})_2$ in CH_2Cl_2 was spin coated onto a Si chip. The solvent evaporated leaving an amorphous film of $Mn(O_2CC_7H_{15})_2$. Other films were prepared in an analogous way.

For multicomponent mixtures a similar procedure was followed. A 2:1:1 mole ratio precursor mixture of lead: zirconium: titanium as prepared as the precursor mixture for $Pb(Zr_{0.5}Ti_{0.5})O_3$. A 0.442 g sample of lead (II) 2-ethylhexanoate (0.89

mmol Pb) was dissolved in CH_2Cl_2. To this solution, 0.217 g (0.45 mmol Ti) of a 75% solution of $Ti(acac)_2(i\text{-}prop)_2$ in isopropanol was added. Then 0.677 g (0.45 mmol Zr) of Zr(IV) 2-ethylhexanoate in mineral spirits (6% Zr).

Photolysis of the Complexes as Thin Films on Silicon Surfaces. The Si (100) surface was used as the substrate in all experiments. A typical experiment will be described. An FTIR spectrum of silicon was obtained and used as a reference throughout the experiment. A thin, amorphous film of $Mn(O_2CC_7H_{15})_2$ was deposited on silicon as described in the previous section. An FTIR spectrum was then obtained (see Figure 1). The sample was placed in the irradiation beam, exposed for 15 min and another FTIR spectrum was obtained. The extent of reaction was monitored by the changes in intensity of the FTIR bands associated with the complex. This procedure was repeated for accumulated photolysis times of 30, 60 min and 24 hrs at which time the FTIR absorptions due to the ligands in the precursor complex decreased to zero intensity. When the intensity of the bands were at the baseline, the sample was then moved to the scanning Auger microprobe for further analysis. Auger spectra were obtained to determine if all the ligands were extruded from the film and to determine the composition of the films. The results are summarized in Tables III and IV.

Lithography. Lithography experiments were performed as follows. The mask was placed in contact with a thin amorphous film of a precursor. The film was irradiated through the mask, and after photolysis, the film was rinsed with ethanol to remove unexposed starting material.

Dosage experiments were conducted in the following manner. Amorphous films of the precursors on Si(100) were made by spin-coating. The precursor films were partially masked with blank silicon chips and irradiated with a 254 nm mercury lamp. The light intensity measured at 4.4 mW/cm^2. After 10 min, the light was removed and the films were rinsed with ethanol. Dosages of 10 sec, 1 min, 5 min, 10 min, 30 min, 2 hrs and 24 hrs were also attempted. The required dosage for a complex is the amount of energy required to produce a pattern that remains on the substrate after rinsing with solvent.

References

1. Tominaga, K.; Sakashita, Y,; Nakashima, H.; Okada, M. J. Cryst. Growth 1994, 145, 219.
2. Hwang, Y. S.; Paek, S. H.; Park, C. S.; Mah, J. P.; Choi, J. S.; Jung, J. K.; Kim, Y. N. J. Mat. Sci. Lett. 1995, 14, 1168.
3. Scott, J. F.; Paz de Araujo, C. A. Science 1989, 246, 1400.
4. Wolf, S., Silicon Processing for the VLSI Era, Vol. 2, Sunset Beach, CA: Lattice Press 1986.
5. Xu, Y., Ferroelectric Materials and Their Applications, Amsterdam: North-Holland 1991.
6. Hill, R. H.; Avey, A. A.; Blair, S. L.; Gao, M.; Palmer, B. J. ICEM-IUMRS '94 Symp. Proc. 1994, 1, 435.
7. Chu, C. W.; Hill, R. H. IUMRS-ICEM'94 Symp. Proc. 1994, 1, 441.

8. Avey, A. A.; Hill, R. H. J. Am. Chem. Soc. 1996, 118, 237.
9. Blair, S. L.; Chu, C. W.; Dammel, R. R.; Hill, R. H. Advances in Resist Technology and Processing XIV, Proceedings of SPIE 1997, 3049, 829.
10. Gao, M.; Hill, R. H. J. Photochem. Photobiol. A: Chem. 1996, 97, 73.
11. Nakamoto, K., Infrared and Raman Spectra of Inorganic and Organometallic Compounds, 4th ed., New York: John Wiley & Sons 1986.
12. Bradley, D. C.; Holloway, C. E. Chem. Commun. 1965, 284.
13. Bradley, D. C.; Holloway, C. E. J. Chem. Soc. A 1965, 282.
14. Kofstad, P., Nonstoichiometry, Diffusion and Electrical Conductivity in Binary Metal Oxides, New York: Wiley-Interscience, 1972.
15. Davis, L. E.; MacDonald, N. C.; Palmberg, P. W.; Riach, G. E.; Weber, R. E., Handbook of Auger Electron Spectroscopy: A reference book of Standard Data for Identification and Interpretation of Auger Electron Spectroscopy Data, 2nd ed., Eden Prairie, MN: Physical Electronics Division Perkin-Elmer Corp. 1976.
16. Yoon, S. G.; Kim, H. G. IEEE Trans. Ultrasonics, Ferroelectrics and Frequency Control 1990, 37, 333.
17. Dana, S. S.; Etzold, K. F.; Clabes, J. J. Appl. Phys. 1991, 69, 4398.
18. Choi, J.-H.; Kim, H.-G. J. Appl. Phys. 1993, 74, 6413.
19. Kakegawa, K.; Mohri, J.; Takahashi, T., Yamamura, H.; Shirasaki, S. Solid State Commun. 1977, 24, 769.
20. JCPDS-ICDD, file number 33-784.
21. Dammel, R., Diazonaphthaquinone-based Resists, Volume TT 11, SPIE-The International Society for Optical Engineering, Washington, USA 1993, pp. 11.

Chapter 6

Molding of Polymeric Microstructures

T. Hanemann, V. Piotter, R. Ruprecht, and J. H. Hausselt

Forschungszentrum Karlsruhe GmbH, Institut für Materialforschung III, Postfach 3640, 76021 Karlsruhe, Germany

Improved molding techniques allow for an economic mass production of micro components with lateral dimensions in the micrometer, structural details in the submicrometer range and high aspect ratios of up to 500. The capabilities of micro injection molding as well as examples of microstructures made from various polymers and related applications like pump housings, cell containers etc., will be demonstrated. Photomolding of UV curable resins as a new reaction injection molding process and related results dealing with optimized curing times and molding properties of different resin compositions will be discussed.

The market success of microcomponents made from various metals, ceramics, polymers or composite materials strongly depends on the development of an economic process technology. For low cost fabrication of macroscopic technical plastic products, different molding techniques, especially injection molding, have been established. Hence, the attempt to adapt this reliable technology for the replication of microcomponents was obvious. Actually the LIGA process developed at the *Forschungszentrum Karlsruhe (FZK),* Germany, (LIGA is the german abbreviation of lithography with synchrotron radiation, electroplating and molding) is one of the most important technologies for the fabrication of microstructures with lateral dimensions in the micrometer, structural details in the submicrometer range, high aspect ratios (height to width ratio) of up to several hundreds and a final average surface roughness of less than 50 nm (*1,2*). The LIGA technique allows the fabrication of mold inserts which contain the inverted surface topology of the desired microcomponents. Alternatively, the mold inserts can be manufactured by various microstructuring techniques like mechanical ultraprecision milling, laser ablation (Laser LIGA), reactive ion etching, and lithography by means of light or particle beams. They are suitable in all relevant micromolding procedures like Thermoplastic Injection Molding (TIM), hot embossing as well as in Reaction Injection Molding (RIM).

In the following discussion, different injection molding techniques and some molded microstructures will be presented.

Thermoplastic Micro Injection Molding

Suitable and reliable molding techniques are basic preconditions for the economic production of microsystem components in large scale series. In contrast to standard molding, micromolding requires modifications of the molding machinery, the molding tool's construction as well as the adaptation of the molding parameters to the small geometrical dimensions and the large flow length to wall thickness ratio (≡ aspect ratio).

Injection molding of LIGA microstructures has been developed at *FZK* since 1986. Basic differences to conventional injection molding machines are the higher precision and the integration of peripheral equipment specific to microtechnology. The microstructures in the LIGA or micromechanically fabricated mold inserts correspond to blind holes. As a consequence the whole mold cavity has to be evacuated prior to the material's injection. Hence the machine periphery comprises a vacuum unit for evacuating the mold's cavity and temperature control units for each half of the molding tool. At *FZK* a full automatically driven two component Ferromatik Milacron K50S2F injection molding machine allows small and middle scale plastic microstructure fabrication (Figure 1). In cooperation with a machine vendor an all-electric injection molding machine has been adapted for the requirements of micromolding. In actuality, small and micro components made from various polymer materials with the following specifications can be molded:

- plate-shaped microparts with microstructures of any lateral form
- volume of the standard substrate base plate: 20 x 60 x 2 mm^3 (width x length x height)
- microstructure height up to 1.6 mm
- smallest wall thickness down to 20 μm
- smallest structural detail 0.2 μm
- aspect ratio up to 30 : 1
- suitable materials: PMMA, PC, PE, PSU, POM, PA12, PEEK, etc.

A typical micromolding cycle starts with closing of the molding tool, evacuation of the cavities and heating up to the injection temperature. At injection temperature, the plastic melt is being injected into the mold. After complete filling the molding tool has to be cooled down to the demolding temperature followed by demolding, while at the same time new plastic material is conveyed and melted. The most critical steps are the molding and the demolding procedures. Thermal shrinkage during the cooling process has to be compensated by a corresponding holding pressure. A reliable demolding without any cracks or flaws is supported by plane parallel structural walls and the excellent surface quality of LIGA microstructures with an average roughness of less than 50 nm. Due to the time consuming heating and cooling steps the cycle times in all micromolding techniques are considerably longer than those for conventional macroscopic molding techniques. As a direct consequence, the reduction of the cycle time is one of the key issues for economic success of microstructured products. In case of micro injection molding, the shortest cycle times reported are 70s using POM and a LIGA mold insert containing microstructures with an aspect ratio of 2.5 (*3*). Using materials with higher viscosities and microstructures with larger aspect ratios the cycle times increase consequently. As an example, injection molding of high performance thermoplastics such as PEEK results in a cycle time of about 2 minutes using micromechanically made mold inserts carrying microstructures with an aspect ratio of up to 5. In case of higher aspect ratios, cycle times often amount up to ten minutes. Figure 2 shows the reduction of the cycle time achieved at *FZK* since October 1994 for LIGA microstructures made from PMMA with aspect ratios of up to 20. The installation of a modified dual cycle temperature equalization followed by the optimization of the molding parameters has resulted in a total cycle time reduction of more than 60 % (*4*). Latest experiments yield cycle times for automated manufacturing

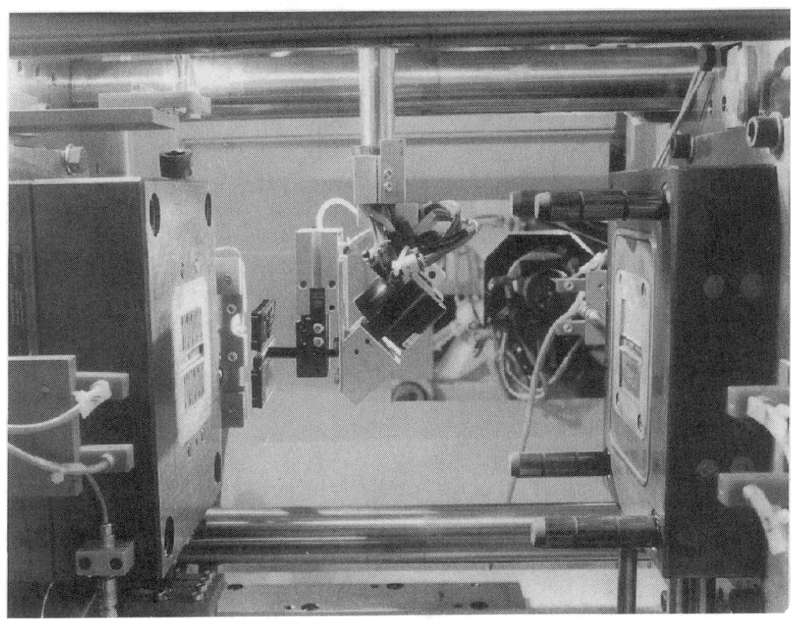

Figure 1. Full automatical driven two component injection molding machine.

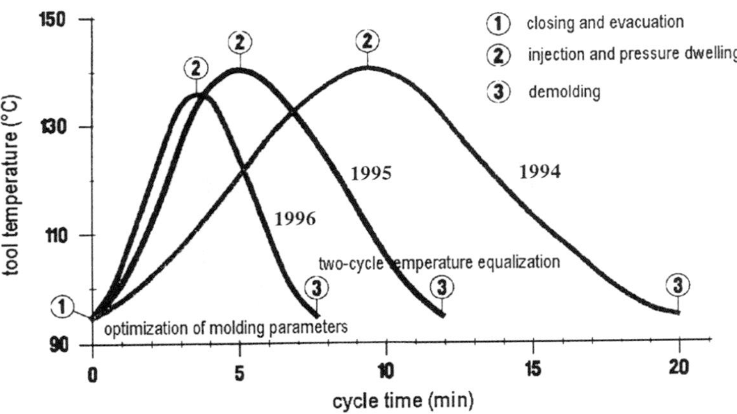

Figure 2. Reduction of the injection molding cycle time since 1994.

using PMMA of less than 5 minutes if the molding tool's geometry, the temperization system, and the molding parameters are being further optimized.

Reaction Injection Molding (RIM). Thermally initiated reaction injection molding is another advanced technique for the fabrication of microstructures. In addition to the described technique, basically only a mixing and metering unit for the cast resins as well as a temperature control for an accurate polymerization temperature has to be included. Due to the low viscosity of the cast resins, a relatively simple and reliable mold filling and excellent molding accuracies are obtained. Thermally curable resins based on acrylates, methacrylates, amides and silicones are practical (*5*). To support demolding, internal mold release agents have to be added which influence the thermomechanical properties of the polymers.

Applications

The molding process represents the link between product development and a middle or large scale production. As mentioned earlier the injection molding of polymer microstructures is determined by the microstructure's aspect ratio as well as by the desired field of application and the resulting material requirements which determine the selection of the polymer. In the following some examples for successfully molded microcomponents are presented:

- Micro injection molding of microoptical benches and a LIGA microspectrometer for the NIR-range (structure height 145 µm, grating step height approx. 1 µm, resulting aspect ratio ≈ 160) using PMMA is under development towards an established molding technique (Figure 3) (*6*). The first commercial breakthrough was the fabrication of a LIGA microspectrometer (spectral range from 380 - 750 nm, structure height 100 µm, grating step height 0.2 µm, resulting aspect ratio of 500 (*7*)) on a middle scale series of approximately 4000 units by hot embossing of thin PMMA foils (*8*).
- Mammal tissue cells can be cultivated if cubic shaped microcontainers (300 x 300 x 310 μm^3) with a wall thickness of 50 µm are used (Figure 4, left side). The injection molded PMMA structures possess small pyramidal rectangular holes (40 x 40 μm^2) with smallest size (6 x 8 μm^2) at the bottom side for the nutritious supply of the cells (*9*).
- Micro membrane pumps with thermo-pneumatic actuators for the conveyance of gases or liquids are mounted in housings made from chemically and thermally resistant polymers (PSU or PEEK) for an extended lifetime (Figure 4, right side).
- Electrostatically driven actuators like microvalves are integrated in housings made from carbon black containing PA12. They are used in process technology and in micro analytical systems. The diameters of the valves are 280 and 300 µm with a smallest wall thickness of 50 µm (Figure 5, left side) (*10*).
- A subsequent electroforming on conducting polymers (here: PA12 with carbon black) allows the fabrication of metal microstructures for e.g. spin nozzles test structures (Figure 5, right side) (*4*).

For special applications micro components with high temperature stability or small thermal expansion coefficient are necessary. Hence, further process developments like micro casting as well as metal and ceramic micro injection molding targeted to the fabrication of metal and ceramic microstructures are under investigation. (*11,12*)

New Process Development: Photomolding of UV Curable Systems

In thermally initiated reaction injection molding the relative slow polymerization starts at elevated temperatures (up to 150 °C) after filling of the microstructured mold inserts with the low viscosity resin. Typical polymerization times necessary for a

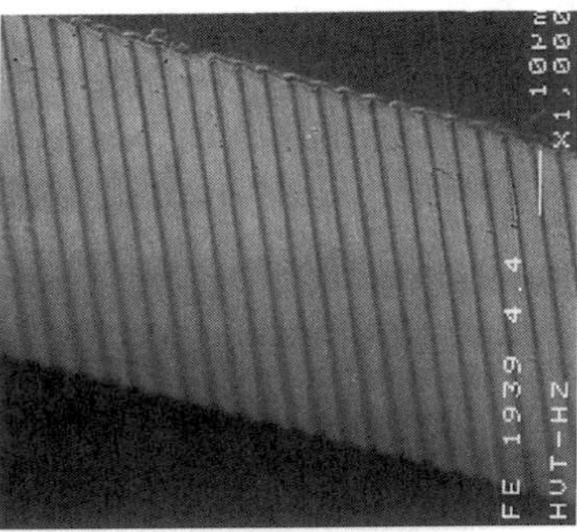

Figure 3. Mold insert and injection molded NIR-microspectrometer (PMMA).

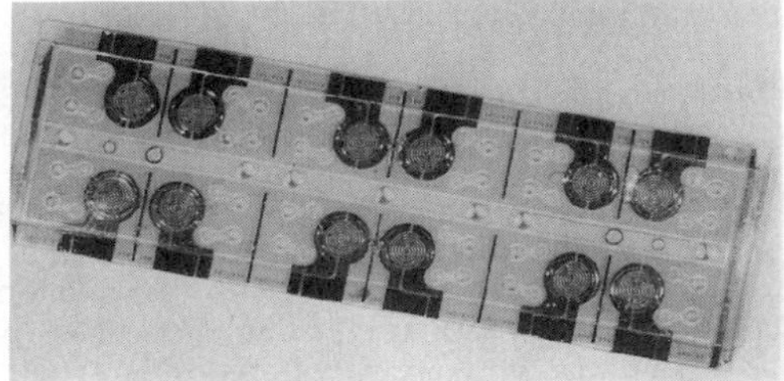

Figure 4. Micro cell container (PMMA) and micro membrane pump housing (PSU).

Figure 5. Microvalve plates, spin nozzles (PA12/carbon black) and electroformed nickel raw structure.

complete hardening of the bulk material can take up to 30 min for a thickness of 3 mm at a curing temperature of 115 °C. After curing the solidified polymer can be demolded at temperatures between 40 and 80 °C in case of PMMA. Our process development now tries to combine the features of reaction injection molding with the advantages of the photopolymerization of methylmethacrylate resins using suitable photoinitiators (*13*). As a direct consequence the whole molding process can be realized at ambient temperature without any heating or cooling steps. Hence the cycling time should be reduced to lower values. The most important process characteristics are described in the following:

- degassing of the resin prior to injection
- variation of the injection pressure from 0.1 up to 5 MPa
- evacuation of the mold insert prior to the resin injection
- use of UV-A and -B transparent glass molding tool
- holding pressure compensates the reaction shrinkage during curing
- demolding by integrated ejector pins
- integration of a small sized powerful UV-source with homogeneous intensity distribution over a large illuminated area (at least 50 x 100 mm²) in the experimental setup.

The first design of the laboratory injection molding equipment has been integrated in a press with a maximum force of 60 kN. In principle the setup consists of two parts: the lower part with the light source and the upper part containing the molding tools. The lower glass molding tool faces the lamp, the upper molding tool carries the mold insert, ejector pins, connectors for vacuum pump and external temperature control as well as the polymer resin reservoir (*14*).

Photopolymerization Times. Commercially available liquid methylmethacrylate resins like Plexit (Roehm GmbH) contain normally approximately 30 - 35 weight% polymer dissolved in the parent monomer. Further increase in the polymer concentration results in a corresponding increase of the viscosity and cannot be used for reaction injection molding. In our experiments we decided to investigate either pure commercial Plexit, dilutions with the monomer (MMA or Diluent30 (stabilized MMA)) or with Ethyltriglycolmethacrylate (ETGM) as a bulky comonomer with respect to the polymerization times and the brittleness of the solidified material (PMMA). Additionally, mixtures containing internal release agents like Butylstearate (SBE) or plastizers like Dibutylphthalate (PSBE, Aldrich) were prepared (*14*).

The photopolymerization time of a polymer resin depends directly on the choice of a suitable photoinitiator. The absorption of a photoinitiator from Ciga Additive GmbH fits with the emission range of an UV A/B-source (F-type). Commercial resins for inks or wood impregnation contain 0.5 - 5.0 weight% photoinitiator, further increase elevates normally the curing times due to a pronounced absorption of the initiator close to the surface. The resulting reduction of the effective light intensity in the bulk material is known as self-quenching process (13). As a consequence, first experiments have to evaluate the proper photoinitiator concentrations. With respect to the relative thick molded substrate base plate (1 - 5 mm) and the desired short polymerization times, a large irradiation density of up to 600 mW/cm² for the photopolymerization of various resin compositions at different photoinitiator contents was applied. All given data were normalized to a thickness of 5 mm (Figure 6). A more detailed investigation on pure Plexit yielded in the fastest polymerization time of 2.5 min @ 3.0 weight% initiator and 600 mW/cm² irradiation density. In comparison to previous experiments with irradiation densities of 250 mW/cm² and 350 mW/cm² a mean reduction of the curing time of 35% could be observed (Figure 7). Stronger dilutions with monomer or other additives extends the hardening process especially at small initiator concentrations. But in all cases the polymerization times are strongly reduced in comparison to thermally initiated reaction injection molding.

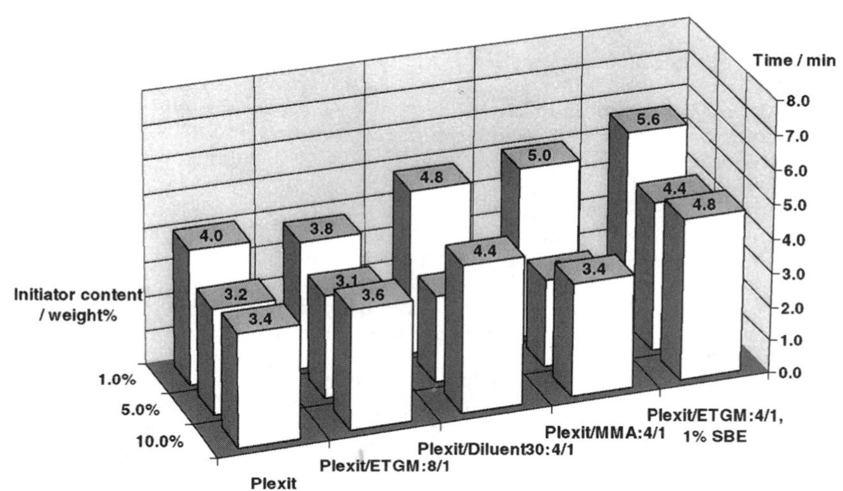

Figure 6. Change of the photopolymerization times with resin composition.

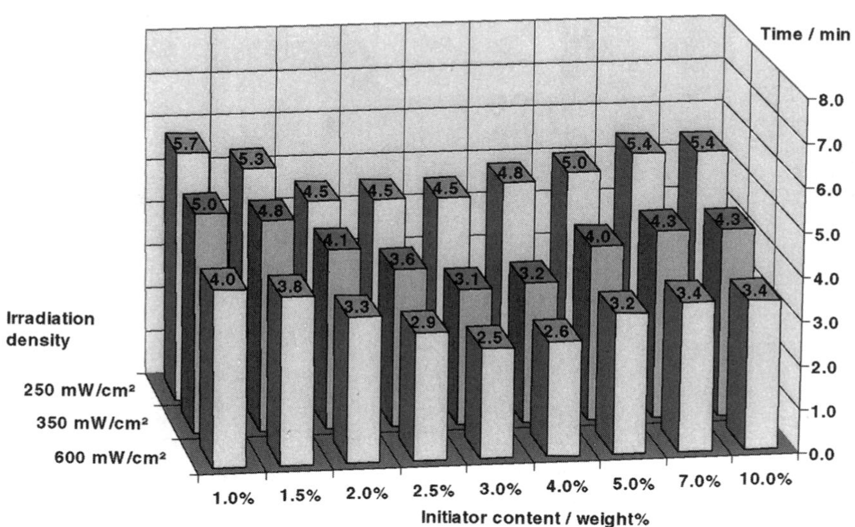

Figure 7. Reduction of the curing times.

First Molding Experiments. In case of micro injection molding of PMMA, the injection and holding pressures are between 50 - 70 MPa at an injection temperature around 230 °C. The photomolding approach reduces the injection and holding pressures to values between 0.5 and 3 MPa at ambient temperature. In micro injection molding, normally, thermal shrinkage of the polymer material during cooling down supports demolding. Here, due to the isothermal process, successful demolding depends strongly on the static friction of the mold insert and the polymer's surface, respectively, as well as on the polymer's elastic behaviour. As a consequence, the surface roughness of the mold insert has to be as low as possible, which is fulfilled if the mold inserts are fabricated either micromechanically or by the LIGA technique. Surprisingly resin compositions containing the above mentioned intrinsic release agents or plastizers (ETGM, PSBE, SBE) show worse demolding properties due to a pronounced brittleness. The best results have been achieved using pure Plexit with 3 weight% photoinitiator. A series of structureless parts with base plate thicknesses between 1 and 5 mm and a rough polymerization time of 1 min/mm thickness could be molded in good quality.

The whole process development of photoinduced reaction injection molding is targeted towards the molding of microstructures. Successful experiments using pure Plexit containing 3 weight% photoinitiator and 1 weight% intrinsic release agent (Wuertz GmbH) allowed the first molding of polymer slabs (width: 300 µm, depth: 600 µm) on a substrate base plate (thickness 2 mm) using a micromechanically fabricated mold insert (Figure 8). It is expected that by the end of the year molding of slab structures with dimensions < 100 µm will be possible in reliable quality. So, a quite elegant method would be available for the fabrication of plastic microstructures e.g. waveguiding elements for future applications in computing or telecommunication.

Figure 8. Mold insert and REM image of photomolded part.

Acknowledgment

We gratefully acknowledge financial support by the Deutsche Forschungsgemeinschaft. Additionally we thank all who contributed successfully to this work.

Literature Cited

1. Becker, E.W.; Ehrfeld, W.; Hagmann, P.; Maner, A.; Münchmeyer, D. *Microelectronic Engineering* **1986**, *4*, 35.
2. Bacher, W.; Menz, W.; Mohr, J. *IEEE Trans. Ind. Electr.* **1995**, *42(5)*, 431.
3. Michaeli, W.; Rogalla, A. *F&M Feinwerktechnik* **1994**, *104*, 641.
4. Piotter, V.; Hanemann, T.; Ruprecht, R.; Haußelt, J. H. *Microsystem Technologies* **1997**, *3(3)*, 129.
5. Hagmann, P.; Ehrfeld, W. *Makromol. Chem. Macromol. Symp.* **1989**, *24*, 241.

6. Heckele, M.; Bacher, W.; Hanemann, T.; Ulrich, H. *Precision Plastic Optics for Optical Storage, Displays, Imaging, and Communications;* SPIE-Proc. 3135; SPIE: San Diego, CA, **1997**; 06-12.
7. Heckele, M.; Bacher, W. *Micromachine Devices* **1997**, *2(2)*, 1.
8. Müller, C.; Mohr, J. *Interdisciplinary Science Reviews* **1993**, *18(3)*, 273.
9. Weibezahn, K.; Knedlitscheck, G.; Dertinger, H.; Schaller, T.; Schubert, K.; Bier, W., **1993**, DP 4132379.3-41.
10. Fahrenberg, J.; Bier, W.; Maas, D.; Menz, W.; Ruprecht, R.; Schomburg, W.K.; *Proc. Micro Mechanics Europe,* Pisa, Italy, **1994**; 17.
11. Piotter, V.; Hanemann, T.; Ruprecht, R.; Thies, A.; Hausselt, J. H. *Micromachining and Microfabrication Technologies III;* SPIE-Proc. 3223; SPIE: Austin, TX, **1997**; 13-22.
12. Ruprecht, R.; Hanemann, T.; Piotter, V.; Haußelt, J.H. *Proc. Micro Materials (Micro Mat '97);* Berlin, Germany, **1997**; 238.
13. Chang, C.-H.; Mar, A.; Tiefenthaler, A.; Wostratzky, D. in *Handbook of Coatings Additives;* Calbo, L. J., Ed.; Marcel Dekker, Inc.: New York, NY, **1992**, Vol. 2; 1-43.
14. Hanemann, T.; Ruprecht, R.; Haußelt, J. H. *Polymeric Materials Science & Engineering* **1997**, *77*, 418.

Microlithographic Chemistry and Processing

Elsa Reichmanis

The intense drive towards designing and fabricating integrated circuits having increasingly more densely packed features of ever shrinking dimensions has placed severe demands on the materials technologies that are used to define the individual circuit elements. The predominant technology today is "conventional photolithography" (i- and g-line photolithography employing 350-450 nm light). Incremental improvements in tool design and performance have allowed the continued use of 350-450 nm light to produce ever smaller features. Additionally, the same basic positive photoresist consisting of a diazonaphthoquinone photoactive compound and a novolac resin has continued to be the resist of choice.

The advent of 248 nm lithography necessitated the first shift in materials chemistry from the traditional novolac based resists to an alternative technology in about 20 years. Chemically amplified resists, first introduced in the 1980 time frame, have been the subject of intense research and development efforts. The concept of chemical amplification employs the photogeneration of an acidic species that catalyzes many subsequent chemical events such as deblocking of a protective group or crosslinking of a matrix resin. The overall quantum efficiency of such reactions is, thus, effectively much higher than that for initial acid generation. The elegant design of an acid sensitive, t-butoxycarbonyl protected poly(hydroxystyrene) introduced by Willson, Ito and Frechet initiated activities related not only to design of alternative approaches, but also to developing fundamental mechanistic understanding of processes such as diffusion phenomena, photochemical mechanisms, etc.

Since the conception of chemical amplification mechanisms almost 2 decades ago, they are now becoming commonplace in the device fabrication arena. As 248 nm lithography becomes the norm in advanced VLSI manufacture, the requisite materials technology must gradually become evaluated and accepted, though further advance in this technology requires continued investigation into materials issues and alternatives. This section describes recent materials design alternatives for robust, manufacturable 248 nm resist technologies. In addition, acid generator chemistries and acid diffusion mechanisms will be discussed along with their possible implications on lithographic performance.

Chapter 7

Acid Labile Cross-Linked Units: A Concept for Improved Positive Deep-UV Photoresists

H.-T. Schacht[1], P. Falcigno[1], N. Münzel[1], R. Schulz[1], and A. Medina[2]

[1]Ciba Specialty Chemicals Inc., Klybeckstrasse 141, CH-4002 Basel, Switzerland
[2]Olin Microelectronic Materials Inc., 200 Massasoit Avenue, East Providence, RI 02914

The transacetalization during the acid catalyzed reaction of vinyl ethers with aliphatic hydroxyl groups leads to a new polymer type which is useful in chemical amplified resist systems. With the transacetalization reaction it is possible to increase the molecular weight through a polymer analogous synthesis via acetal bridges. First, the principle is evaluated with polymers with aliphatic alcohol units attached to the backbone, and then applied to polyhydroxystyrene. In the second case bifunctional alcohols are used for the crosslinking units. Two-component resists containing polymers with acid labile crosslinking units and a photo acid generator show higher thermal flow stability and increased resolution depending on their increased molecular weight distribution. Gel permeation chromatography shows that the crosslinked units are completely cleaved in the exposed area, like the monofunctional acetal blocking groups. The resists exhibit good stability against air borne and substrate borne contamination.

Going down to quarter and sub quarter micron features it was very soon understood that the diazonaphthoquinone/novolak system is not able to fulfill the requirements of a lithographic process at 248 nm (*1-4*). The main issue is the low transparency of the diazonaphthoquinone photo acid generator and the novolak binder at this wavelength (*5*).

One approach to provide resists with high sensitivity and contrast involves the principle of chemical amplification (CA) (*6-8*). The amplification effect is achieved by employing a photo generated acid as a catalyst to carry out a cascade of chemical reactions in the resist film. Catalytic chain lengths of >1000 have been reported in the literature (*9,10*). However, the high catalytic chain lengths also enhance the resist's sensitivity to airborne basic contaminants and basic moieties on the substrate (*11,12*). In a positive tone resist such deactivation of the photoacid by airborne or substrate bound basic contaminants results in "T-tops" or "foot" formation respectively.

The influence of the airborne contamination is especially critical for the pattern profiles during the time between exposure and post exposure bake, due to the diffusion of the base into the resist. Several attempts to prevent the deactivation of the acid in the resist are described in the literature. The problem of airborne contamination could be solved to some extent by air filtration (*11*), applying a protective topcoat (*13,14*), incorporation of additives into the resist formulation (*15,16*) or annealing the resist (*17,18*). In addition, it was demonstrated that resists based on low activation deprotection chemistry are very robust not only towards airborne basic contaminates, but also towards basic contamination from the substrate. In this case, the deprotection reaction proceeds rapidly after the exposure at room temperature. Therefore, the neutralization of the photogenerated acid by basic contaminants between exposure and post exposure bake affects the deblocking reaction much less. Examples for a binder resin using low energy deprotection systems are poly-(4-hydroxystyrene) (PHS) partially protected with silyl ethers (*19*), ketals (*20-24*) and acetals (*24-31*). The higher thermal stability of the acetal protected polymers compared to ketal protected resins (*24,32*) has led us to focus on this type of protecting group.

The chemical structure of the acetal influences the thermal properties of the resin such as glass transition temperature, decomposition temperature and thermal flow stability (*32-35*). This paper describes the improved thermal properties by increasing the molecular weight via a transacetalization reaction. The polymers containing such crosslinking units were evaluated in two-component positive photoresists.

Experimental

Materials. Poly-(4-hydroxystyrene/4-vinylcyclohexanol) (PHSVC) was obtained by partial hydrogenation of commercially available PHS as described in the literature (*36,37*). The vinylcyclohexanol content of the polymers was analyzed by ^{13}C-NMR measurements and controlled between 5-15%.

The acid catalyzed reactions of PHSVC and PHS with ethyl propenylether, ethyl vinyl ether, n-propyl vinyl ether, cyclohexyl vinyl ether, and n-butyl, i-butyl and t-butyl vinyl ether were run as previously described (*38*). In the case of the PHSVC the crosslinking degree was controlled by adjusting the ratio of the cyclohexanol units in the polymer. For PHS the crosslinking degree was controlled by the addition of bifunctional aliphatic alcohol such as ethylene glycol, 1,4-butanediol, 1,6-hexanediol, 1,8-octanediol, 1,12-dodecanediol, 1,4-cyclohexanediol, diethyleneglycol, 4,8-bis(hydroxymethyl)-tricyclo(5,2,1,0/2,6)decane, 4,4'-isopropylidenedicyclohexanol and 1,3-dihydroxy-1,2,3,4-tetrahydronaphtalene (0.5 to 20 weight percent relative to PHS) to the reaction mixture. The polymers were isolated by precipitation in water. All polymers are soluble in typical resist solvents like propylene glycol monomethyl ether acetate or methyl-3-methoxypropionate.

The ratio of the cyclohexanol units in the starting material (PHSVC), the acetal blocking level of the phenol groups and the amount of the crosslinked units were determined by inverse gated ^{13}C-NMR decoupling experiments. Molecular weight was determined by gel permeation chromatography (GPC) in THF utilizing PHS standards.

Lithography. For the lithographic evaluations the polymer was dissolved in propylene glycol monomethyl ether acetate along with a standardized amount of a proprietary photo acid generator, which forms a sulfonic acid after irradiation. All solutions were filtered through 0.2 mm Teflon membrane filters. The photoresist solutions were spin coated onto silicon substrates, soft baked for 60s at 130°C, exposed, post exposure baked for 60 s at 110°C, and developed using 0.262 N aqueous tetramethylammonium hydroxide in a single puddle process. Thermal flow experiments were conducted on a hot plate, where the patterned substrates (20 μm pads) were baked for 3 minutes. For the post exposure delay (PED) experiments, the resist was compared lithographically side-by-side with a resist containing poly-(4-acetoxystyrene-4-t-butoxy-carbonyloxystyrene-sulfone) as the polymeric binder.

Results and Discussion

The acid catalyzed addition of hydroxy compounds to enol ether (I) is a general method to synthesize acetals as shown in Scheme 1(*39*). Especially for the synthesis of mixed o/o-acetals (II) this route is widely used. Under certain conditions (amount of the alcohol R'''-OH, substituent R) the reaction proceeds further to yield symmetric acetals III (*40,41*).

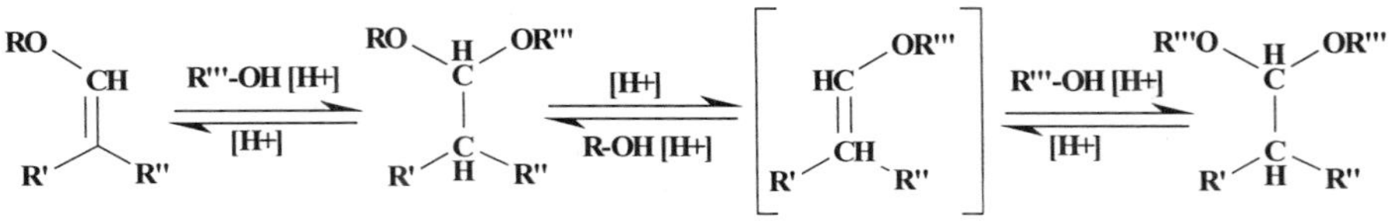

(I) (II) (III)

Scheme 1. Acetalization and Transacetalization of vinyl ether.

Acetal Protected PHS. Scheme 2 illustrates the possible equilibrium reactions of PHS (IV) with vinyl ethers. Under acidic conditions the phenolic hydroxyl groups of IV are converted into acetal groups (VI). The conversion degree depends on the reaction conditions (temperature, catalyst, time, solvent).

The stability of the intermediate oxonium ions V and VII determine which of the two possible routes leading either to the phenolic species IV, or to the aromatic vinyl ether species VIII is preferred. In the case of an aliphatic group R, the oxonium ion V, stabilized by an aromatic system, is preferred over the aliphatic oxonium ion VII. This favors the cleavage to the phenolic compound IV but not the formation of the vinyl ether VIII. This vinyl ether reacts further to the symmetrical species III as depicted in Scheme 1. High catalyst concentration or high temperature causes the formation of the aromatic vinylether VIII to compete with the reaction leading to IV (*42*). In this case an intermolecular crosslinking between two polymer chains occurs.

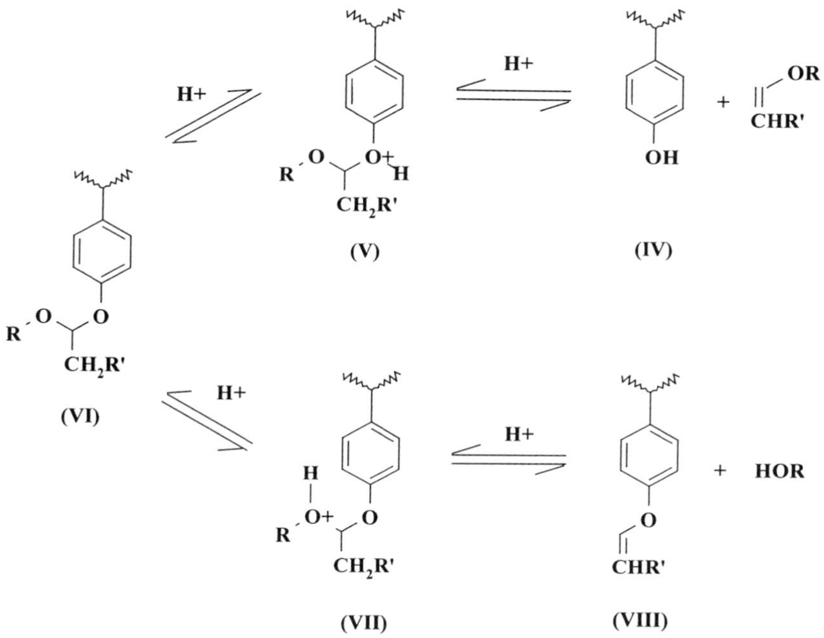

Scheme 2: Acidolysis of acetal protected polyhydroxystyrene

PHSCV as Backbone. If PHS is replaced by PHSCV in the same reaction scheme, the molecular weight distribution after the acetalization is shifted significantly towards higher molecular weight, except when dihydropyrane (DHP) is used as vinyl ether. In parallel, a strong increase of the polydispersivity (M_w/M_n) is observed. The increase of the molecular weight depends on the type of vinyl ether used in the protection reaction. Reacting PHSVC ($M_w = 5000$, $M_w/M_n = 2.1$, cyclohexanol content in the polymer of 7%) with ethyl vinyl ether (EVE), ethyl propenylether (EPE), i-butyl (i-BVE) and n-butyl (n-BVE) vinyl ether resulted in polymers with a M_w of 10000 to 11000 and a polydispersity (M_w/M_n) around 3. The reaction with cyclohexyl vinyl ether yielded a polymer with M_w of 16000 and M_w/M_n of 4.2. In the case of t-butyl vinyl ether M_w increased even more from 5000 to 40000 and M_w/M_n from 2.1 to 7.3.

Table 1 shows the relative concentration of the possible acetal moieties of the PHSVC reaction with various vinyl ethers. Besides the acetal protected OH groups B and C (scheme 3), two different structural units D and E were identified by ^{13}C-NMR analysis (*42*). The structural unit E is only observed with t-butyl vinyl ether. However, no aromatic to aromatic intra and inter chain crosslinking (F in scheme 3) was observed in either the PHS or PHSVC system, unless high catalyst concentration or high temperature was used. Not only the type of vinyl ether units influenced the molecular weight but also the amount of the vinylcyclohexanol comonomer ratio in PHSCV.

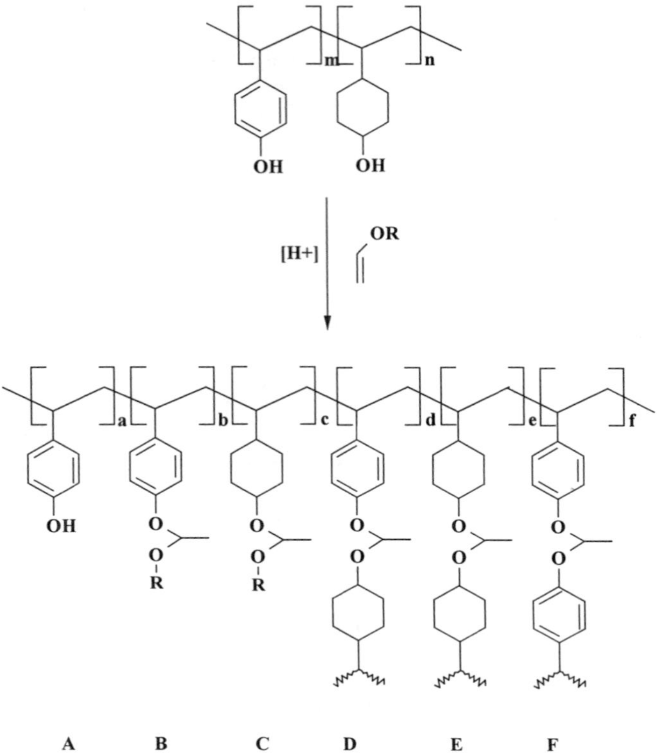

Scheme 3. Acid catalyzed reaction of PHSVC and vinyl ether

Table 1. Molecular weight distributions and relative concentrations of acetal moieties of protected PHSVC

Vinyl ether	GPC		Relative Concentration Measured ^{13}C-NMR [%]				
	M_w	M_w/M_n	B	C	D	E	F
DHP	7000	1.8	85	15	0	0	0
i-BVE	10000	2.8	85	11	4	0	0
n-BVE	11000	2.9	86	10	4	0	0
EPE	11000	3.3	84	12	4	0	0
EVE	11000	3.0	87	9	4	0	0
CHVE	16000	4.2	B+D=96	C+E=4	-	-	0
t-BVE	40000	7.3	70	18	5	3	0
t-BVE*	28000	4.3	82	13	4	1	0

* starting polymer had a cyclohexanol unit content of 5%

The investigated vinyl ethers resulted in different molecular weights, thus showing different reactivity in the formation of crosslinking units (D and E shown in Scheme 3). Higher M_w and wider polydispersivity is obtained in the order, primary < secondary < tertiary alkyl group attached to the vinyl ether oxygen atom. If DHP is used, the six membered ring formation is preferred over the open chain of the oxonium ion intermediate (*39,42*).

Crosslinked PHS. The concept of transacetalization can also be applied to PHS. However, since PHS does not contain aliphatic alcohol moieties, which are necessary for the effective crosslinking reaction, we have added aliphatic diols (HO-Z-OH) to the reaction mixture. Examples for suitable alcohols are ethylene glycol, 1,4-butanediol, 1,6-hexanediol, 1,8-octanediol, 1,12-dodecanediol, 1,4-cyclohexanediol, diethylene-glycol, 4,4'-isopropylidenedicyclohexanol and 1,3-dihydroxy-1,2,3,4-tetrahydronaphtalene. The molecular weight of the resulting crosslinked polymer strongly depends on the amount and the type of the added diol.

According to corresponding ^{1}H- and ^{13}C-NMR data we propose the structure depicted in Scheme 4 for the resulting polymers. The polymer chains are linked via the bifunctional acetal moieties. However, there is no evidence for crosslinking units between two phenolic groups (F in Scheme 3). ^{1}H-NMR investigations exclude any type of side groups containing vinyl ether moieties. ^{13}C-NMR spectra show no evidence for other acetal protecting groups as described in Scheme 3. The same crosslinking unit can be synthesized by the reaction of ethyleneglycol divinyl ether or hexanediol divinyl ether. Because of the limited availability of divinyl ethers, the diol route is preferred.

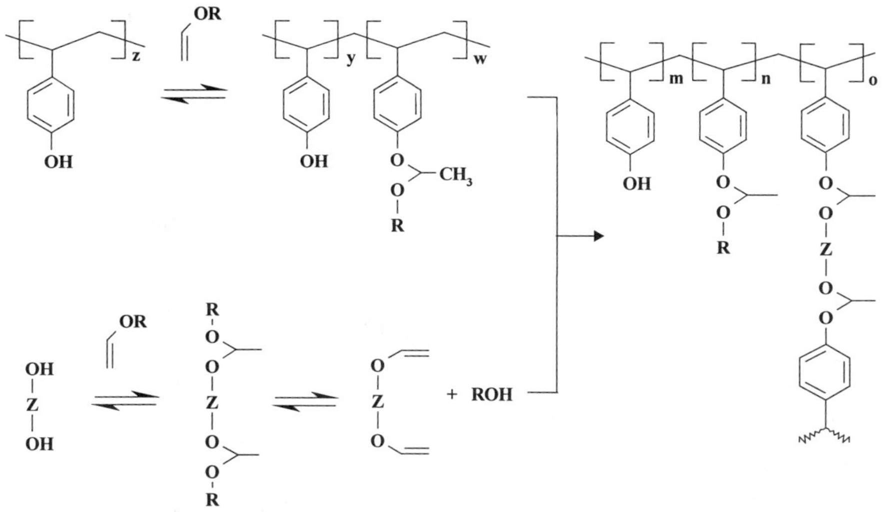

Scheme 4. Acid catalyzed reaction of PHS with diols and vinyl ethers.

This transacetalization allowed us to synthesize a wide variety of polymers with different blocking levels and various molecular weights. For the comparison of the different crosslinked and linear acetal protected polymers we defined the crosslinking degree X in equation 1:

$$X = [M_w(cross)/M_w(lin)] - 1 \quad (1)$$

M_w(cross) is the weight average molecular weight of the crosslinked polymer and M_w(lin) is the weight average molecular weight of the corresponding linear polymer with the same acetal blocking level under the provision that M_w before the acetalization reactions is the same for both.

Figure 1 shows the linear correlation between the feed ratio of the bifunctional alcohol and the crosslinking level X. Since the molecular weight is proportional to the crosslinking level, M_w can be controlled by the diol feed ratio, if the reaction conditions (temperature, time, amount of catalyst, solvent etc.) are kept constant.

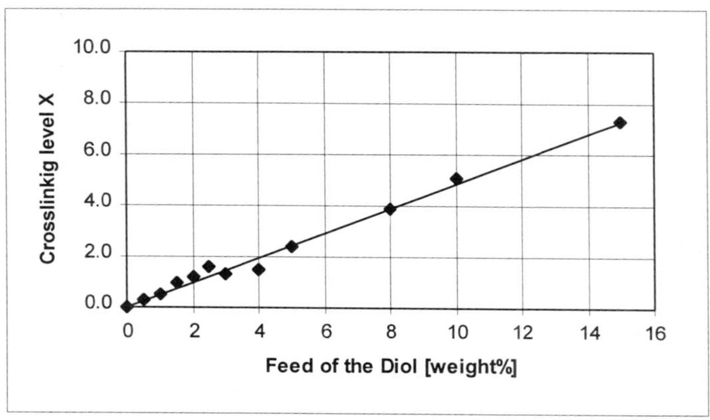

Figure 1: Crosslinking degree as function of diol feed ratio.

Table 2 shows the influence of the diol chain length on the crosslinking level under the same reaction conditions. In general the crosslinking level increased with the distance of the two alcohol groups in the diol. The crosslinking level is independent on whether the diol is a straight chain or a more rigid moiety (e.g. 1,3-dihydroxy-1,2,3,4-tetrahydronaphtalene). According to our ^{1}H- and ^{13}C-NMR data we see no evidence of free vinyl ether or alcohol moieties. The differences in the molecular weights can be explained in terms of the preference of diols with a short distance between the hydroxyl groups to undergo intramolecular versus intermolecular crosslinking. Only the intermolecular reactions lead to an increase in M_w. If we compare diols with the same number of atoms between the hydroxyl groups, it was observed that diols with secondary alcohol groups lead to higher crosslinking than diols with primary alcohol groups due to their different reactivity.

Table II. Molecular weight distributions of PHS crosslinked with various diols (feed ratio: 2 mol%)

Diol	C-Atoms between OH	Type of alcohol	GPC M_w	M_w/M_n	X
Ethylene Gylcol	2	primary	15000	1.1	0.2
1,4-Butanediol	4	primary	18000	1.2	0.4
1,4-Cyclohexanediol	4	secondary	24000	1.4	0.9
1,6-Hexanediol	6	primary	24000	1.3	0.9
1,3-Dihydroxy-1,2,3,4-Tetrahydronaphtalene	7	primary	24000	1.4	0.9
4,4'-Isopropylidene-Dicyclohexanol	10	Secondary	40000	1.9	1.9
1,12-Dodekanediol	12	Primary	30000	1.6	1.3

Acid Labile Crosslinked Units. To investigate if the crosslinking units cleave with a different or similar rate as monofunctional acetals during a typical DUV process, we measured the molecular weight distribution of the resist before soft bake, after soft bake (SB; 130°C, 60 sec) and after post exposure bake (PEB; 115°C, 60 sec). Figure 2 shows the GPC traces of an acetal protected PHS crosslinked via 4,4'-isopropylidene-dicyclohexanol. The molecular weight distributions before and after soft bake are identical. Thus, the polymer is stable under the bake conditions. The GPC trace of the resist after exposure indicates that the acidolysis of the crosslinking units in the presence of the photo generated acid proceeds to completion even at room temperature within several minutes. The resulting narrow molecular weight distribution is essentially identical to that of the PHS used as starting material. After PEB there is no change in the molecular weight. An additional PEB is applied in order to complete the deblocking reaction independent of the post exposure time delay and to improve the resolution.

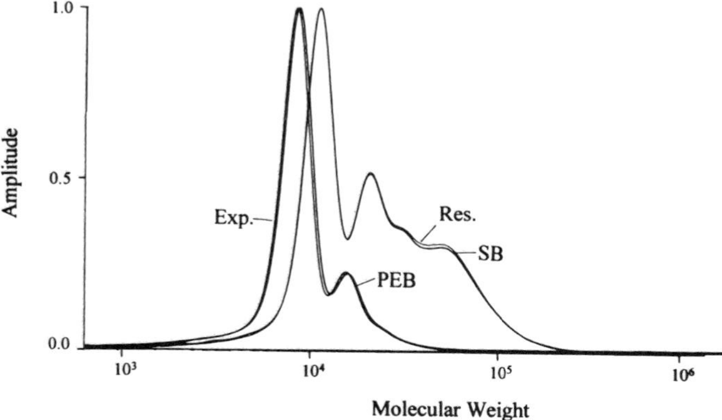

Figure 2: GPC traces of a resist before soft bake (Res.), after soft bake (SB), after exposure (Exp.) and after post exposure bake (PEB).

Thermal Flow Stability. The thermal flow stability of resists containing polymers with different X-values were evaluated using the same polymer backbone and the same acetal protecting group. Generally the thermal flow stability increases with the X-value. Figure 3 visualizes this result. The main effect already occurs at a low crosslinking level (X = 0.8). Here, the resist flow temperature is improved by more than 10°C. Higher crosslinking degree yields only a slight increase of the thermal stability (about 5°) of the pattern.

Resolution. Generally, there is a trade-off between resolution limit and molecular weight of the polymer, because of the lower dissolution rate of higher molecular weight resins (*49*). However, in the case of the polymers with acid labile crosslinked units a contrary effect is seen. In this case low molecular weight PHS is formed after exposure and PEB with good solubility in standard TMAH developer. Scheme 5 illustrates the decrease of the molecular weight of the crosslinked polymer after exposure and post exposure bake.

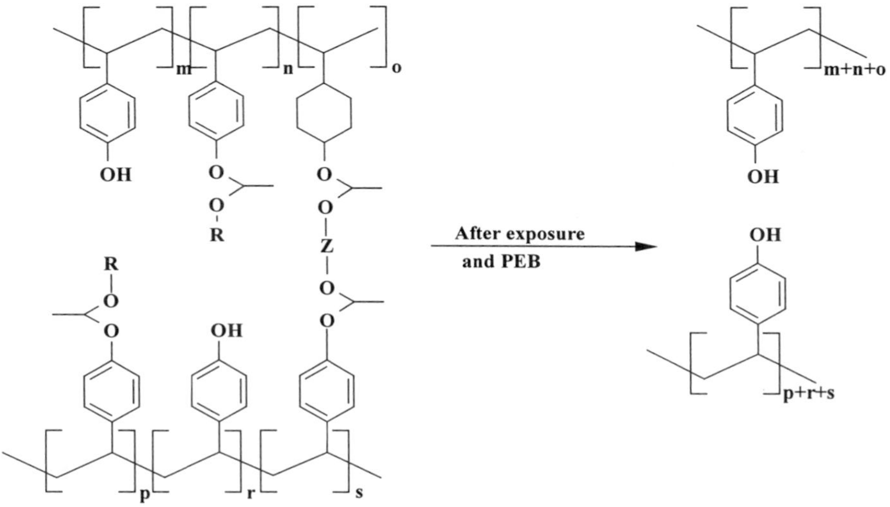

After exposure and PEB

Scheme 5. Cleavage of acetal units during exposure and PEB

The dissolution rate curves in Figure 4 show that the crosslinked (X = 2.2) and the linear (X = 0) acetal protected PHS have about same R_{max} value. In the unexposed areas, crosslinking can significantly improve inhibition. Thus, we can combine the benefits of maintaining high M_w in the unexposed areas and low M_w in the exposed areas. In the curves in Figure 4 the R_{min} is reduced from 2.0 Å/sec to less than 1Å/sec. Figure 5 demonstrates that resolution is improved by 0.04 μm with increased crosslinking level (X = 3.9). Figure 6 shows the resolution capability down to 0.21 μm contact holes.

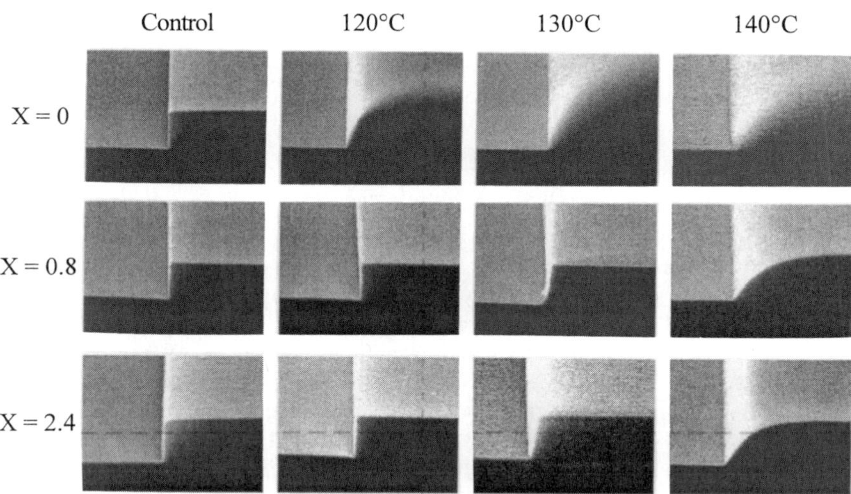

Figure 3. SEM micrographs illustrating the effect of the crosslinking degree X on thermal flow stability

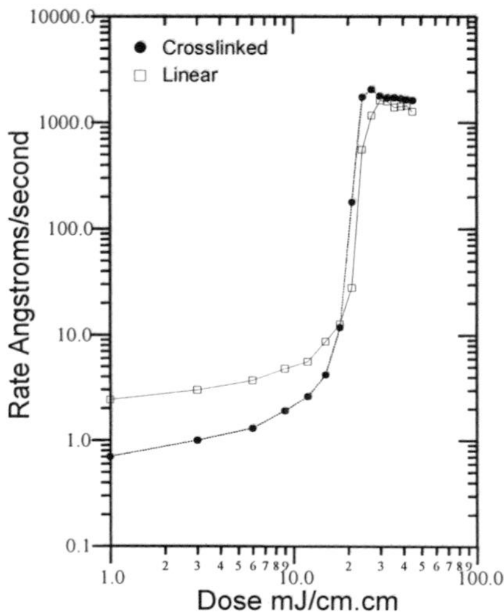

Figure 4. Dissolution rate vs. exposure dose curves for resists. Reproduced with permission from reference 44.

Stability against Base Contamination. A photoresist with a partially protected PHS crosslinked with 4,4'-isopropylidenedicyclohexanol as polymeric binder was lithographically compared side-by-side with a polymer partially protected with t-butoxycarbonyl (t-BOC) (*45-48*) in a non purified cleanroom atmosphere. When postbaked 30 min after exposure, the formulation containing t-BOC protected resin, as a typical high deprotection energy system, produced images with severe T-top formation. Processed under the same conditions partially crosslinked acetal protected PHS shows no capping after 3 hours, as illustrated in Figure 7.

Figure 8 is depicting the resist insensitivity to TiN substrates. The resist based on a partially crosslinked acetal protected PHS as binder resin was processed on TiN, one of the most critical substrates in this respect (*48-50*). In contrast to the t-BOC protected polymer based formulations no footing was observed.

Conclusion

The concept of acid labile crosslinked units in polymers for positive DUV resist was evaluated with poly-(4-hydroxystyrene/4-vinylcyclohexanol) and implemented on a poly-4-hydroxystyrene backbone. The principle to increase the molecular weight by a transacetalization reaction was used to improve acetal protected phenolic resins for advanced chemically amplified resists. The polymer chains were linked via diols with acetal bridges. This type of linkage can be applied to a wide range of phenolic polymer binders. Resolution and thermal stability of the photoresist patterns were improved. Gel permeation chromatography and dissolution rate measurements prove that the crosslinked units remain essentially intact in the unexposed areas, thus enhancing

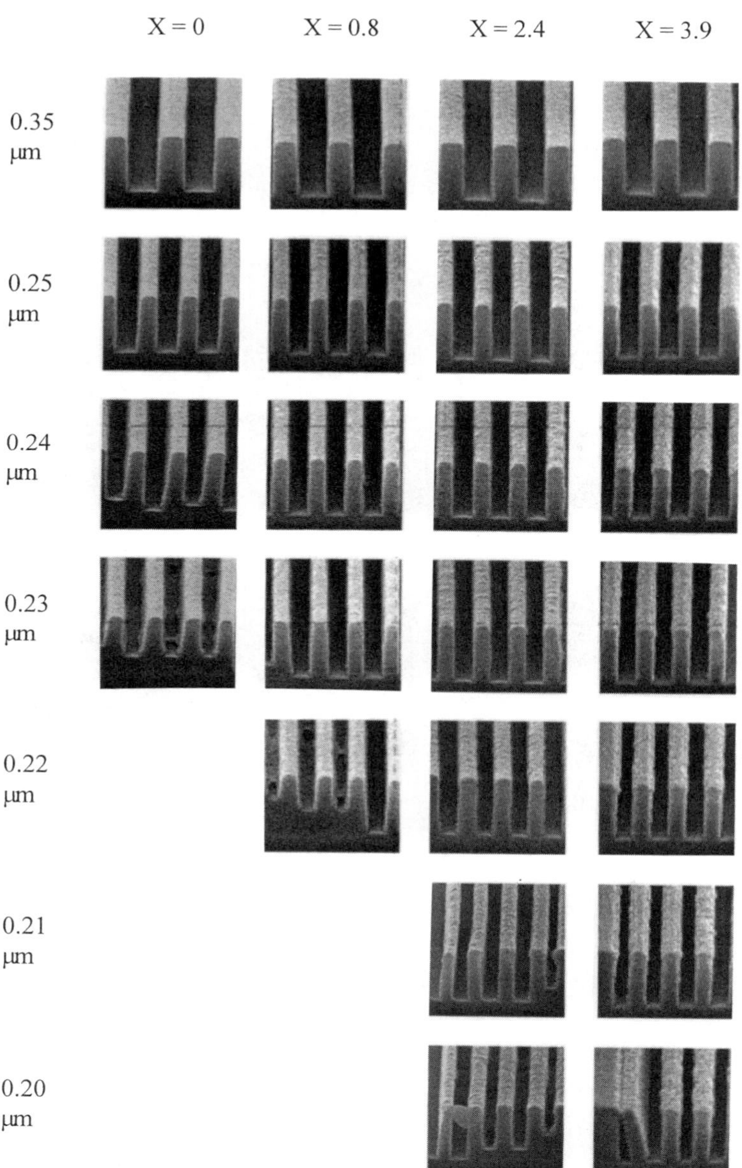

Figure 5. SEM micrographs showing the effect of the crosslinking degree X on

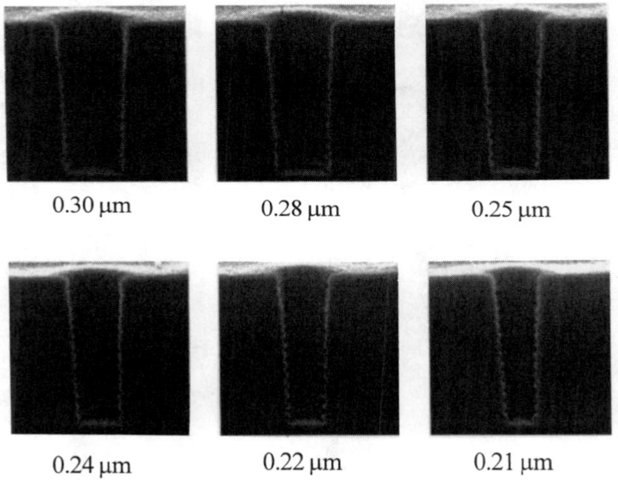

Figure 6. SEM micrographs showing resolution of contact holes. Reproduced with permission from reference 44.

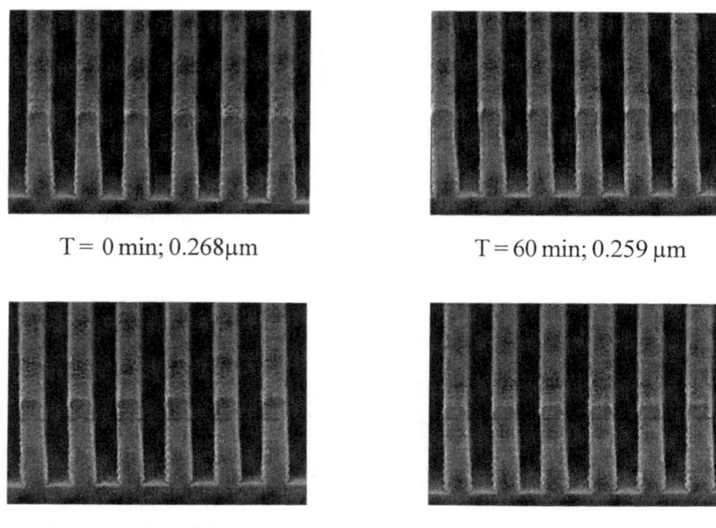

Figure 7. SEM micrographs showing 0.25 μm line/space patterns with 3 h, 2 h, 1 h and no time delay between exposure and PEB.

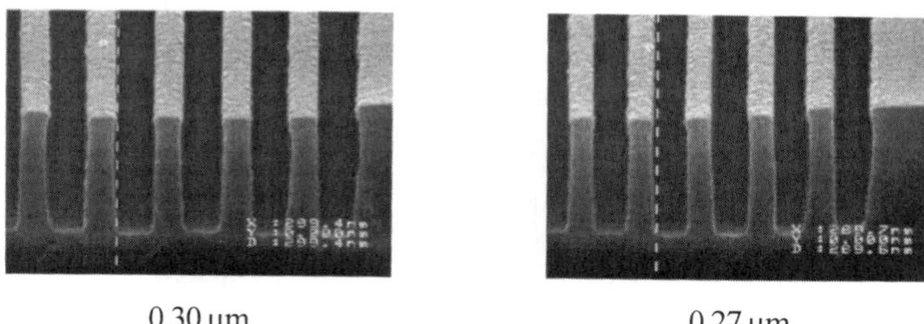

Figure 8. SEM micrographs illustrating resolution for dense lines over TiN.

the thermal stability of the patterns and providing improved dissolution inhibition. In the exposed areas, the cleavage of the blocking and crosslinking units is nearly completed after exposure at room temperature. The resist based on these materials exhibited good stability against basic contamination.

Acknowledgments

The authors wish to thank B. Nathal, H.-J. Kirner, N. Reichlin, K. Petschel and especially A. Zettler for their efforts in the synthesis of the materials. We also wish to thank Heinz Holzwarth, Christoph de Leo, A. Foltzer, T. Tschan, J.-P. Unterreiner, U. Pfeiffer, J. Zenhäusern and K. Eisenblätter for the lithographic work, H. Stephan and J. Schneider for their NMR support and C. Mertesdorf , L. Ferreira (OMM), T. Sarubbi (OMM) and N. R. Bantu (OMM) for their helpful discussions. In addition we want to thank O. Nalamasu, J. Kometani, F. M. Houlihan and A. G. Timko (all Lucent Technology, Bell Labs) for the fruitful collaboration.

Literature Cited

1. Sugita, K.; Ueno, N. *Prog. Polym. Sci.* **1992**, *17*, 319.
2. Willson, C. G.; Miller, R. D.; Pederson, L. A.; Regitz, M. *Proc. SPIE , Adv. Resist Technol. Process. IV* **1987**, *771*, 2.
3. Hayase, R.; Onishi, Y.; Niki, H.; Oyasato, N.; Hayase, S. *J. Electrochem. Soc.* **1994**, *141*, 3141.
4. Orvek, K. J.; Palmer, S. R.; Garza, C. M.; Fuller, G. E. *Proc. SPIE , Adv. Resist Technol. Process. III* **1986**, *631*, 83.
5. "*Principles, Practices and Materials*", Semiconductor Lithography, Moreau, W. M.; Plenum New York, **1988**, 372.
6. Ito, H.; Willson, C. G. *Technical Papers of SPE Regional Technical Conference on Photopolymers* **1982**, 331.
7. Ito, H.; Willson, C.G. in *"Polymers in Electronics"*; Davidson, T., Ed.; Symposium Series 242, American Chemical Society: Washington, D. C., **1984**, p. 11.
8. Reichmanis, E.; Houlihan, F. M.; Nalamasu, O.; Neenan, T. X. *Chem. Mater.* **1991**, *3*, 394.
9. McKean, D. R.; Schädeli, U. P.; MacDonald, S. A. in *"Polymers in Microlithography"*; Reichmanis, E.; MacDonald, S. A., Ed.; Symposium Series 412, American Chemical Society: Washington, D. C., **1991**, p. 27.
10. Houlihan, F. M.; Neenan, T. X.; Reichmanis, E.; Kometani, J. M.; Chin, E. *Chem. Mater.* **1991**, *3*, 462.
11. MacDonald, S. A.; Clecak, N. J.; Wendt, H. R.; Willson, C. D.; Snyder, C. D.; Knors, C. J.; Deyoe, N. B. Maltabes, J. G.; Morrow, J. R.; McGuire, A. E.; Holmes, S. J. *Proc. SPIE , Adv. Resist Technol. Process. VIII* **1991**, *1466*, 2.
12. Nalamasu, O.; Reichmanis, E.; Cheng, M.; Kometani, J. M.; Houlihan, F. M.; Neenan, T. X.; Bohere, M. P.; Dixen, D. A.; Thompson, L. F.; Takemoto, C. H. *Proc. SPIE , Adv. Resist Technol. Process. VIII* **1991**, *1466*, 13.
13. Nalamasu, O.; Cheng, M.; Timko, A. G.; Pol, V.; Reichmanis, E.; Thompson, L. *J. Photopolymer. Sci. Technol.*, **1991**, *4*, 299.

14. Kumada, T.; Tanaka, Y.; Ueyama, A.; Kubota, S.; Koezuka, H.; Hanawa, T.; Morimoto, H. *Proc. SPIE, Adv. Resist Technol. Process. X* **1993**, *1925*, 31.
15. Funhoff, D. J. H.; Binder, H.; Schwalm, R. *Proc. SPIE, Adv. Resist Technol. Process. IX* **1992**, *1672*, 46.
16. Przybilla, K. J.; Kinoshita, Y.; Kubo, T.; Masuda, S.; Okazaki, H.; Padmanaban, M.; Pawlowski, G.; Roeschert, H.; Spiess, W.; Suehiro, N. *Proc. SPIE, Adv. Resist Technol. Process. X* **1993**, *1925*, 76.
17. Ito, H.; Breyta, G.; Hofer, D.; Sooriyakumaran, R.; Seeger, D. J. *J. Photopolym. Sci. Technol.* **1994**, *7*, 433.
18. Ito, H.; Breyta, G.; Hofer, D.; Sooriyakumaran, R. in *"Polymers for Advanced Imaging and Packaging"*; Reichmanis, E.; Ober, C. K.; MacDonald, S. A.; Iwayanagi, T.; Nishikubo, T., Ed.; Symposium Series 614, American Chemical Society: Washington, D. C., **1995**, p. 21.
19. Crivello, J. V.; Colon, D. A.; Lee, J. L. *Polym. Mater. Sci. Eng.* **1989**, *61*, 422.
20. Huang, W. S.; Kwong, R.; Katnani A. D.; Khojasteh, M. *Proc. SPIE, Adv. Resist Technol. Process. XI* **1994**, *2195*, 37.
21. Huang, W. S.; Kwong, R; Katnami; A. D.; Khojasteh, M.; Lee; K. Y. *Mat. Res. Soc. Symp. Proc.* **1994**, *324*, 493.
22. Lee, K. Y.; Huang, W. S. *J. Vac. Sci. Technol. B* **1993**, *11(6)*, 2807.
23. Huang, W. S.; Katnani A. D.; Yang, D.; Brunsvold, B.; Bantu, N. R.; Kohjasteh, M.; Sooriyakumaran, R.; Kwong, R.; Lee, K. Y.; Hefferon, G.; Orsula, G.; Cameron J.; Denison M.; Sinta R.; Thackeray J. *J. Photopolym. Sci. Technol.* **1995**, *8*, 525.
24. Mertesdorf C.; Münzel, N.; Holzwarth, H.; Falcigno, P.; Schacht, H.-T.; Rhode, O.; Schulz, R.; Slater, S.; Nalamasu, O.; Timko, A. G.; Neenan, T. X. *Proc. SPIE, Adv. Resist Technol. Process. XII* **1995**, *2438*, 84.
25. Hesp, S. A. M.; Hayashi N.; Ueno, T. *J. Appl. Polym. Sci.* **1991**, *42*, 877.
26. Hattori, T.; Schlegel, L.; Imai, A.; Hayashi, N.; Ueno, T. *J. Photopolym. Sci. Technol.* **1993**, *6*, 497.
27. Hattori, T.; Schlegel, L.; Imai, A.; Hayashi, N.; Ueno, T. *Proc. SPIE , Adv. Resist Technol. Process. X* **1993**, *1993*, 146.
28. Funhoff, D. J. H.; Binder; H.; Goethals, M.; Reuhman-Huisken, M. E.; Schwalm, R.; Van Driesche, V.; Vinet, F. *J. Inf. Rec. Mat.* **1994**, *21*, 311.
29. Mertesdorf, C.; Falcigno, P.; Münzel, N.; Nathal, B.; Schacht, H.-T., Zettler, A. *Proced. ACS Div. Polymeric Materials, Science and Engineering* **1995**, *72*, 147.
30. Hattori, T.; Imai, A.; Yamanaka, R.; Ueno, T.; Shiraishi, H. *J. Photopolym. Sci. Technol.* **1996**, *9*, 611.
31. Houlihan, F. M.; Nalamasu, O.; Reichmanis, E.; Timko, A. G.; Varlemann, U.; Wallow, T.; Bantu, N. R.; Biafore J.; Sarubbi, T.; Falcigno P.; Kirner, H. J.; Münzel, N.; Petschel, K., Schacht, H.-T., Schulz, R. *Proc. SPIE, Adv. Resist Technol. Process. XIV* **1997**, *3049*, 466.
32. Mertesdorf, C.; Münzel, N.; Falcigno P.; Kirner, H. J.; Nathal, B., Schacht, H.-T.; Schulz, R.; Slater, S.; Zettler, A. in *"Polymers for Advanced Imaging and Packaging"*; Reichmanis, E.; Ober, C. K.; MacDonald, S. A.; Iwayanagi, T.; Nishikubo, T., Ed.; Symposium Series 614, American Chemical Society: Washington, D. C., **1995**, p. 35.

33. Paniez, P. J.; Rosilio, C.; Mouanda, B.; Vinet, F. *Proc. SPIE, Adv. Resist Technol. Process. XI* **1994**, *2195*, 14.
34. Vinet, F.; Buffet, N.; Fanton P.; Pain, L.; Paniez, P. J. *Proc. SPIE, Adv. Resist Technol. Process. XII* **1995**, *2438*, 202.
35. Ito, H. *J. Polym. Sci., Part A* 1986, *24*, 2971.
36. Maruzen Oil Co., Ltd. Jpn. Kokai Tokkyo Koho, JP 88-137091; *Chem. Abstr.* **1989**, *111*, 195665m.
37. Hitachi Chemical Co., Ltd. Jpn. Kokai Tokkyo Koho, JP 87-289336; *Chem. Abstr.* **1989**, *111*, 215560d.
38. Shiraishi, H.; Hayashi, N.; Ueno, T.; Sakamizu, T.; Murai, F.; *J. Vac. Sci. Technol. B* **1991**, *9*, 3343.
39. *Methoden der organischen Chemie*; Houben-Weyl; 4. Edition, Stuttgart, **1991**, Vol. E14a/1, p. 323.
40. Sandler, S. R.; Karo, W.; *Organic Functional Group Preparations*; Academic Press, New York, **1972**, Vol. III, p. 1.
41. Effenberger, F. *Angew. Chem.* **1969**, *81*, 374.
42. Schacht, H.-T.; Falcigno, P.; Münzel, N.; Holzwarth, H.; Schneider, J. *J. Photopolym. Sci. Technol. Process.* **1996**, *9*, 573.
43. Itani, T.; Iwasaki, H.; Fujimoto, M.; Kasama, K. *Proc. SPIE, Adv. Resist Technol. Process. XI* **1994**, *2195*, 126.
44. Bantu, N.; Maxwell, B.; Medina, A.; Sarubbi, T.; Toukhy, M.; Schacht, H.-T.; Falcigno, P.; Münzel, N.; Petschel, K.; Houlihan, F. M.; Nalamasu, O.; Timko, A.G. *Proc. SPIE, Adv. Resist Technol. Process. XIV* **1997**, *3049*, 324.
45. Nalamasu, O.; Timko, A. G.; Cheng, M., Kometani, J. M.; Galvin, M.; Heffner, S.; Slater, S. G.; Blakeney, A. J.; Münzel, N.; Schulz, R.; Holzwarth, H.; Mertesdorf, C.; Schacht, H.-T. *Proc. SPIE , Adv. Resist Technol. Process. X* **1993**, *1993*, 155.
46. Münzel, N.; Holzwarth, H.; Falcigno, P.; Schacht, H.-T.; Schulz, R.; Nalamasu, O.; Timko, A. G.; Reichmanis, E.; Stone, D. R.; Slater, S. G.; Blakeney A. J. *Proc. SPIE, Adv. Resist Technol. Process. XI* **1994**, *2195*, 47.
47. Nalamasu, O.; Timko, A. G.; Reichmanis, E.; Houlihan F. M.; Novembre, A. E.; Tarascon, R.; Münzel, N.; Slater, S. G. in *"Polymers for Advanced Imaging and Packaging"*; Reichmanis, E.; Ober, C. K.; MacDonald, S. A.; Iwayanagi, T.; Nishikubo, T., Ed.; Symposium Series 614, American Chemical Society: Washington, D. C., **1995**, p. 4.
48. Nalamasu, O.; Kometani, J.; Cheng, M.; Timko, A. G.; Reichmanis, E.; Slater S.G.; Blakeney, A. J. *J. Vac. Sci. Technol. B* **1992**, *10*, 2536.
49. Dean, K. R.; Carpio, R. A.; Rich, G. K. *Proc. SPIE, Adv. Resist Technol. Process. XII* **1995**, *2438*, 514.
50. Yabe, S.; Watanabe, M.; Satou, I.; Taguchi, T. *J. Photopolym. Sci. Technol. Process.* **1997**, *10*, 465.

Chapter 8

Chemistry of Ketal Resist System and Its Lithographic Performance

Wu-Song Huang[1], Kim Y. Lee[2], Rao Bantu[3], Ranee Kwong[1], Ahmad Katnani[1], Mahmoud Khojasteh[1], William Brunsvold[1], Steven Holmes[1], Ronald Nunes[1], Tsuyoshi Shibata[1,4], George Orsula[5], James Cameron[5], Dominic Yang[5], and Roger Sinta[5]

[1]IBM Microelectronics, Hopewell Junction, NY 12533
[2]ETEC System Incorporated, Hayward, CA 94545
[3]Olin Microelectronic Materials, Providence, RI 02914
[4]Toshiba Corporation, Alliance Partner
[5]Shipley Company, 455 Forset Street, Marlboro, MA 01752

Since the introduction of chemically amplified resist systems to DUV technology, the environmental stability and bake latitudes have been the major concern of this type of chemistry. Ketal resist systems have been very robust towards these issues. The methoxypropene protected polyhydroxystyrene resist is our first initial work on ketal system. This resist, after optimization, has 0 nm /°C PEB sensitivity; no environmental concern; minimal slimming; 0.263N TMAH developer compatible and 5-6 months storage shelf life at room temperature. This paper will discuss its chemistry and its lithographic performance.

Recently, a significant shift from I-line to Deep-UV lithography has occurred in the semiconductor industries due to the requirement of printing 250 nm images or below. This migration has spurred many resist companies to develop a high performance chemically amplified DUV resist system within a very short time, in order to grip some market share in its early stage. In this highly competitive resist business, many different chemistry platforms have been investigated and various resist products are developed. Although majority of the resist compositions have an acid labile protecting group on a polymer backbone, the activation energy of the protecting group seems to dictate the resist property and lithography process. In general, low activation energy systems have better environmental stability and bake latitudes, but also have the tendency of line width slimming. On the other hand, high activation energy systems usually have poor environmental stability and bake latitudes, but very few reported literature has shown that there is line width slimming. To achieve the environmental stability in high activation energy system, an annealing concept has been introduced to a hydroxystyrene-t-butyl acrylate copolymer based resist (*6-10*). This system requires higher post apply bake (PAB) and post exposure bake (PEB) temperatures. Ketal resist system (KRS) has an extremely low activation energy, which will deprotect during exposure, therefore is extremely robust towards base contamination (*1-5*). We will discuss the chemistry and lithographic performance of KRS, which contains a base resin of polyhydroxystyrene protected with methoxypropanyl (MOP) in this chapter.

Result and Discussion

Synthesis of Ketal Protected Polymer. A low cost process to synthesize ketal protected polyvinylphenols (PVP) has been developed. As shown in Scheme I, the corresponding alkoxyalkenes were added to a solution of PVP with catalytic amount of acids that would then produce the corresponding ketal protected PVP. The extent of protection is controlled by the alkoxyalkene loadings. The blocking level may deviate from the loading depending on the moisture content of the polymer and the reaction solvent. One mole of water will consume two moles of alkoxyalkene. The amount of protection can be determined by C13 NMR as shown in Figure 1. The mole percent protection is calculated by dividing the area integration of peak at 121 ppm to the combination of peaks at 121 ppm and 115 ppm. The protection level can also be qualitatively identified by the two peaks at 1065 cm^{-1} and 1130 cm^{-1} in IR spectrum (Figure 2).

The acid can catalyze the deprotection of the MOP group as shown in Scheme II. The water reacts with the deprotected cation to form acetone and methanol. Both are very volatile species and will not condense on the lenses of the stepper. However, the triflic acid generated in most of our formulations can be a concern and should be replaced for the large production environment despite the minuscule amount of acid detected in the lab experiments. With insufficient amount of water, the methanol will react with the deprotected cation to form dimethoxypropane as intermediate. In the absence of water, during thermal deprotection, the deprotected cation regenerates the methoxypropene group which has been proven by trapping and analyzing the gas generated during deprotection by NMR.

Manufacture of KRS Resist. Figure 3 shows the typical dissolution rate of the MOP protected PVP's which were synthesized using p-toluenesulfonic acid as catalyst. This figure shows a sharp change in dissolution rate from 0% to 10% and starts to level off around 20% protection. This suggests that to achieve a good lithographic performance, it is desirable to have at least 20% protection. The resist sensitivity for partially TBOC protected PVP system is sensitive to the blocking level. For KRS, the resist is very insensitive to the blocking level of the polymer(*10*). With blocking level of MOP between 10% to 45%, the lithographic dose is essentially the same using 0.14 N tetramethylammonium hydroxide (TMAH) developer. When the protecting level becomes too high, wetting the resist surface becomes difficult, and the dose drifts slightly higher. Table I shows the dose to clear (E_0) vs. the protection level. The E_0 changes with the thickness as shown in the swing curve (Figure 4) but not with the protection level. Only when it reaches 50% protection, the E_0 starts to drift to a higher value. With excess methoxypropene in the reaction, the highest protection level achieved on PVP in the lab is 77%, but there was no effort spent to find the highest protection level on PVP. When the protection

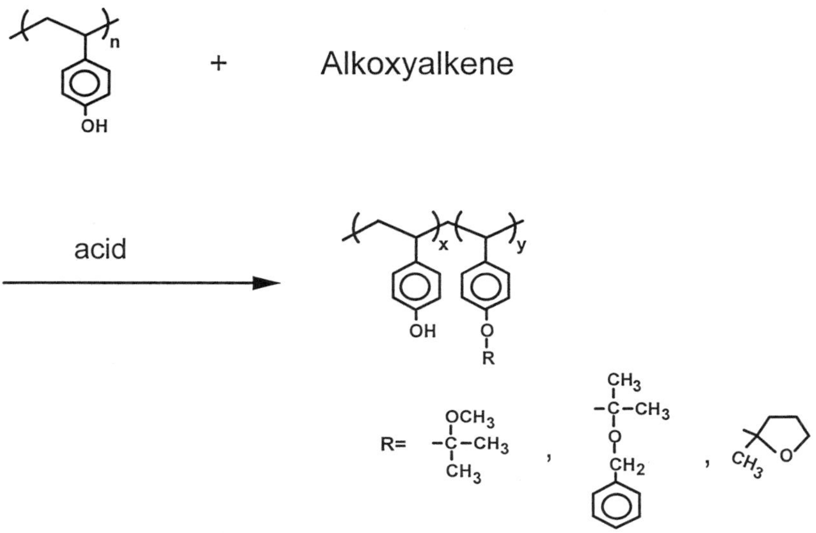

+ Alkoxyalkene

acid

OH

R= $-C(OCH_3)(CH_3)-CH_3$, $-C(CH_3)_2-O-CH_2-C_6H_5$, 2-methyltetrahydrofuran-2-yl

Scheme I. Synthetic scheme for partially ketal protected PVP.

Protection = a / a+b

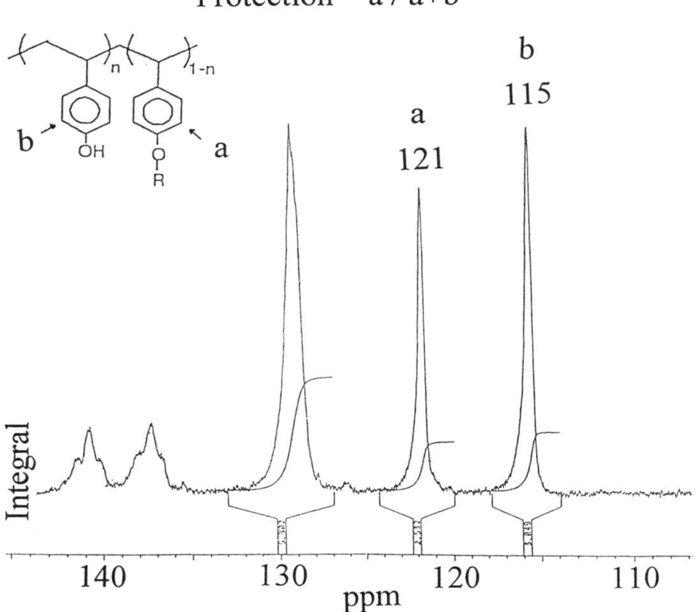

Figure 1. NMR spectrum of partially ketal protected PVP.

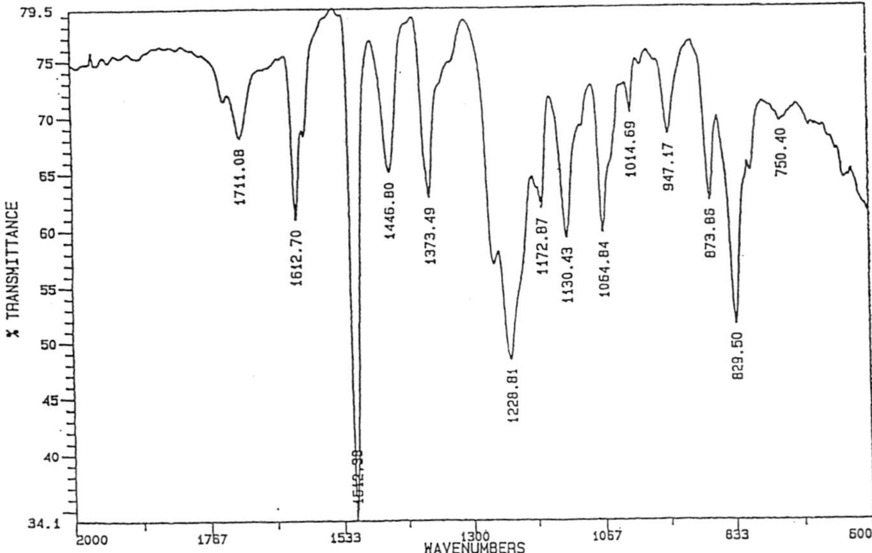

Figure 2. IR spectrum of partially methoxypropene protected PVP.

Ketals

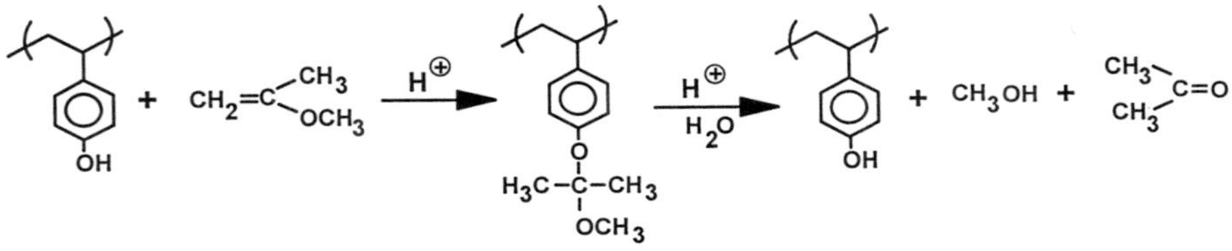

OH + $CH_2=C(CH_3)OCH_3$ $\xrightarrow{H^{\oplus}}$ $O-C(CH_3)_2-OCH_3$ $\xrightarrow[H_2O]{H^{\oplus}}$ OH + CH_3OH + $(CH_3)_2C=O$

SchemeII. Protection and deprotection schemes of methoxypropene to PVP.

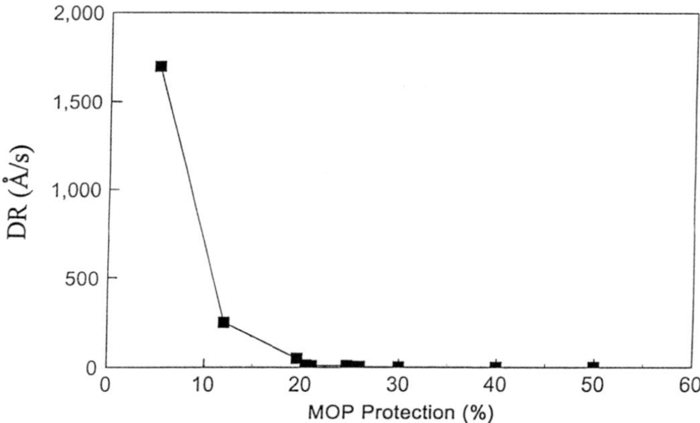

Figure 3. Dissolution rates of the polymer vs. MOP protection level in 0.263N TMAH developer.

level is too low, there is no change in lithographic dose but excessive film loss on resist after development.

Table I. Mole percent protection of MOP vs. E_0 with different resist thickness

Protection (%)	Thickness (Å)	E_0 (mJ/cm^2)
14	9,241	7
23	9,261	7
20.6	9,261	11
31.8	8,755	11
19.3	8,275	13
50.8*	8,307	15

*Developed with 0.15N for 60s instead of 0.14N for 80s, PAB is 100°C for 90s instead of 100°C for 120s

The amount of photoacid generator (PAG) in the resist determines the amount of acid generated during photo process, thus determining the resist sensitivity. As was reported in KRS system (*2*), lithographic dose of the resist is linearly proportional to the reciprocal of PAG loading at fixed amount of basic additive (Figure 5). If we maintain the same ratio between PAG and base and vary the PAG loading, the lithographic dose is essentially the same as shown in Figure 6. In other words, if we can control the precise ratio of PAG to base, one can put any amount of polymer to the solvent to prepare the resist solution.

Shelf Life of KRS. Traditionally, we define the shelf life of a resist based on the lithographic dose drifting out of the required spec of ± 3% of the original dose. With this definition, one will not be able to determine shelf life of KRS. If the same resist thickness is maintained, KRS does not change lithographic dose. As described in the previous section, the MOP blocking level on PVP does not affect the lithographic dose. In the aging process, the polymer slowly loses a small portion of the MOP and decreases its blocking level. Changing blocking level reduces polymers' molecular weight which in turn reduces the resist solution's viscosity, and then decreases the coating thickness. With improved synthetic method for the polymer with any trace amount of acids removed from the resist solution, the thickness does not change at all in two months (at 20°C and 40°C) as shown in Figure 7. The observation in the laboratory indicates that the thickness did not decrease more than 30Å within 5-6 months period at room temperature.

Line width Slimming. Experimental findings have shown that KRS has a line width slimming effect. The amount of slimming ranges from 0-50 nm depending on tool and location and is independent of feature size in most cases. As shown in Figure 8, the slimming occurred very fast in the first 15 minutes and then leveled off after 40-50 minutes. The line width does not change afterwards, even up to 24 hours and beyond. Some of the experiments showed slimming occurring only on the bottom of the resist images and the top did not change. The cause of this slimming is

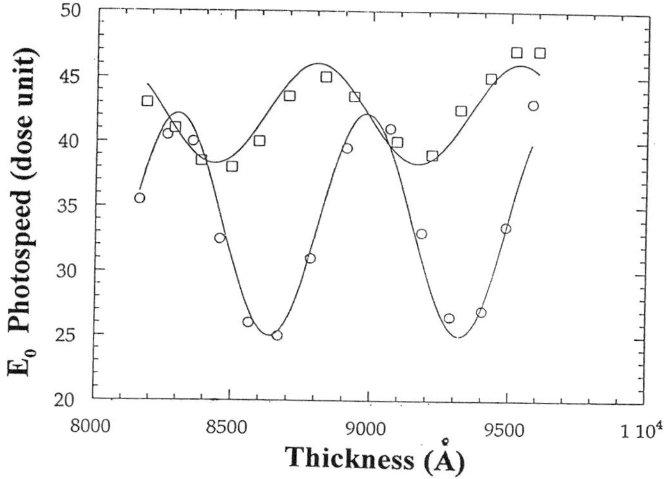

Figure 4. Swing curves of KRS on Si (O) and on BARL (□) using 0.45NA GCA stepper.

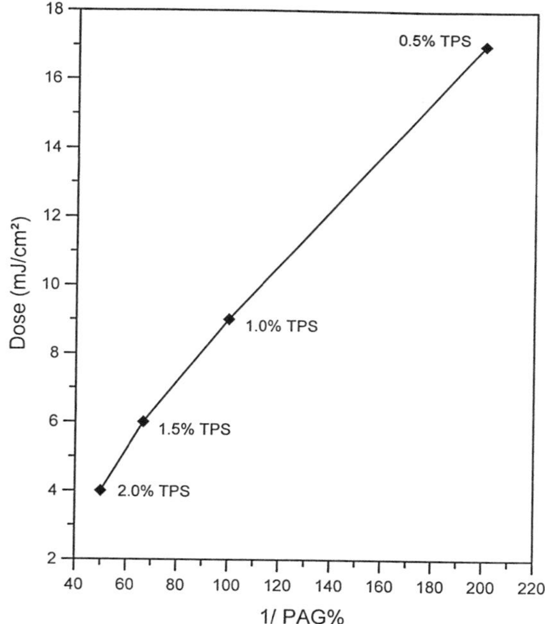

Figure 5. Doses vs. 1/PAG for KRS. TPS loadings are 2, 1.5, 1 and 0.5% of the total solid content with fixed base amount.

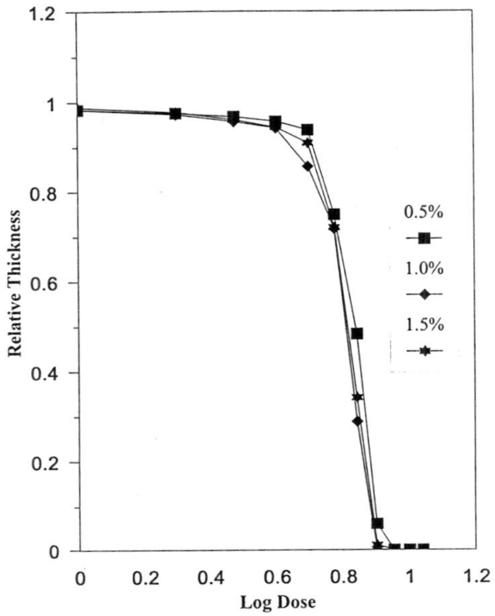

Figure 6. The contrast curves of KRS with various PAG loadings from 0.5% to 1.5% using the same PAG to base ratio.

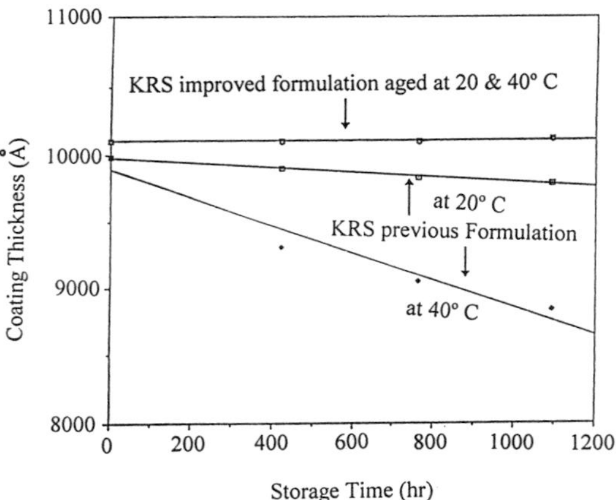

Figure 7. KRS aging data. The coating thickness change of KRS resist formulation stored at 20 and 40° C for 1100 hours.

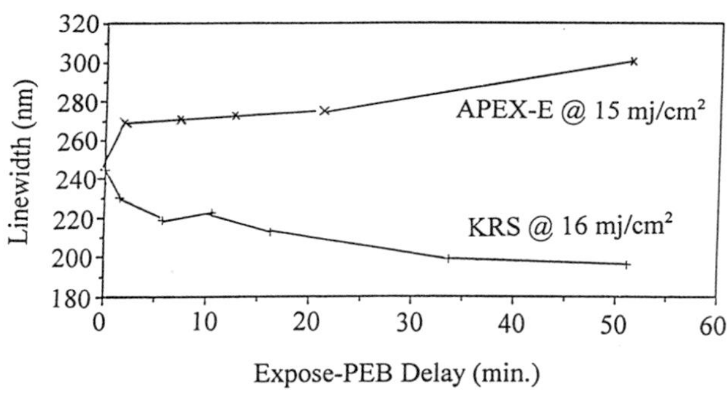

Figure 8. Linewidth vs. exposure-PEB delay time for KRS and APEX-E.

not very clear, and the common assumption is due to the effect of the acid diffusion. Slimming effect can be reduced by using bulkier and weaker acid, higher PAB temperature, higher base loading, and stronger developer. With the new KRS formulation for 0.263N developer, the line width slimming is not detectable on the 250 and 225 nm images after one hour delay, as shown in Figure 9.

Standing Waves. Most of the commercial DUV resists have some standing waves, with APEX-E type TBOC system having less. The cause of standing waves is the lack of acid diffusion in KRS resist. Since the reaction is complete during the exposure step, the PEB step is not efficient in removing the standing waves of KRS. Some success was achieved with adding plasticizer to lower the Tg of the resist. This approach reduces the resist's thermal flow resistance and no extensive study was done on its lithographic performance. Since the resist is compatible to most organic Anti-Reflective Coatings (ARC's), the best approach is to use ARC in the resist process if standing waves is a concern.

Bake latitudes and Environmental Stability. For low activation systems, the bake latitudes are usually very good with PEB sensitivity less than 4 nm/°C (*11*). KRS is extremely insensitive to PAB and PEB temperatures. The PEB sensitivity is 0 nm/°C as shown in Figure 10 for the 250 nm resist images. The low PAB sensitivity is common in most of the resist systems except in a few special cases.

As reported earlier (*1-5*), KRS is extremely environmentally stable. No matter where this resist is processed, there is no observation of any contamination effect so far.

0.263N TMAH KRS. Most of the early work on KRS was processed with 0.14N TMAH developer. The PAG used was triphenylsulfonium triflate. By changing the PAG to a higher inhibiting triarysulfonium PAG, we have been able to process the resist with 0.263N TMAH developer. The lithographic performance is superior to the 0.14N TMAH version with linear resolution down to 225 nm using 0.50 NA and 0.60NA tool (Figure 11). The dose latitude for 250 nm is greater than 20% and focus latitude is around 1.2 μm (Figure 12).

Conclusion

In this paper we have described a very viable resist system that has great potential in meeting many challenges of the aggressive semiconductor industries for their insatiable demand for highly performing resist. The resist is environmentally stable, insensitive to bake temperatures, easy to manufacture, compatible with 0.263N developer, and has a reasonable shelf life. The most important thing is that the resist can be adapted for particular uses by choosing different protecting groups, which gives an unlimited future for this type of resist.

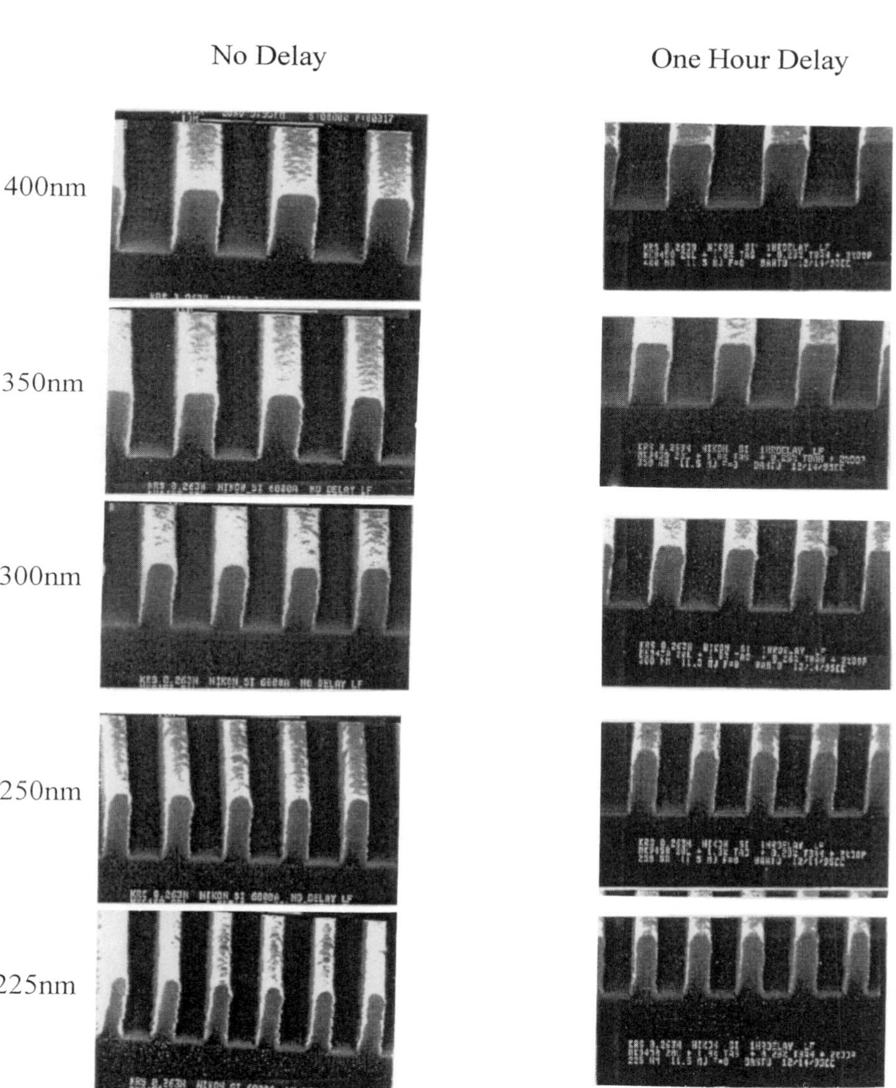

Figure 9.KRS L/S images from 400nm to 225nm with and without PEB delay.

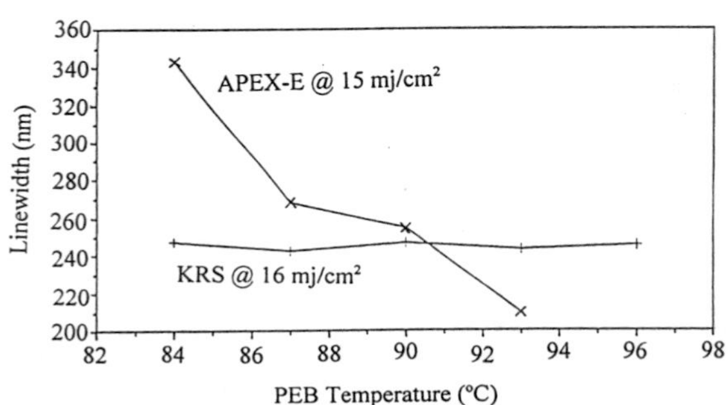

Figure 10. Linewidth variation against PEB temperatures for APEX-E and KRS using MSII stepper (0.50 NA) stepper.

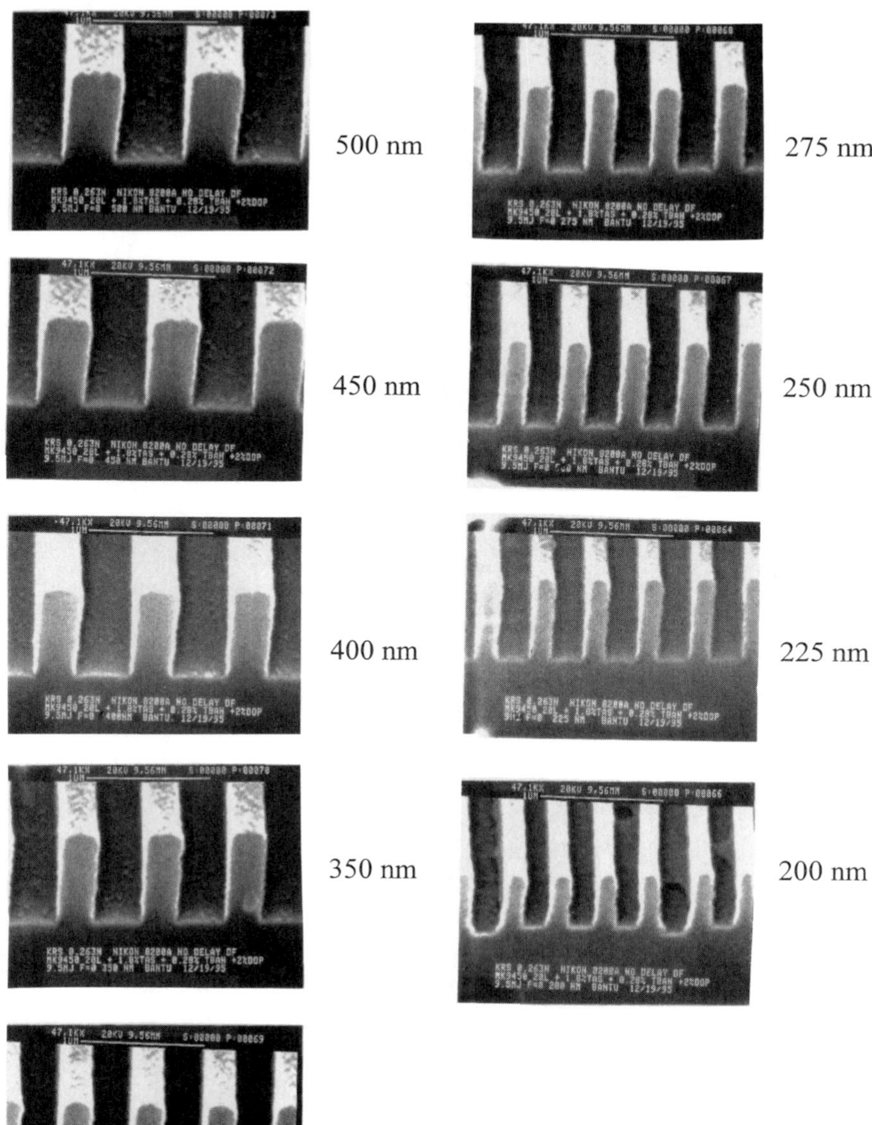

Figure 11. KRS L/S images for 820nm thick resist using Nikon EXX 0.60NA stepper.

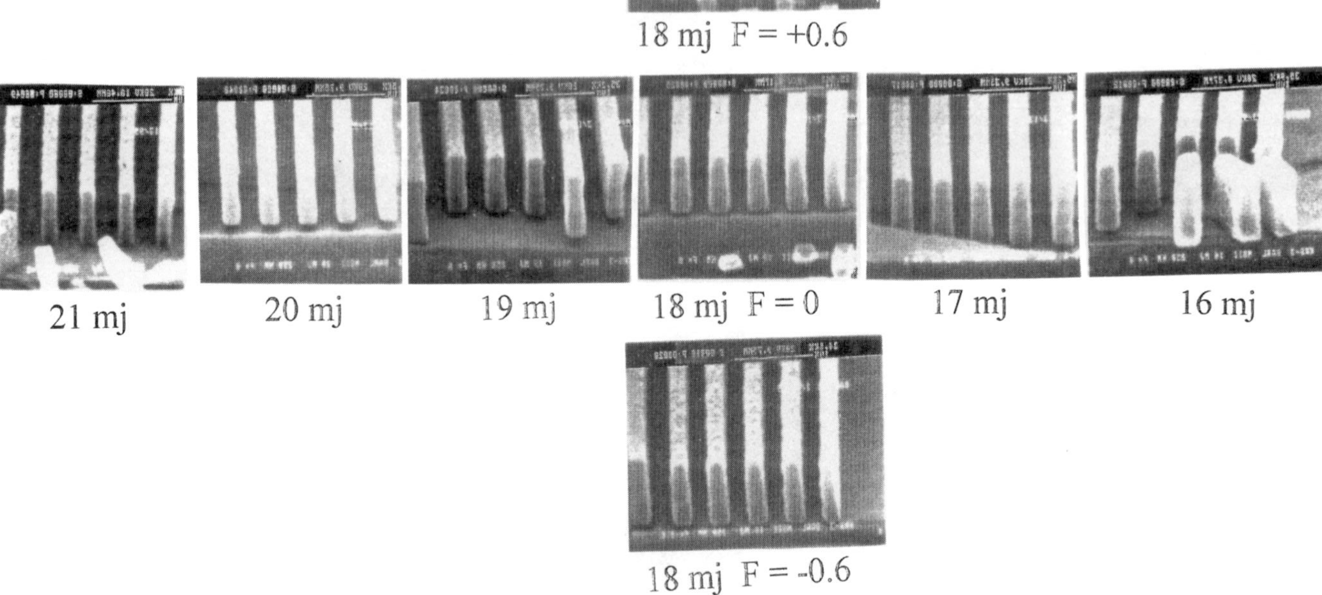

Figure 12. KRS 250nm L/S images from 16 to 21 mj/cm^2 with 1.2 micron focus latitude.

Acknowledgments

The authors would like to extend their gratitude to G. Hefferon, P. Jaganathan, K. Petrillo, R. Lang, G. Spinillo, W. Moreau, J. Fahey, R. Sooriyakamuran, J. Sturtevant, M. Denison, and J. Thackeray for their invaluable discussions and technical support.

Literature Cited

1. Huang,W.S.; Kwong, R.; Katnani, A.D.; Khojasteh, M. and Lee, K.Y., *Mat. Res. Soc. Symp. W. Proc.*, **1994**, *324*, 493
2. Huang,W.S.; Kwong, R., Katnani, A.D.; Khojasteh, M., *Proc. SPIE*, **1994**, *2195*, 37
3. Lee, K.Y. and Huang, W.S., *J. Vac. Sci. Tecnnol. B*, **1993**, *11(6)*, 2807
4. Huang, W.S., Kwong, R., Katnani, A.D.; Khojasteh, M.; and Lee, K.Y., *Photopolymers SPE Conference*, **1994**, 96
5. Huang, W.S., Katnani, A.D.; Yang, D.; Brunsvold, B.; Bantu, R.; Khojasteh, M; Sooriyakumaran, R.; Kwong, R.; Lee, K. Y.; Hefferon, G.; Orsula, G.; Cameron, J.; Denison, M.; Sinta, R. and Thackeray, J., *J. Photopolym. Sci. Technol.*, **1995**, *8 (4)*, 525
6. Ito, H.; England, W.P.; Clecak, N.J.; Breyta, G.; Lee, H.; Yoon, D.Y.; Sooriyakumaran, R. and Hinsberg, W.D., *Proc. SPIE*, **1993**, *1925*, 65
7. Ito, H.; England, W.P.; Sooriyakumaran, R.; Clecak, N.J.; Breyta, G.; Hinsberg, W.D.; Lee, H.; Yoon, D.Y., *J. Photopolym. Sci. Technol.*, **1993**, *6(4)*, 547
8. Ito, H.; Breyta, G.; Hofer, D.; Sooriyakumaran, R., *Polym. Mater. Sci. Eng.*, **1995**, *72*, 144
9. Ito, H.; Breyta, G.; Sooriyakumaran, R. and Hofer, D., *J. Photopolym. Sci. Technol.*, **1995**, *8(4)*, 505
10. Conley, W.; Breyta, G.; Brunsvold, B.; DePietro, R.; Hofer, D.; Holms, S.; Ito, H.; Nunes, R.; Ficftl, G.; Hagerty, P.; Thackeray, J., *Proc. SPIE*, **1996**, *2724*, 34
11. Mertesdorf, C.; Munzel, N.; Holzwarth, H.; Falcigno, P.; Schacht, H.-T.; Rohde, O.; Schulz, R.; Slater, S.G.; Frey, D.; Nalamasu, O.; Timko, A.G.; Neenan,T.X., *Proc. SPIE*, **1995**, *2438*, 84

Chapter 9

Photoacid Diffusion in Chemically Amplified DUV Resists

Toshiro Itani, Hiroshi Yoshino, Shuichi Hashimoto, Mitsuharu Yamana, Norihiko Samoto, and Kunihiko Kasama

NEC Corporation, 1120 Shimokuzawa, Sagamihara, Kanagawa 229-11, Japan

In order to clarify the effects of acid diffusion on lithographic performance, the acid diffusion behavior in chemically amplified positive KrF resists was studied. The resists consisted of tert-butoxycarbonyl (t-BOC) protected polyhydroxystyrene and a benzenesulfonic acid derivative as a photoacid generator (PAG). Acid diffusion coefficient and diffusion length were obtained using Fick's diffusion law by analyzing the amount of generated acid and ion conductivity of a resist film. As a result, it was confirmed that the acid diffusion is ruled by only one mechanism, and two diffusion paths, which correspond to the remaining solvent in the resist film and hydrophilic OH sites in the base resin, existed. Moreover, the acid diffusion length was decreased by increasing photoacid bulkiness. Furthermore, it was found that additional base component not only quenched photo-generated acid but also suppressed the acid diffusion. Based on the experimental analysis, the acid diffusion behavior in the resist film was clarified and the relationship between acid diffusion and resist performance was obtained.

A chemically amplified resist based on acid catalysis is the most promising technology for next generation lithography such as deep-ultraviolet, electron beam and X-ray lithography. Many efforts to improve inherent resist material as well as the resist processing have been reported.[1-18] For improving resist performance, it is very important to understand the role of each component in resist formulation in lithographic performance (such as resolution capability, focus margin, and standing wave effect), as well as in the process stability such as exposure dose margin and delay time stability between exposure and the post-exposure bake (PEB). Therefore, inherent resist characteristics such as acid generation characteristics, acid diffusion behavior, deblocking reaction, and dissolution characteristics have been investigated. In particular, the influences of photoacid generators (PAGs) and their diffusion behavior are considered to be very large, and many articles have been published on this subject.[3-11] In this article, the acid diffusion behavior in a resist film is investigated in chemically amplified positive KrF resists for various PEB conditions, blocking levels of base resin, prebake temperatures, molecular weight dispersions, and base loading. As a result, the acid diffusion behavior in resist film has been clarified and the effects of acid diffusion on lithographic performance have been revealed.

Experimental

Materials and Processing. Chemically amplified positive KrF resists, consisting of tert-butoxycarbonyl (t-BOC) protected polyhydroxystyrene and a benzenesulfonic acid derivative PAG (5 wt%), were used. The casting solvent is propyreneglycol-monomethyletheracetate (PGMEA). Through the irradiation of this PAG with KrF excimer laser light, 2,4-dimethylbenzenesulfonic acid is generated. The resist samples were coated on silicon substrates primed with hexamethyldisilazane to a thickness of 0.7 μm and then prebaked at 90°C for 90 s (as a standard condition). The samples were exposed by a KrF excimer laser stepper with a 0.50-NA lens. Then, PEB was carried out at 100°C for 90 s (as a standard condition) on a hot plate within 5 min after exposure to minimize airborne contamination.

Measurement of Acid Diffusion. Acid diffusion coefficient D and diffusion length L were obtained from Fick's diffusion law by using the following equations:

$$D = \sigma kT/[H]q^2,$$
$$L = (2Dt)^{1/2},$$

where σ, k, T, $[H]$, q, and t are ion conductivity, the Boltzmann constant, diffusion temperature, the amount of acid, ionic charge, and diffusion time, respectively. The amount of acid was determined by the following. The resist samples were coated on silicon substrates with a thickness of 0.7 μm, and then prebaked at 90°C for 90 s. These wafers were exposed at various exposure doses of 0 – 100 mJ/cm^2. The resist film on each wafer was collected by dissolving in acetone, and tetrabromophenolblue as an indicator dye was added. The solution was diluted to a certain amount in an acid-free treated messflask. The amount acid was determined by measuring the absorption of the solution to a light of 619 nm wave length. The ion conductivity of the resist film was determined by the following. The resist samples were coated on quartz substrate that has an arched electrode, with a thickness of 0.7 μm, and then prebaked at 90°C for 90 s. These wafers were exposed at various exposure doses of 0 – 100 mJ/cm^2, and PEB was carried out at 100°C for 90 s. The ion current was measured by electronic probe on arched electrode, and the ion conductivity was calculated.[3,5,6,11]

Results and Discussion

Acid Diffusion Reaction Mechanism. We evaluated the acid diffusion behavior under several PEB conditions (60-100°C for 90 s). Figure 1 shows the amount of generated acid [H] as a function of exposure dose. The amount of generated acid increases exponentially with increasing exposure dose. Figure 2 shows the ion conductivity σ of the resist film as a function of exposure dose. The σ value also increases exponentially with increasing exposure dose for each PEB temperature, and a higher PEB temperature induces higher ion conductivity. By using these parameters and the equations which originated from the diffusion law, the diffusion coefficient and the acid diffusion length were obtained. Figure 3 shows plots of diffusion coefficient versus exposure dose at several PEB temperatures. A higher PEB temperature gave rise to higher diffusion coefficient. At low PEB temperatures (≤90°C), the diffusion coefficient increased with increasing exposure dose and then saturated to a constant value. On the other hand, at high PEB temperatures (>90°C), the diffusion coefficient decreased with increasing exposure dose and then saturated to a constant value. It is considered that the diffusion coefficient at high PEB temperatures and lower exposure dose, was affected by the acid concentration reduction, originating from the acid disappearance via evaporation from the resist film surface during high PEB treatment (>100°C). Figure 4 shows acid diffusion length as a function of PEB time for several PEB temperatures. The acid diffusion length increased with increasing PEB time, and higher PEB temperature brought about longer diffusion length.

In order to analyze the activation energy (Ea) for acid diffusion, Arrhenius plots were obtained in the lower temperature range (60-100°C) where the effect of acid disappearance is minimized. Figure 5 shows Arrhenius plots of diffusion coefficient. Each plot shows a straight line for each exposure dose. This fact indicates that only one mechanism dominates the acid diffuison. Figure 6 shows the activation energy of the acid diffusion, which was estimated from the slope of the Arrhenius plots. The activation energy decreases with increasing exposure dose, and tends to saturate to a constant value. It is considered that hydrophilic OH groups of the base resin, generated by deprotection of hydrophobic t-BOC groups, became one of the diffusion paths. Therefore, the acid diffusion rate is increased at higher OH group concentrations.

Analysis of Acid Diffusion Paths. To clarify the acid diffusion path in the resist film, we evaluated the acid diffusion behavior for various prebake temperatures (90-150°C, 90s) and t-BOC blocking levels (0-60%), focusing our attention on the concentration of remaining solvent in the resist film. Figure 7 shows resist film thickness as a function of prebake temperature. The resist film thickness decreased with increasing prebake temperature. Also, a higher blocking level induced a larger loss of film thickness. This indicates that the t-BOC groups were decomposed and volatilized from the resist film; therefore, the resist film volume decreased to a greater extent at a higher blocking level than at a lower one. Figure 8 shows the concentration of the remaining solvent in the resist film as a function of prebake temperature. The concentration of remaining solvent was determined by gas

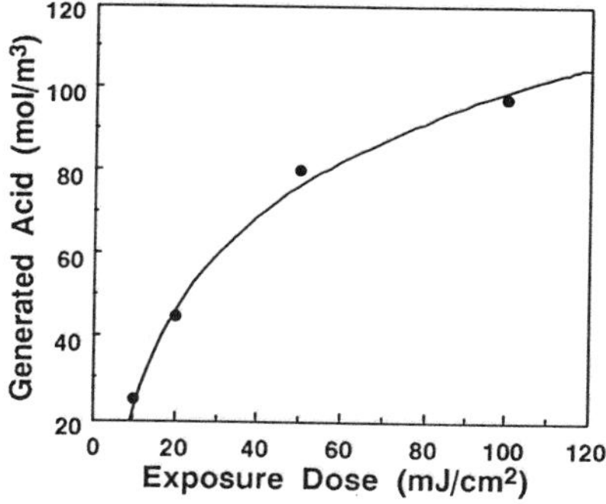

Figure 1. Amount of generated acid as a function of exposure dose.

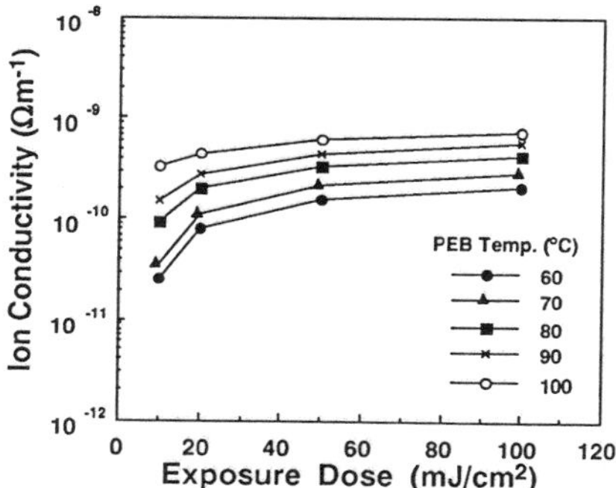

Figure 2. Ion conductivity of resist film as a function of exposure dose at various PEB temperatures.

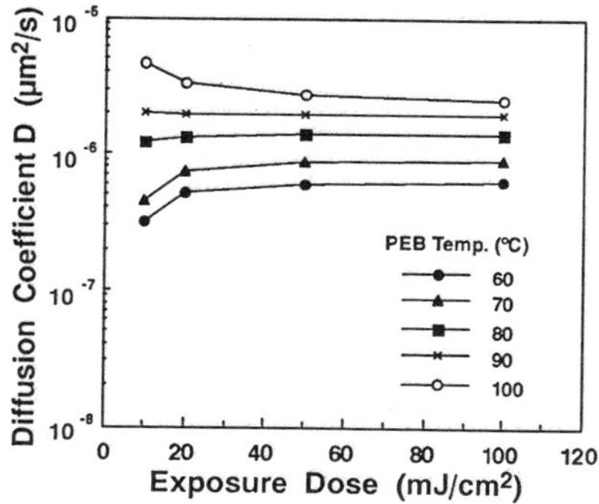

Figure 3. Diffusion Coefficient as a function of exposure dose at various PEB temperatures.

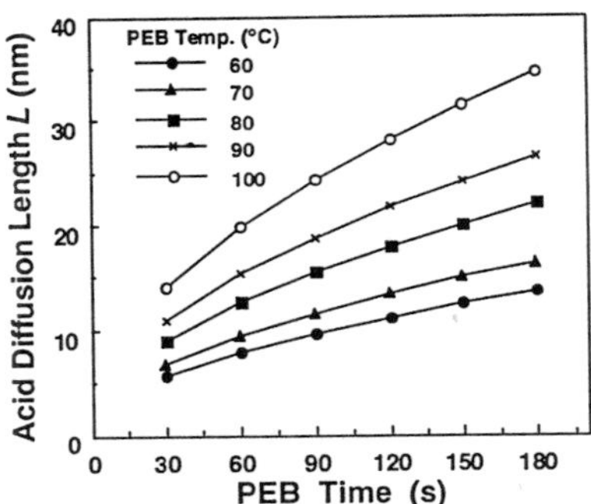

Figure 4. Acid diffusion length as a function of PEB time at various PEB temperatures.

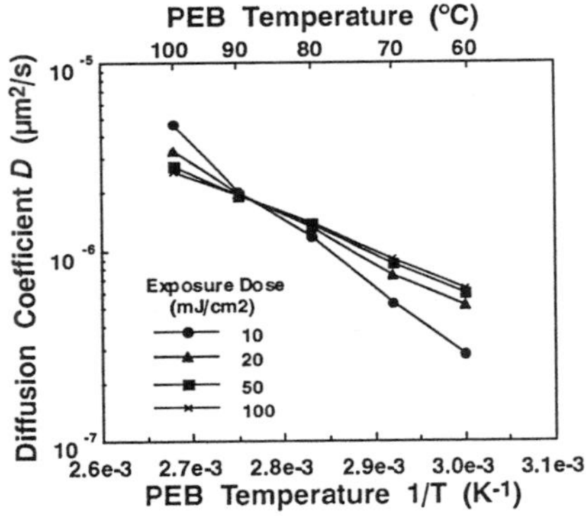

Figure 5. Arrhenius plots of diffusion coefficient.

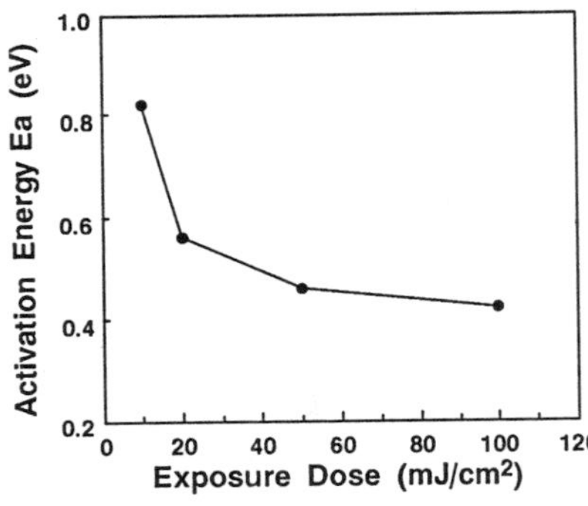

Figure 6. Activation energy of acid diffusion.

chromatograph-mass spectrometer. The amount of remaining solvent decreased with increasing prebake temperature. Moreover, the remaining solvent was lower in a higher blocking level resist than in a lower one.

A diffusion coefficient was estimated from the amount of generated acid and the ion conductivity. Figures 9 and 10 show plots of diffusion coefficient versus exposure dose for various prebake temperatures and blocking levels. The diffusion coefficient decreased with increasing prebake temperature or blocking level for each exposure dose. Based on the these results, the acid diffusion length for several prebake temperatures and blocking levels were calculated for a diffuison time of 90 s. Figure 11 shows plots of acid diffusion length versus prebake temperature, at a 50-mJ/cm2 exposure dose, for several blocking levels. The acid diffusion length decreased almost linearly with increasing prebake temperature, and a lower blocking level was associated with a longer acid diffusion length. The resist with a blocking level of 0% showed a relatively higher acid diffusion property. Figure 12 shows the relationship between acid diffusion length and remaining solvent, at 50-mJ/cm^2 exposure dose, obtained by combining Figures 8 and 11. It was found that the acid diffusion length increased with increasing remaining solvent and that the use of lower blocking level resulted in longer acid diffusion length, even if the remaining solvent concentration remained constant. This indicates that the remaining solvent corresponds to one of the diffusion channels within the resist film. In a previous section, we found that the activation energy of the acid diffusion decreased with increasing exposure dose and had a tendency to saturate at a constant value. It is considered that hydrophilic OH groups of the base resin constitutes another diffusion path. Figure 13 shows a schematic diagram of the acid diffusion behavior within the resist film. The acid (H^+) diffuses via hydrophilic OH sites in the remaining solvent field.

In order to study the above results, actual lithographic performance was evaluated for various prebake temperatures and blocking levels. The resist pattern could not be obtained using a blocking level of 20% and a 130°C or 150°C prebake temperature. This is because the blocking groups decomposed during the prebake in these cases. The resolution capabilities are summarized in Table I. A resolution of 0.22-μm lines and spaces (L&S) was obtained at a blocking level of 40% and a 90°C or 110°C prebake temperature. It should be noted that optimum prebake temperature and blocking level can produce a better resolution capability. Figure 14 shows scanning electron microscope (SEM) micrographs of a 0.25-μm L&S pattern. A higher prebake temperature or higher blocking level is associated with the occurrence of a strong standing-wave effect at the pattern sidewall. It was confirmed that the strong standing-wave effect is brought about due to a shorter acid diffusion length at higher prebake temperature or higher blocking level and that an optimum diffusion length exists from the viewpoint of pattern profile.

In summary, the existence of a direct relationship between remaining solvent and acid diffusion length was revealed, and the existence of two diffusion paths, i.e., the remaining solvent in the resist film and hydrophilic OH sites of the base resin, was confirmed. Moreover, it was found that the change of acid diffusion length corresponds directly to the lithographic performance.

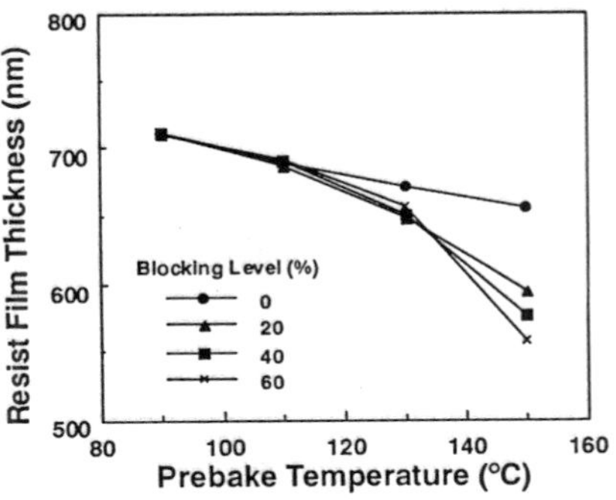

Figure 7. Resist film thickness as a function of prebake temperature for various blocking levels.

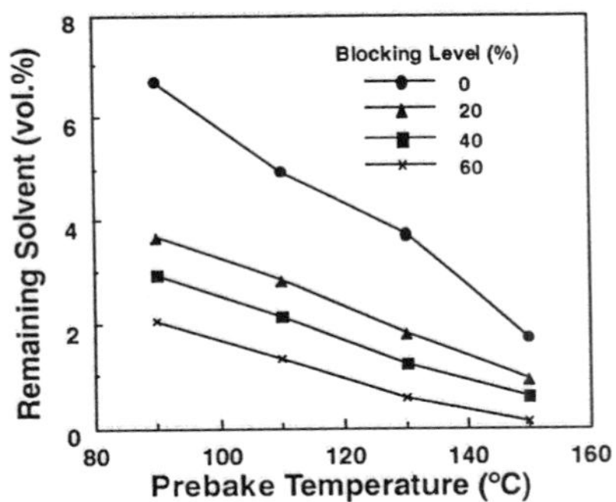

Figure 8. Concentration of remaining solvent in the resist film as a function of prebake temperature.

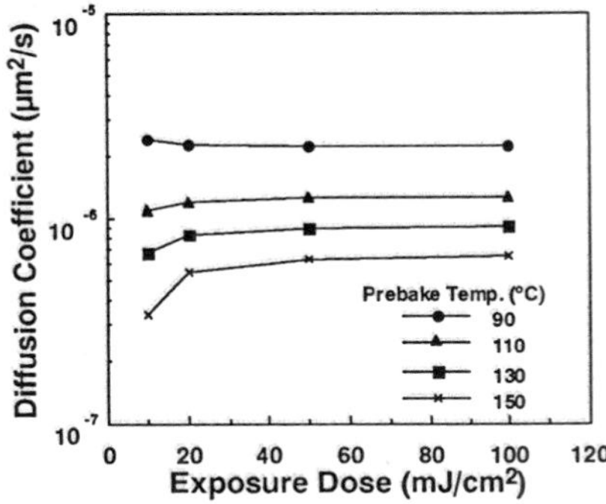

Figure 9. Diffusion coefficient as a function of exposure dose for various prebake temperatures.

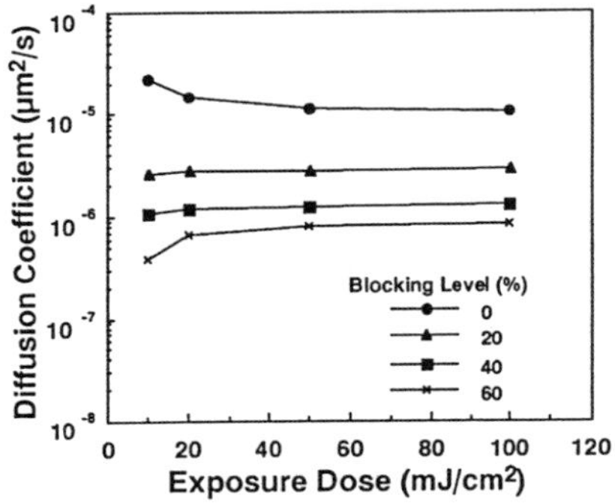

Figure 10. Diffusion coefficient as a function of exposure dose for various t-BOC blocking levels.

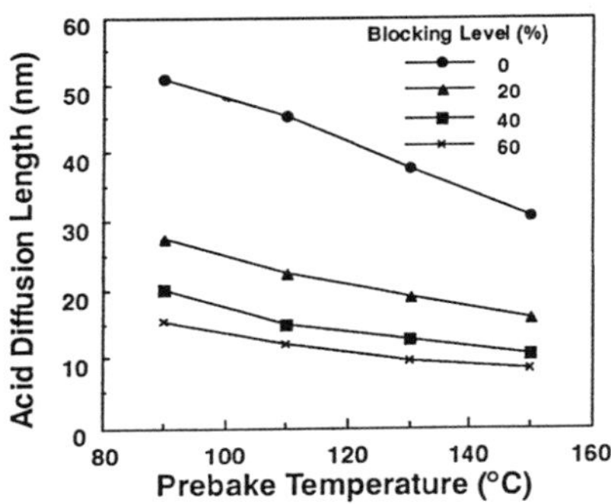

Figure 11. Acid diffusion length as a function of prebake temperature for various t-BOC blocking levels.

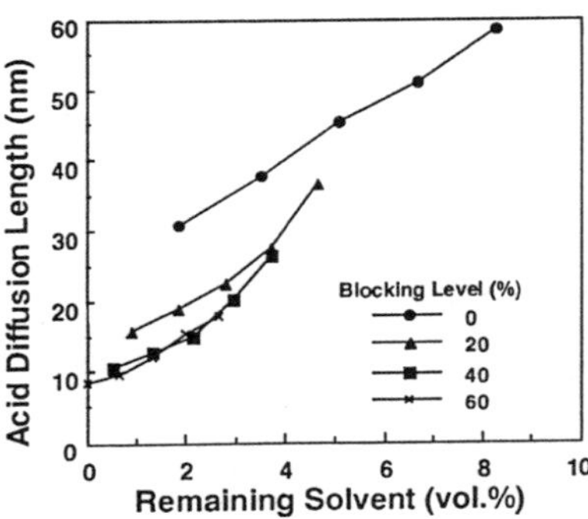

Figure 12. Acid diffusion length as a function of remaining solvent in resist film at various blocking levels.

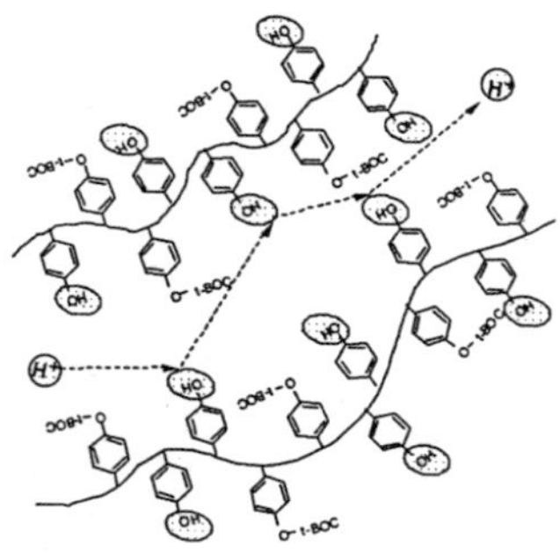

Figure 13. Scheme of acid diffusion behavior within the resist film.

Table I. Resolution capability of resist for various prebake temperatures and t-BOC blocking levels

(μm L&S)

Blocking level	Prebake temperature (℃)			
(%)	90	110	130	150
20	0.23	0.23	no pattern	no pattern
40	0.22	0.22	0.24	no pattern
60	0.24	0.23	0.24	no pattern

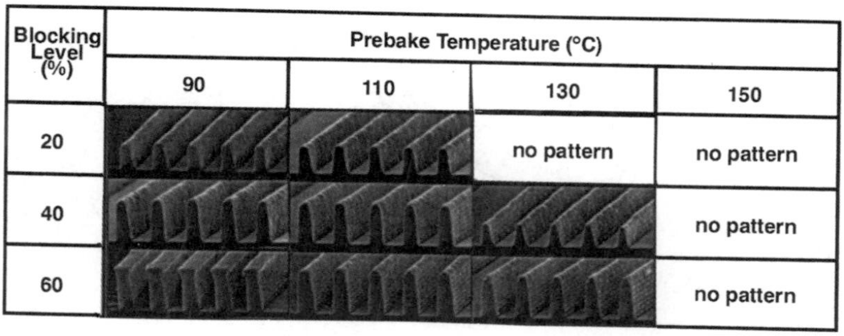

Blocking Level (%)	Prebake Temperature (°C)			
	90	110	130	150
20			no pattern	no pattern
40				no pattern
60				no pattern

Figure 14. SEM micrograph of 0.25 μm pattern for various prebake temperatures and t-BOC blocking levels.

Effect of Molecular Weight Dispersion. The influences of molecular weight dispersion (Mw/Mn) of the base resin were evaluated. Molecular weight dispersion of the base resin was 1.2, 4.0 or 9.0. These were prepared by mixing higher Mw/Mn resin and lower Mw/Mn resin. Molecular weight was constant to be 25000 among the samples. Acid diffusion parameters under an optimum exposure dose and actual acid diffusion length under 90-s PEB time are listed in Table II. The amount of generated acid under optimum exposure dose was much the same among these Mw/Mn cases. Figure 15 shows the exposure dose dependence of diffusion coefficient for various Mw/Mn. The exposure dose dependence was small for every case, but lower Mw/Mn brought about higher acid diffusion coefficient. The effect of Mw/Mn was small, however, lower Mw/Mn brought about higher acid diffusion. It is considered that acid diffusion reaction is promoted and acid diffusion length becomes uniform for lower Mw/Mn. Based on the above discussion, the acid diffusion model was considered. Figure 16 shows the acid diffusion model. In the previous section, we described that photo-generated acid diffuses via both remaining solvent or hydrophilic OH site within the resist film. In this figure, hatched circles represent photo-generated acid within a polymer matrix (white blocks). The reason why average acid diffusion length becomes longer for lower Mw/Mn (a), compared to that for higher Mw/Mn (b), is that photo-generated acid apparently diffuses more uniformly and smoothly in a homogeneous polymer matrix (a).

Effect of Acid Structure. Next, we evaluated the effect of photoacid bulkiness on acid diffusion. For the benzenesulfonic acid PAG, four types of substituents, 4-fluoro (F), 4-chloro (Cl), 2,4-dimethyl (diMe), and 4-tert-butyl (tertBu), were studied. Acid strength of these PAGs was almost the same (pKa~2.5 in H_2O at 20°C). The transmittance of resist films, which affects resolution and profile, was almost constant at 49~51%/0.7 μm. Therefore the resist performance is expected to depend only on the substituent type. The acid diffusion parameters and acid diffusion length of these PAGs are listed in Table III. The acid diffusion length of the diMe substituent was comparatively smaller than those of the F and Cl substituents, and acid diffusion length of the tertBu substituent was fairly small compared with the others. This fact indicates that the acid diffusion in the resist film is decreased by increasing photoacid bulkiness. It is believed that the acid catalytic reaction number decreased for the shorter acid diffusion length. Figure 17 shows a SEM micrograph of the resist pattern profile and resolution capability for various levels of photoacid bulkiness. Resolution capability was much the same, (0.25-μm L&S), regardless of photoacid bulkiness. However, the resist pattern profiles, especially its top rectangularity, improved with increasing photoacid bulkiness and T-topping profiles were observed in the tertBu substituent. It is considered that one of the reasons for this top profile change is shorter acid diffusion length in the -tertBu substituent. In a chemically amplified resist, acid loss at the resist film surface is caused by acid volatilization or quench with airborne base contamination, and this acid loss should be compensated for by acid diffusion from the resist bulk region and catalytic reaction. However, it is difficult to compensate for the acid loss by using a shorter acid-diffusion length. Therefore, a T-topping profile is observed in the tertBu substituent.

Table II. Diffusion parameters and acid diffusion lengths for various Mw/Mn

Mw/Mn	1.2	4.0	9.0
Amount of acid ($10^{25}m^{-3}$)	3.0	3.1	3.1
Ion conductivity ($10^{-10}\,\Omega\,m^{-1}$)	5.0	4.5	3.5
Diffusion coefficient ($10^{-6}\mu m^2/s$)	3.3	3.0	2.4
Diffusion length (nm)	24	23	21

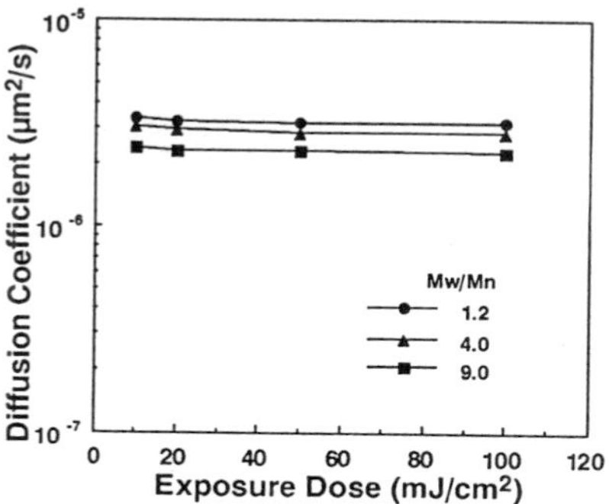

Figure 15. Diffusion coefficient as a function of exposure dose for various Mw/Mn.

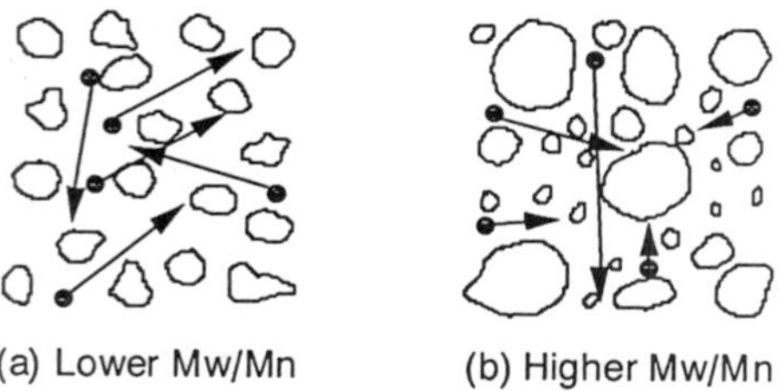

Figure 16. Scheme of acid diffusion model; (a) lower Mw/Mn, (b) higher Mw/Mn.

Table III. Diffusion parameters and acid diffusion lengths for various photoacid substituents

Substituent of PAG	F	Cl	diMe	tertBu
Amount of acid ($10^{25}m^{-3}$)	1.8	2.5	2.8	3.6
Ion conductivity ($10^{-10}\,\Omega\,m^{-1}$)	3.0	4.5	3.5	3.3
Diffusion coefficient ($10^{-6}\mu m^2/s$)	3.4	3.6	2.5	1.8
Diffusion length (nm)	25	26	21	18

Effect of Additional Base Component. In order to evaluate the influence of a base additive on acid diffusion, N-methylpyrrolidone (NMP) was used as an additional base component (0.1, 0.2, 0.3 wt%). Figure 18 shows the amount of generated acid or base in several samples, which have PAG, and base loading as a function of exposure dose. For PAG-containing resists, a higher base concentration brought about a smaller amount of acid. It was confirmed that the additional base component quenched some of the generated acid. The ion conductivity of the resist film also increased exponentially with increasing exposure dose for each base concentration of PAG-containing resists, and higher base concentration induced lower ion conductivity. The ion conductivity was also suppressed by base additive. Figure 19 shows exposure dose dependence of diffusion coefficient for various base loadings. For PAG-containing resists, it was found that the exposure dose dependence was small and that higher base concentration brought about lower acid-diffusion coefficient. It was confirmed that additional base component not only quenched photo-generated acid but also suppressed the acid diffusion within the resist film. For no-PAG-containing samples, the diffusion coefficient of the base becomes smaller with increasing exposure dose and higher base concentration leads to lower diffusion coefficient. It is considered that ionization efficiency might be reduced in a lower NMP concentration that NMP combines with decomposed blocking groups directly generated by KrF exposure, and that ion current is suppressed in higher exposure dose. Figure 20 shows plots of diffusion lengths of acid or base versus base concentration at 50-mJ/cm^2 for 90-s diffusion time. The clear relationship between acid diffusion length and base concentration is obtained. This fact indicates that optimum acid-diffusion length can be adjusted by the addition of a base component.

In order to analyze the above results, actual lithographic performance was evaluated. The resist sensitivity and resolution limit are listed in Table VI. The sensitivity was degraded with increasing a base concentration because the generated acid was quenched and its diffusion length was suppressed by the additional base. The resolution of 0.23-μm L&S was obtained with 0.1-wt% and 0.2-wt% base concentrations. Figure 21 shows a SEM micrograph of the 0.25-μm L&S pattern. T-topping profiles were observed in no-additional-base resist, because of airborne contamination even if delay time was less than 1 min. T-topping profiles were suppressed with increasing base concentration, and higher base concentration induced a mild standing-wave effect on the pattern side-wall. It was confirmed that the additional base is effective for suppressing the T-topping profile. Moreover, it was found that the optimum base concentration which gives the optimum acid diffusion length was estimated to be 0.2-wt% for better resolution capability and pattern profile. In this case, an optimum acid-diffusion length of 14 nm was obtained. This value was smaller than that obtained in the previous section. It is considered that the optimum diffusion length is different in the base-containing situation. Therefore, in order to find the most suitable diffusion length, the other parameters, such as combination between acid and base structure should be investigated.

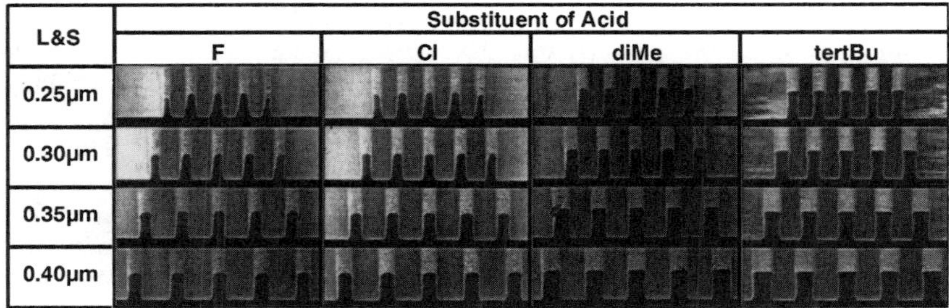

Figure 17. SEM micrograph of 0.25 μm pattern for various levels of photoacid bulkiness.

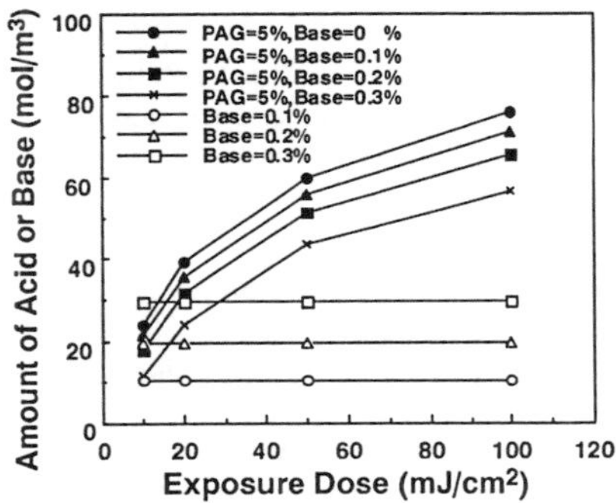

Figure 18. Amount of generated acid or base as a function of exposure dose for various PAG or base loadings.

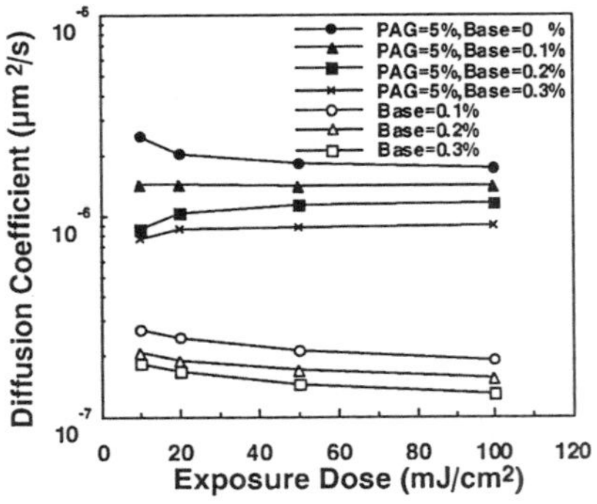

Figure 19. Diffusion coefficient as a function of exposure dose for various base loadings.

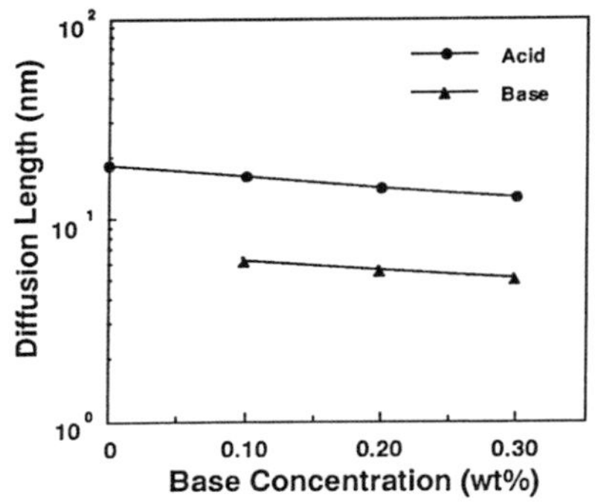

Figure 20. Acid or base diffusion length as a function of base concentration.

Table IV. Resist sensitivity and resolution capability for various NMP loadings

NMP (wt%)	Sensitivity (mJ/cm^2)	Resolution (μm L&S)
0	3.0	3.1
0.1	5.0	4.5
0.2	3.3	3.0
0.3	24	23

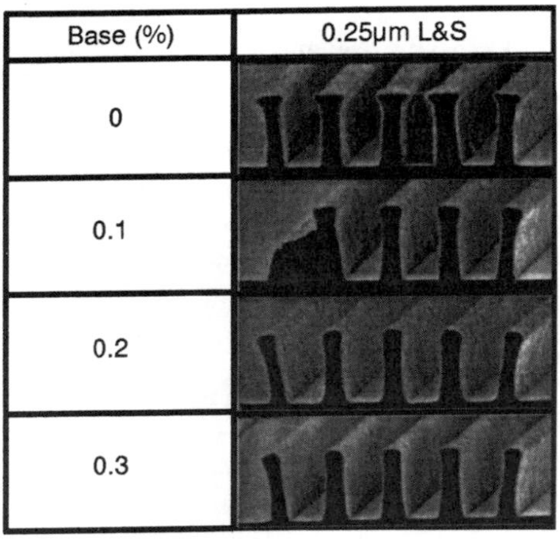

Base (%)	0.25μm L&S
0	
0.1	
0.2	
0.3	

Figure 21. SEM micrograph of 0.25 μm pattern for various NMP loadings.

Conclusions

Photoacid diffusion behavior in t-BOC-blocked chemically amplified positive DUV resists under various conditions was studied. Based on the experimental results, it was confirmed that only one mechanism dominated the acid diffusion in the resist film, and two diffusion paths, i.e., the remaining solvent in the resist film and hydrophilic OH sites of base phenolic resin, existed. Moreover, the effects of molecular weight dispersion, acid structure, and additional base component on both acid-diffusion behavior and lithographic performance were revealed. Finally, the acid diffusion behavior in the resist film was clarified and the acid diffusion length that affected the resist performance could be controlled.

Acknowledgments

The authors would like to thank Dr. O. Mizuno and Dr. N. Endo for their helpful suggestions and encouragement.

References

1. Itani, T.; Yoshino, H.; Hashimoto, S.; Yamana, M.; Samoto, N.; Kasama, K. *J. Vac. Sci. Technol.* **1996,** *B14*, 4226.
2. Itani, T., Yoshino, H.; Hashimoto, S.; Yamana, M.; Samoto, N.; Kasama, K. *Jpn. J. Appl. Phys.* **1996,** *35*, 6501.
3. Itani, T.; Yoshino, H.; Fujimoto, M.; Kasama, K. *J. Vac. Sci. Technol.* **1995,** *B13*, 3026.
4. Itani, T.; Yoshino, H.; Hashimoto, S.; Yamana, M.; Samoto, N.; Kasama, K. *Microel. Eng.* **1996,** *35*, 149.
5. Houlihan, F.M.; Chin, E.; Nalamasu, O.; Kometani, J.M.; Harley, R. In *Microelectronics Technology*; Reichmanis, E.; Ober, C.K.; MacDonald, S.A.; Iwayanagi, T.; Nishikubo, T., Eds.; Symposium Series 614; American Chemical Society: Washington, D. C., 1995, pp 84-109.
6. McKean, D.R.; Schaedeli, U.; MacDonald, S.A. In *Polymers in Microlithography*; Reichmanis, E.; Macdonald, S.A.; Iwayanagi, T., Eds.; Symposium Series 412; American Chemical Society: Washington, D. C., 1989, pp27-38.
7. Deguchi, K.; Ishiyama, T.; Horiuchi, T.; Yoshikawa, A. *Jpn. J. Appl. Phys.* **1990,** *29*, 2207.
8. Schlegel, L.; Ueno, T.; Shiraishi, H.; Hayashi, N.; Hesp, S.; Iwayanagi, T. *Proc. MicroProcess Conference,* **1989,** 54.
9. Schlegel, L.; Ueno, T.; Hayashi, N.; Iwayanagi, T. *J. Vac. Sci. Technol.* **1991,** *B9*, 278.
10. Nakamura, J.; Ban, H.; Deguchi, K.; Tanaka, A. *Jpn. J. Appl. Phys.* **1991,** *30*, 2619.

11. McKean, D.R.; Allen, R.D.; Kasai, P.H.; Schaedeli, U.; MacDonald, S.A. *Proc. SPIE.* **1992,** *1672*, 94.
12. Yoshimura, T.; Shiraishi, H.; Okazaki, S. *Jpn. J. Appl. Phys.* **1995,** *34*, 6786.
13. Asakawa, K.; Ushirogouchi, T.; Nakase, M. *J. Photopolymer Sci. Technol.* **1994,** *7 (3)*, 497.
14. Fedynyshyn, T.H.; Thackeray, J.W.; Georger, J.H.; Denison, M.D. *J. Vac. Sci. Technol.* **1994,** *B12*, 3888.
15. Raptis, I.; Grella, L.; Argitis, P.; Gentili, M.; Glezos, N.; Petrocco, G. *Microel. Eng.* **1995,** *30*, 295.
16. Kawai, Y.; Otaka, A.; Tanaka, A.; Matsuda, T. *Jpn. J. Appl. Phys.* **1994,** *33*, 7023.
17. Hashimoto, S.; Itani, T.; Yoshino, H.; Yamana, M.; Samoto, N.; Kasama, K. *J. Photopolymer Sci. Technol.* **1996,** *9 (4)*, 591.
18. Iwasaki, H.; Itani, T.; Fujimoto, M.; Kasama, K. *Proc. SPIE.* **1994,** *2195*, 164.

Chapter 10

Highly Photosensitive Diazo Compounds as Photoacid Generators for Chemically Amplified Resists

Kieko Harada[1], Masahito Kushida[1], Kyoichi Saito[1], Kazuyuki Sugita[1], and Hirotada Iida[2]

[1]**Department of Specialty Materials, Chiba University, 1-33 Yayoi-cho, Inage-ku, Chiba 263, Japan**

[2]**Toyo Goshei Kogyo Company, Research Center, 563 Komegasaki, Funabashi-shi, Chiba 273, Japan**

Diazo compounds are the photosensitive components for the contrast enhanced lithography (CEL) materials and the photoacid generator(PAG) for PS printing plates based on thermal crosslinking. This paper reports on PAG performance of p-benzoylamino-2,5-ethoxybenzenediazonium triflate (BTf) for microlithography resists. The rate of acid generation from BTf was larger than that of diphenyliodonium triflate(ITf). Deprotection of poly(tert-butyloxycarbonyloxystyrene) (tBOCHS) with BTf was 3 times faster than that with ITf after postexposure bake at same temperature. BTf with higher sensitivity and thermal stability may be expected to be PAG of diazo compounds applicable to microlithography resists.

Diazo compounds are the photosensitive components for copying of the diazo-type and vesicular processes, the contrast enhancement lithography (CEL) materials and the photoacid generator (PAG) for thermal-crosslinking type PS printing plates (1). Diazo compounds substituted with chlorine atom were used for thermal ring-opening polymerization of tetrahydrofuran (2). 2,5-Diethoxy-4-tolylthiobenzenediazonium ion coupled with various Lewis acids (SbF_6^-, PF_6^-, BF_4^- etc.) was used for ring-opening photopolymerization of 1,2-epoxypropane (3). 4-(4-Methoxyanilino)-benzenediazonium hexafluorophosphate with an absorption at long wavelengths was an initiator for cationic polymerization of poly(2,3-epoxypropyl methacrylate) by argon ion laser (4). Diazo compounds substituted with dialkylamino groups showed high photosensitivity and thermal stability, but they can not be used as PAG (5). The acid generated was trapped by coodination to lone-pair electrons on the nitrogen of the dimethyl amino group (6), but the coodinated acid was reported to dissociate from the nitrogen by heating (7).

Diazo compounds substituted with a benzoylamino group have been used for wet diazo-type copying process, which offer high sensitivity and thermal stability. This paper will report on PAG performance of these compounds for microlithography resists.

Experimental

Materials. 4-Benzoylamino-2,5-ethoxybenzenediazonium hexafluorophosphate (BP), was obtained by adding ammonium hexafluorophosphate to an aqueous solution of 4-benzoylamino -2,5-ethoxybenzenediazonium zinc chloride. The crude solids were recrystallized from methanol. 4-Benzoylamino-2,5-ethoxybenzenediazonium trifluoromethanesulfonate (triflate) (BTf) was prepared by diazotizing the corresponding amino compound with isopentyl nitrite in acetic acid containing trifluoromethanesulfonic acid (4). After completion of diazotization, the precipitated crystals were filtered and dried. p-Phenylaminobenzenediazonium hexafluoro phosphate (DPD) and p-dimethylaminobenzenediazonium hexafluorophosphate (DED) were prepared similarly to BP. The matrix polymers, poly(p-hydroxystyrene)(PHS) and PHS partially protected with a t-butoxycarbonyl (tBOC) group (20% protected, Mw. = 8,000) (tBOCHS), were purchased from Aldrich Chemical Co. Inc. and Iweiss Chemical Co. respectively. Diphenyliodonium hexafluorophosphate (IP) and diphenyliodonium triflate(ITf), were obtained from Midori Chemical Co. Tetrabromophenol Blue (TBPB) indicator was obtained from Aldrich Co. Inc. Diethylene glycol dimethyl ether (DGDE) was used as the coating solvent.

The PHS film (thickness : 4 μ m) was formed on a glass plate by spin-coating a polymer solution containing a matrix polymer PHS, TBPB indicator, acid generator, and DGDE, (0.5g: 0.03g: $5.8\text{X}10^{-5}$M: 1.0 ml). The tBOCHS film was formed from a polymer solution containing tBOCHS, acid generator and DGDE (0.5g: $5.8\text{X}10^{-5}$M : 1.0 ml).

Measurements. The BP and BTf in methanol were exposed to a super-high pressure mercury lamp, Toshiba SHL-100-UV-2 (100W) with a Matsuda UV-D2 (Transparency:300-400 nm) filter. The polymer layers containinng BP or BTf were exposed to a super-high pressure mercury lamp, Toshiba SHL-100-UV-2 (100W). The resist films containing IP or ITf were exposed to a low pressure mercury lamp (LPL), Toshiba GL-8 (8W). The incident energies of the high and low pressure mercury lamps were 3.83 and 0.42 mW/cm^2, respectively. Absorption spectra were measured with a Shimazu UV-180 spectrophotometer. Thermal analysis were performed by Mac Science 001 Thermal Analyzing System at a heating rate of 5℃/min for TGA and DTA under N_2 atmosphere. FTIR spectra of the films on an ITO glass were measured with a Perkin Elmer 1600 spectrophotometer. IR spectra of the film on a NaCl plate were measured with a Hitachi I-2000 spectrophotometer.

Results and Discussion

Photolysis and Thermolysis. The absorption maxima of diazo compounds BTf and BP in methanol were at 396 nm and 392 nm as shown in Table 1.

After UV irradiation of the methanol solution, the concentrations of the remaining diazo compounds were calculated from the absorbance at λ max and are plotted against irradiation time as shown in Fig.1 The absorbed photon numbers ($4.36\text{X}10^{-7}$ einstein/cm^2) were measured by potassium ferrioxalate actinometry. The quantum yields for photodecomposition , Φ, are calculated from Fig.1 and shown in Table 1.

Table 1

Name	Chemical Structure	λ max (nm)	Φ [a]	Td [b] (℃)
BP	C_6H_5-C(=O)-N(H)-$C_6H_2(OC_2H_5)_2$-N_2PF_6 (structure with OC_2H_5 and OC2H5 substituents)	392	0.58	159.0
BTf	C_6H_5-C(=O)-N(H)-$C_6H_2(OC_2H_5)_2$-$N_2CF_3SO_3$	396	0.58	147.5
DPD	C_6H_5-N(H)-C_6H_4-N_2PF_6	379	0.56	149.0
DED	$(C_2H_5)_2N$-C_6H_4-N_2PF_6	382	0.63	143.0
BD	C_6H_5-N_2BF_4	264	0.38(9)	77.0
ClBD	Cl-C_6H_4-N_2BF_4	283	0.29(9)	117.5
IP	$(C_6H_5)_2I\ PF_6$	280	0.20(10)	235.3
ITf	$(C_6H_5)_2I\ CF_3SO_3$	280	____	172.9

a) Quantum yield in methanol
b) Thermal decomposition temperature by TGA analysis

The values of BP and BTf were smaller than that of diazo compounds with diethylamino substituent (DED) and close to that of p-diazodiphenylamine (DPD) which is used as the photosensitive emulsion for screen printing plates. The value with BP is three times as large as that of diphenyliodonium hexafluorophosphate (IP).

TGA and DTA curves of BTf and ITf are shown in Fig. 2. The degradation temperature Td of BP was the highest of the diazo compounds studied as shown in Table 1. In the case of common diazonium cation , Td of BP is higher than that of BTf. The thermal stability of diazo compounds depended on the counter anion. Td of IP was higher than that of ITf, which is in the same correlation to BP and BTf. Td of the diazo compounds was lower than that of the iodonium salts with the same counter anion.

BP and BTf were found to be highly sensitive and thermally stable, and the performance as acid generators for resists was studied next.

Photo-acid Generation. After exposing the PHS film, the absorption change at 620 nm of the indicator, which coresponds to the amount of the generated acid was measured, and photo-acid generation efficiency of PAG in polymers was calculated (8). The relationship of exposure dose and generated acid equivalent is shown in Fig. 3. Since the acid generated from BTf does not coordinate to lone pair electrons on nitrogen of the benzoyl amino group, it was able to be determined in this process. The rate of acid generation was determined by measuring the generated acid equivalent after irradiation (10 mJ/cm^2). The rate of acid generation from BTf was one point five times larger than that of ITf.

Acid-catalyzed Deprotection of tBOCHS. Acid-catalyzed conversion of tBOCHS to PHS was measured as follows: After exposing the film, the conversion was measured by IR absorption changes at 1755 cm^{-1} and 1150 cm^{-1} (correspond to ν (C=O) and ν (C-O-C), respectively) as shown in Fig.4. There were few spectral changes at 3200-3600 cm^{-1} by IR transmittance spectrum.

The relationship between the remaining ν (C-O-C) bond (%) by FTIR and the irradiation dose is shown in Fig.5. The deprotection rate was compared by the dose at the half of remaining of ν (C-O-C). The deprotection rate of tBOCHS with BTf was three times larger than that with ITf after postexposure bake (PEB) for 10 min at 90℃. The relative rate was same as the ratio of the quantum yields in the solution. After PEB for 1 min at 100℃, tBOCHS with BTf was not deprotected. It is considered the generated acid from BTf was coodinated to photoproducts, probably azo dyes. Deprotection rate of tBOCHS with ITf after PEB for 1 min at 100℃ was five times larger than that with BTf after PEB for 10 min at 90℃. The tBOCHS films containing ITf coincided to those of the PHS films by irradiation of 40 mJ/cm^2 and PEB for 1 min at 100℃.

Conclusion

The acid generated from diazo compounds substituted with a benzoylamino group was not trapped by coodination to lone pair electrons on the nitrogen and worked as the catalysis for deprotection of t-BOC groups. The rate of acid generation from BTf was

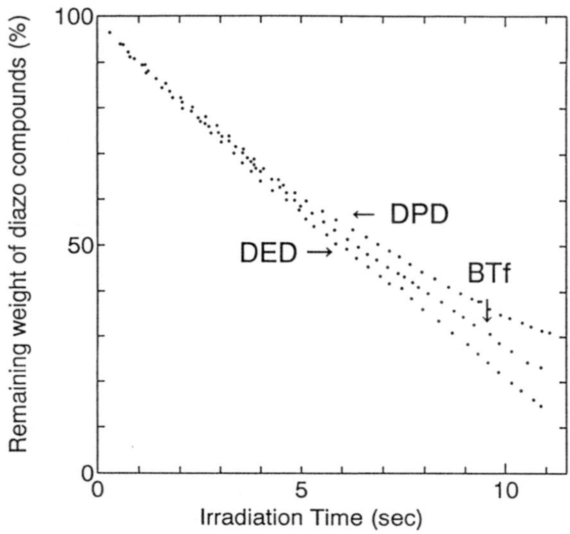

Fig.1 Photolysis of diazo compounds, BTf, DED and DPD.

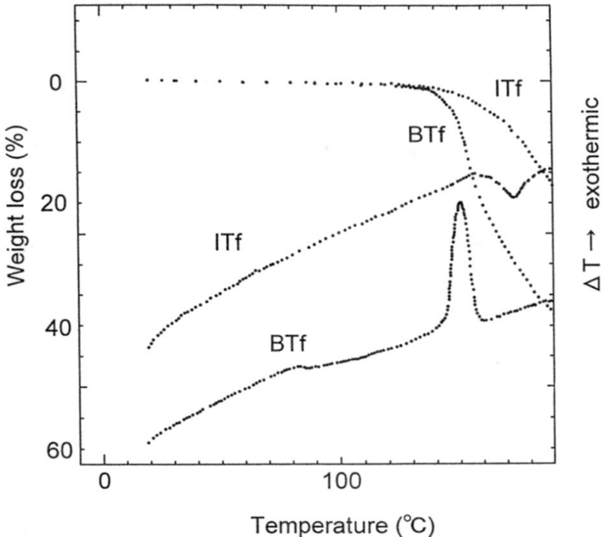

Fig.2 TGA-DTA of BTf and ITf.

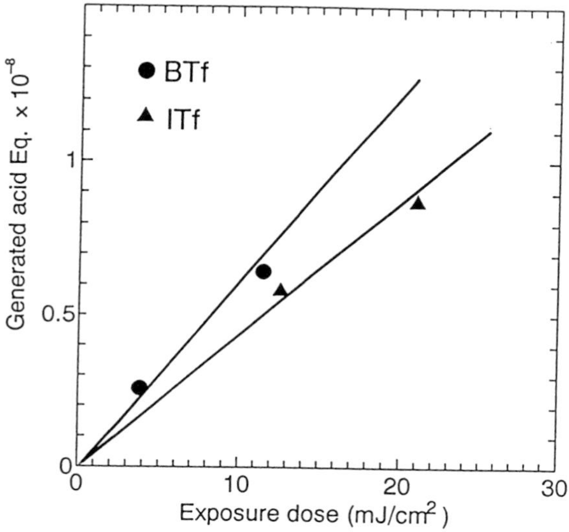

Fig.3 Relationship between generated acid equivalent and exposure dose.

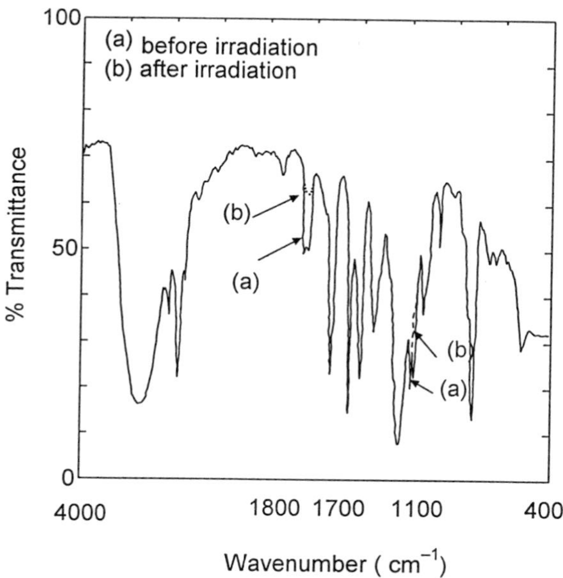

Fig.4 IR spectra of tBOCHS film containing ITf.

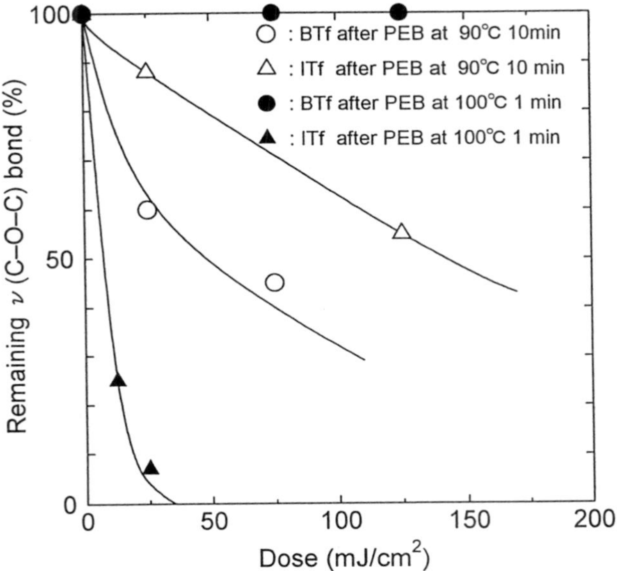

Fig.5 Deprotection of tBOCHS with BTf or ITf by irradiation and postexposure bake.

higher than that from ITf at the same temperature. Deprotection of tBOCHS with BTf was three times larger than that with ITf after PEB for 10 min at 90℃. However the rate was one fifth of that with ITf after PEB for 1 min at 100℃. BTf with high sensitivity and thermal stability may be expected to be a photoacid generator of diazo compounds applicable to microlithography resists.

Acknowledgment

The authors thank Profs. Tsuguo Yamaoka, Fumihiko Akutsu, Norihisa Kobayashi and Shigeru Takahara of Chiba University and Dr. Hideo Kikuchi of Toyo Gosei Co. Ltd. for their helpful suggestions.

References

1. Kodak Co., Japan Patent Publication,1995-20629
2. Dreyfuss, M.P.; Dreyfuss, P., J.Polym.Sci.,Part A-1, **1966**, 7, 2179.
3. Bal, T.S.; Cox, A.; Kemp, T.J.; Moira, P.P, Polymer, **1980**, 21, 423.
4. Koseki, K.; Kawabata, M.; Arai, M.; Yamaoka, T., J.Chem.Soc.Japan, **1984**, (2) 329.
5. Uchino, S.; Hashimoto, M.; Iwayanagi, "Polymers in Microlithography" Reichmanis, E.; MacDonald, S.M.; Iwayanagi,T.; eds., ACS Symp. Ser.412 Amer.Chem.Soc., Washington, D.C., **1989**, pp.319-331.
6. Harada, K.; Ueno, N.; Sugita, K.; Suzuki, S., J. Soc. Photogr. Sci. Technol.Japan **1984**, 7, 351.
7. Harada, K.; Ueno, N.; Sugita, K., Kobunshi Ronbunshu, **1988**, 47, 295.
8. Hatta, T.; Yamaoka, T., Polymers for Advanced Technol. **1994**, 5, 90.
9. Tsunoda, T.; Yamaoka, T., J.Soc.Photogr.Sci.Technol. Japan, **1966**, 29, 197.
10. Crivello, J.V.; Lam, J. H. W., Macromolecules, **1977**,10, 1307.

Chapter 11

Exploration of Chemically Amplified Resist Mechanisms and Performance at Small Linewidths

James W. Taylor, Paul M. Dentinger, Steven J. Rhyner, and Geoffrey W. Reynolds

Department of Chemistry, Center for X-ray Lithography, University of Wisconsin at Madison, Madison, WI 53589

When exposed to X-rays, chemically-amplified resists show very high resolution at the required sensitivity, but fundamental quantitative questions about the reaction and performance remain. What and how much chemical change is necessary to sufficiently decrease the dissolution rate? How many cycles does the acid undergo during the post-exposure bake? What is the sidewall roughness? For Shipley SAL 605 negative-tone resist, quantitative measurements show that under lithographic conditions: 5 x10^{-06} moles/cm^3 of acid are produced, the acid cycles about 26 times, 2% of the phenols are protected, a cross-linking reaction is not necessary for the dissolution rate to be sufficiently changed, and the sidewall roughness is on the order of 5.2 nm. The mechanistic implications of these quantitative observations on resist performance will be discussed.

Semiconductor manufacturing applications, using 1.0 nm X-rays as the exposure source, require resist sensitivities on the order of 50-100 mJ/cm^2 in order to meet the desired throughput. Conventional resists, where an interaction between the exposing radiation and resist directly defines sensitivity, have not been able to meet this need. Attention has turned to chemically-amplified resists where the exposure creates a species which catalyzes multiple chemical events during the post-exposure bake(PEB). There are a variety of resist systems that can demonstrate chemical amplification, but Shipley SAL 605 - a negative-tone resist showing sensitivities on the order of 100 mJ/cm^2 - is the system chosen for this quantitative study.

The mechanism for this resist has been described as the creation of an acid from the interaction of the exposing radiation with a photoacid generator(PAG) within the film matrix, a rate-limiting reaction between the acid and hexamethoxymethylmelamine(HMMM), and a cross-linking reaction between the cationic intermediate on the HMMM and the novolac matrix (*1*). The presumed mechanism is shown in Figure 1 where one molecule of HMMM is shown coupling with one hydroxy on a novolac. O-alkylation has been shown to occur quantitatively over C-alkylation (*2*) as shown in the Figure. Evidence of the coupling reaction is found at 988 cm^{-1} in the IR, (*3*) and corresponds to the formation of an ether bond between the iminium ion and the oxygen from the novolac polymer. The HMMM has

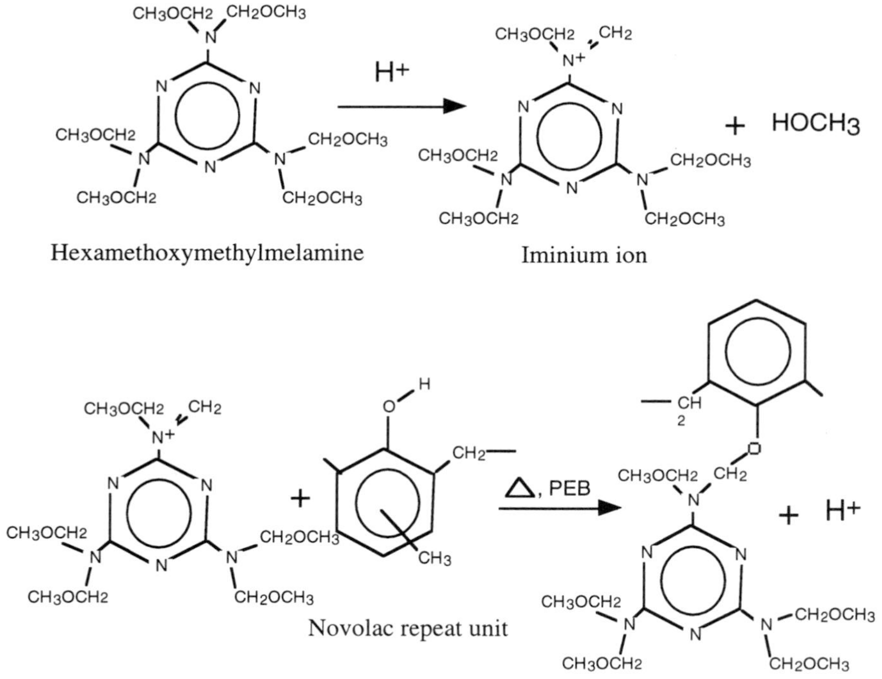
H+
+ HOCH3
Hexamethoxymethylmelamine
Iminium ion
Δ, PEB
+ H+
Novolac repeat unit

Figure 1. Schematic of mechanism of resist reaction. (Adapted from ref. 5)

six potential sites for reaction, and this suggests that cross-linking could occur if more than one novolac oligomer reacted to a single HMMM.

Quantitative Measurements of Resist Performance

In this paper, there are a number of questions about this resist that we address. First, how is the differential dissolution rate created from the chemical-amplification reaction such that the exposed and reacted material withstands the development step with aqueous tetramethylammonium hydroxide(TMAH)? In order to answer this first major question it was necessary to address quantitatively several issues. How much acid is created during exposure? How much methanol product remains in the film after PEB? (This is of concern because of its possible effect on the kinetics of the reaction.) How many cycles of catalytic reaction does the acid undergo before the desired differential dissolution rate is reached? The second question to be addressed quantitatively is what are the top surface and the sidewall roughness after the resist is developed?

Analytical Techniques Used for the Quantitation. The specific details of the analytical techniques used in these studies and of the optimized processing conditions are described in a previous publications (*4-5*) and will only be reviewed here. The X-ray exposures were done on a negative-tone resist, SAL 605 (The Shipley Co.). Films of 0.5 μm thickness were exposed at 125 mJ/cm^2 to a blank Si:N membrane. This dose and other resist processing parameters were consistent with statistically-optimized conditions for printing 0.215 μm features.

Results of the Quantitative Measurements. The moles of HMMM reacted were measured with gel permeation chromatography(GPC). If the cross-linking reaction coupled two novolac chains, one might expect the molecular weight to at least double. What was observed from a series of chromatograms for various baking times at 125 mJ/cm^2 exposure dose to SAL 605 was that the peak corresponding to unreacted HMMM was decreased while a corresponding increase in absorbance was observed throughout the novolac fraction. There did not appear to be the creation of species that were substantially higher in molecular weight. This suggests that a linking reaction had occurred, but cross-linking events were not prominent. Because unreacted HMMM is separated from the novolac resin in the chromatogram, it was possible to quantify the amount of the HMMM reacted by calibrating the peak height of the GPC against a standard of Cymel 300 (American Cyanamid, Wayne, NJ). The moles of HMMM reacted were then measured by subtracting the free HMMM in the reacted films from the free HMMM in the unreacted films. This allowed the amount of HMMM that was reacted to be quantified, but the question of how many reactions occurred per HMMM molecule remained. This was approached by measuring the reaction product, methanol.

Because the methanol could be either evolved from the resist film or retained in the film, the methanol left in the film was first quantified by GPC followed by gas chromatography(GC). This latter experiment revealed that there was no methanol left in the film after a PEB of 108°C for 60 sec - the optimized conditions used for dose to print. The amount of methanol evolved from the film could be measured by inserting a wafer in a sealed bomb, heating the system in an oven for the amount of time needed for PEB, and then analyzing the gas by GC. The amount of methanol produced from a given area of wafer at 0.5 μm thickness and detected in the gas could then be related to the amount of HMMM reacted on the identical wafer, as determined by GPC. This is shown in Figure 2. The slope of the line was used to calculate that 1.27 ± 0.24 moles of methanol were produced/mole of HMMM reacted. In addition,

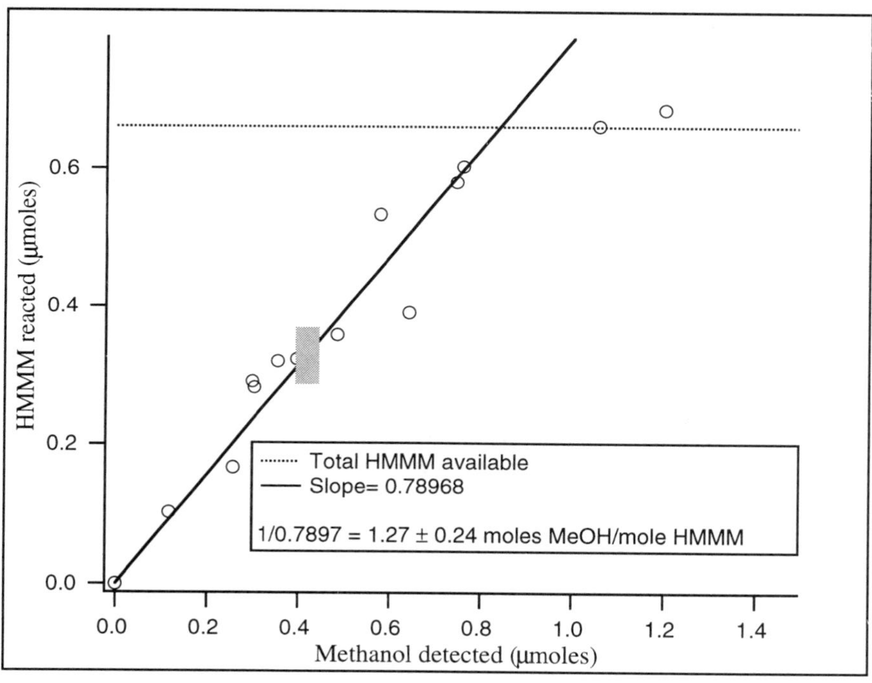

Figure 2. Comparison of moles measured for constant volume of resist, methanol:HMMM. The gray box represents the normal lithographic conditions for this resist. (Adapted from ref. 5)

the amount of reaction produced during optimized conditions of 125 mJ/cm^2 and a PEB on vacuum hot plate is shown as the gray box in Figure 2. At bake times producing about double the reaction necessary for lithography, all of the HMMM is reacted approximately once, but the methanol is continually produced due to the ability for the HMMM to make multiple links per molecule. If the predominant mechanism were a cross-linking reaction, one would have expected at least 2.0 moles of methanol for each mole of HMMM reacted for this system. The fact that considerably less methanol was produced than expected for a cross-linking reaction suggests, along with the GPC, that the differential dissolution rate between the exposed and unexposed resist is created by another process rather than simple cross-linking. Using this method of quantitation, we have not yet assumed O-alkylation vs. C-alkylation. C-alkylation would also produce methanol, and would also result in the correct amount of measured HMMM molecules reacted in the GPC experiment. However, O-alkylation has been shown to occur quantitatively over C-alkylation (*2*) and will only be considered in the FTIR calibration.

In a previous study, the amount of acid produced from 125 mJ/cm^2 exposure of the resist film was determined (*5*). The value was 5 x 10^{-6} moles/cm^3. If the amount of ether produced per acid during a standard hot-plate PEB could be determined, it would be possible to determine how many cycles the acid reacted before it created the required differential solubility. Although the formation of the ether peak could be monitored with FTIR at 988 cm^{-1},[3] the molar absorptivity of the peak was not known. However, the ether IR peak height can be calibrated by measuring the methanol produced by GC and relating the IR absorbance on the same wafer. Figure 3 shows a plot of FTIR peak height at 988 cm^{-1} vs. methanol concentration produced from the film. One methanol is produced for each ether formed, and the x-axis is the concentration of ethers times the path length, 0.5 μm. From this graph, we can obtain a molar absorptivity of the ether peak of 3.1 x 10^5 cm^2/mole. By knowing the molar absorptivity, we can process a wafer under standard conditions on a hot plate and measure the actual ether peaks by IR. From the known acid concentration with dose, the number of reactions per acid molecule is found to be 26 ± 8. In addition, because the total number of reactions have been quantified, one can calculate the percentage of phenolic groups protected by the HMMM reactions under standard lithographic conditions. The percentage of phenols reacted is approximately 2%.

Conclusions about the Mechanism from the Quantitative Measurements. Although the above quantitation steps did reveal a substantial amount of information about the reactivity of SAL 605, the remaining question is why the reaction of essentially one molecule of HMMM with one phenolic group on every other oligomer had such a profound effect on the solubility of the film. The previous suggestion of an increase of the molecular weight upon exposure and PEB clearly does not fit the data. One suggestion, advanced by others (*6*), is that disruption of the hydrogen bonding around the hydroxyl group might have a significant effect on dissolution rate, or that the insertion of the HMMM into the polymer matrix creates a steric hindrance to the TMAH in attacking other hydroxyl groups in the polymer. At this time, we do not have further data to distinguish between these and other possibilities, but we do believe that the concept of a cross-linking reaction is not applicable to this system.

The Problem of Top Surface and Sidewall Roughness

Because the resist is used in patterning, the edge roughness of the resist image is important. The importance of the roughness is derived from the error that is permitted to the resist for control of the critical dimension (CD). For a 100 nm line, the permissible error assigned to the resist is ±7% or ± 7 nm. For this CD, a sidewall roughness greater than 3 nm could contribute substantially to the error.

Two sources of the roughness are: 1) the mask itself and its method of generation; and 2) the developed surface layer arising from the interaction between the developer and the linked resist system. Surface roughness can be measured with an Atomic Force Microscope (AFM). Measuring the sidewall, however, is much more difficult because one has to locate and trace the sidewall of the resist line with the AFM, and the normal configuration of standard AFMs have feedback monitoring only in the Z direction. It is not possible, then, to obtain meaningful data on surfaces whose surface normal is perpendicular to the Z direction unless one employs a system for both the Z and X directions as suggested by Martin et al.(*7*). We have overcome the Z direction problem by creating a series of long lines at high resolution with SAL 605, cleaving the wafer parallel to the line pattern, tipping the cleaved wafer on its edge, and scanning the sidewall of the resist in a similar way to a scan of the surface. These measurements were made with a Topometrix Voyager AFM (Santa Clara, CA) operated in the non-contact mode with high resonance frequency silicon tips. Both the top surface and the sidewall measurements are convoluted with tip shape. Because the process described above tilts the surface normal of the resist line sidewall parallel to the Z axis, this permits measurement of the sidewall using the same tip that is used for top surface.

In these measurements there was serious concern over what roughness value to report. Peak-to-peak roughness would be one measure, but if the wafers were subsequently processed by etching, the peaks might be eroded rapidly and not contribute substantially to a change in critical dimension (CD) control on the wafer. Another measure would be power spectral density. This approach can provide spatial frequency information, but this did not appear critical to the processing of the resists. The root mean square roughness (rms) appeared to be more meaningful for resist processing, and these are the measurements reported here. Here the rms roughness is defined as in equation 1 where n is the total number of data points, Z is the average of all height

$$R_{rms} = \left[\frac{1}{n} \sum_{i}^{n} (Z_i - Z)^2 \right]^{1/2} \tag{1}$$

data, and Z_i is the height of each data point in the image. All images taken were 300x300 points.

Results from Top Surface and Sidewall Roughness Measurements. For Shipley SAL 605, we observed a top surface and a sidewall roughness on the order of 7 ± 1 nm rms for the top surface and 5.2 ± 0.5 nm rms for the sidewall roughness under normal processing conditions of 70 mJ/cm^2, a post apply bake of 116°C for 50 seconds, a PEB of 108°C for 59 seconds, and development in 0.254 N tetramethylammonium hydroxide (TMAH) (Shipley MF-320) by immersion for 80 seconds. The top surface roughness depended on processing conditions and as described later. Typical AFM images are shown as Figure 4 for both the top surface measurement and the sidewall. We expect the mask to contribute on the order of 1.5 to 2 nm rms to the roughness because this appeared to be the limiting value of the roughness with a wide variety of processing conditions. The contributions from the mask to the sidewall roughness, however, are still under investigation.

Figure 5 illustrates our observations that the top surface becomes more rough with an increase in developer concentration. In this experiment, three wafers, each with a dose array from 40 to 250 mJ/cm^2, were developed with solutions of diluted MF-312, a TMAH developer, until the unexposed resist cleared from the wafer.

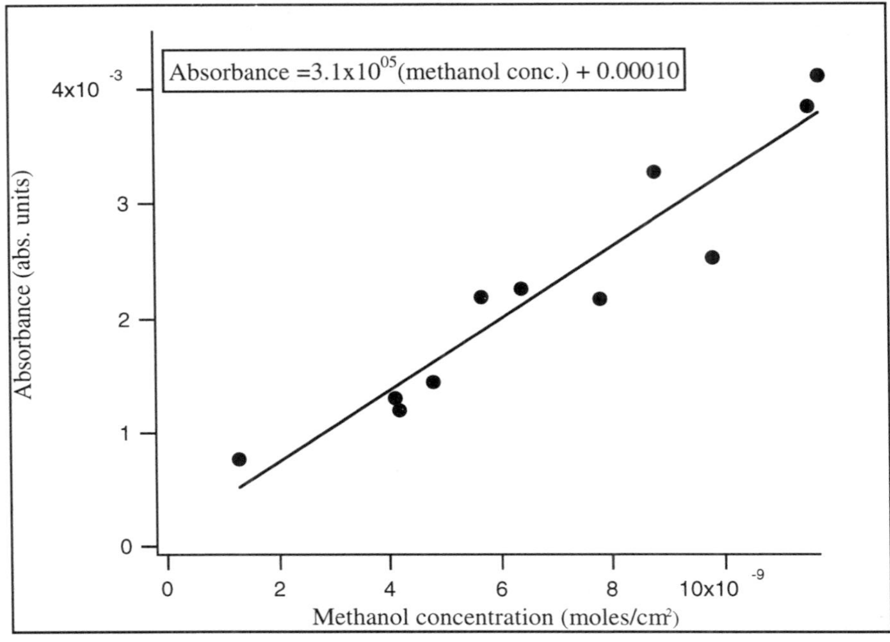

Figure 3. Comparative plot of IR ether peak intensity vs. methanol produced. (Adapted from ref. 5)

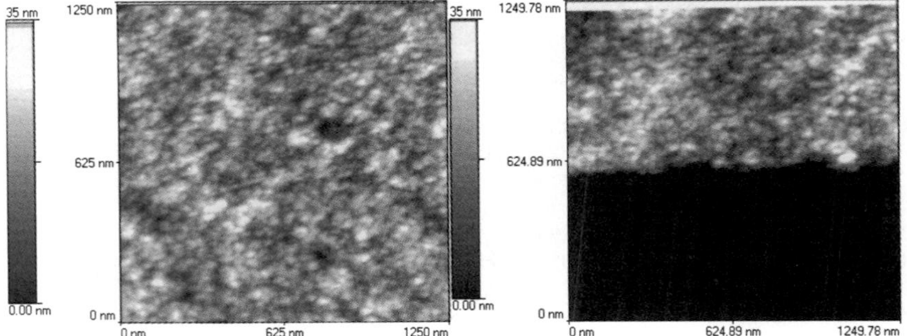

Figure 4. Typical top surface (left) and sidewall (right) images of Shipley SAL 605 negative-tone, chemically-amplified resist. The color bar at the left is representative of the peak-to-peak height difference in the images.

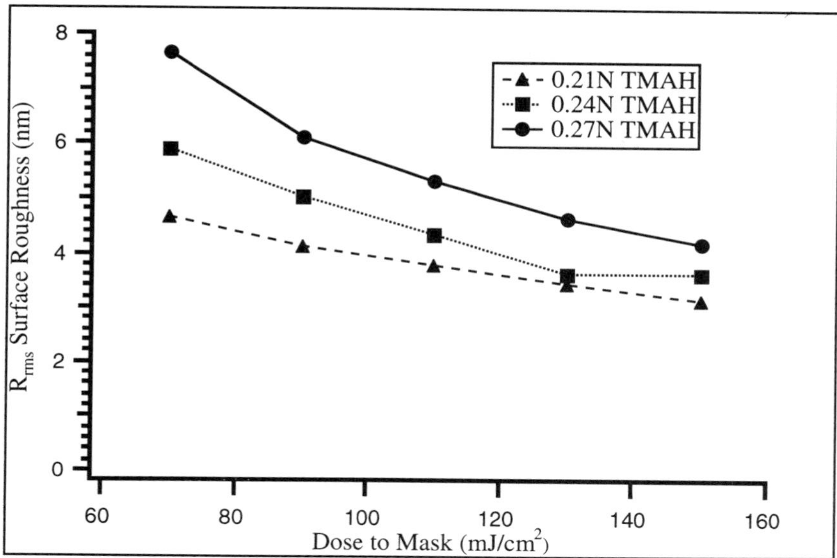

Figure 5. Effect of developer concentration and dose on top surface roughness of Shipley SAL 605. The developer MF-312, a tetramethylammonium hydroxide developer, was diluted with water to 0.27N, 0.24N, and 0.21N concentrations.

From the figure it can be seen that for a constant dose of 90 mJ/cm^2 and a developer concentration of 0.270 N, the top surface roughness was 6.4 nm; at 0.240 N it was 5.2 nm; and at 0.210 N it was 4.2 nm. The general trends are that the top surface roughness decreased with increasing dose and decreasing developer concentration. In another experiment to measure sidewall roughness with 0.25 N and 0.22 N TMAH, the sidewall roughness was essentially constant at 5.2 ± 0.5 nm and was independent with respect to developer concentration and doses ranging from 60 to 140 mJ/cm^2.

Figure 5 showed a dependence of top surface roughness on dose. In order to study the effect of developer additives on the top surface roughness without convoluting the observed roughness with dose, a series of experiments were performed on unexposed wafers. The wafers, however, were subject to the same thermal history (*i.e.* same post apply bake and same "post exposure bake") as normally exposed wafers would have been. The wafers were developed for a measured time in developers of either NaOH or TMAH with various salts added. The development process was arrested by rinsing the wafer in deionized water. Remaining film thickness measurements provided a means to calculate an approximate dissolution rate. Similar studies with exposed resist also demonstrated the general trends that are presented below.

The effects of adding various salts with either NaOH or TMAH developer produced different results for the surface roughness. These results are summarized in Figure 6. With 0.175 N NaOH (top, Figure 6) the surface roughness was constant at 3.75 nm, and no effect was seen with LiCl, KCl, RbCl, NaBr, or NaI for added salt concentrations spanning from ≈4 mMolar to 1.3 Molar for each salt. The dissolution rate increased for all the salts up to a concentration of 1.3 Molar, as has been observed by others (*8*). The analogous experiment was performed with TMAH

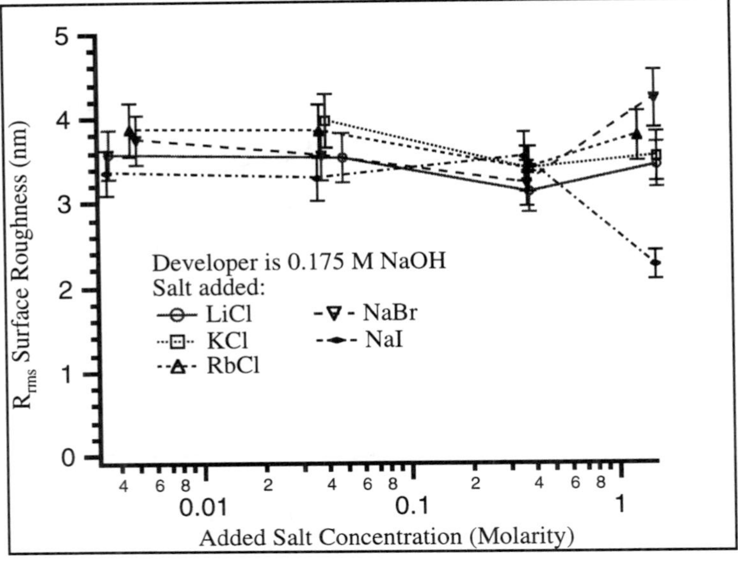

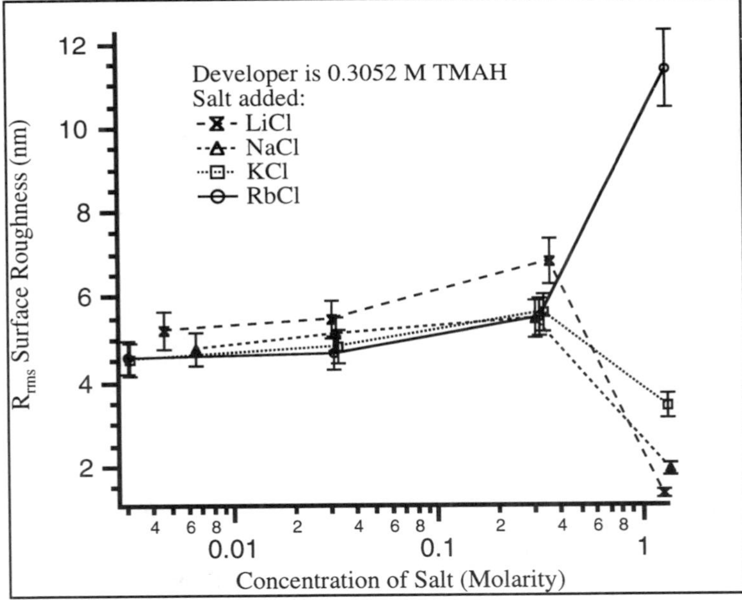

Figure 6. Effect of developer additives on top surface roughness. Top: Developer is 0.175N NaOH with added alkaline chloride salts. Bottom: Developer is 0.305N TMAH with added alkaline chloride salts.

developers at 0.305 N as shown in the bottom of Figure 6. Again, the added salts were LiCl, NaCl, KCl, and RbCl, each in concentrations spanning from 4 mMolar to 1.3 Molar. The top surface roughness was constant at about 5 nm but decreased at 1.3 Molar salt concentration for all salts except RbCl. The dissolution rate in the TMAH with the added salts, however, fell monotonically from values around 35 nm/sec without added salt to approximately 5 nm/sec with the 1.3 M concentration. Thus, there seemed to be a trade off between a reasonable dissolution rate and a smooth surface.

Conclusions from the Roughness Measurements to Date. The measured sidewall roughness of SAL 605 was 5.2 ± 0.5 nm rms and is close to the error limit for 100 nm CD. It is, however, unknown at present if etching will reduce the roughness of the feature sidewall. Studies are in progress to determine this. We are also exploring the possibility of a combination of developer conditions and changes in ionic strength that will produce the smallest sidewall roughness and retain an acceptable dissolution rate. At present, this has been an illusive goal, but if a correlation of the results of the sidewall measurements with various theories of the dissolution mechanism for this type of chemically-amplified resist could be accomplished, a reasonable path to reduce roughness could be suggested.

Acknowledgments. The work was supported by contract 96-LP-452 from the Semiconductor Research Corporation and by DARPA/ONR grant N-00014-96-1-0588. The Synchrotron Radiation Center is supported by the National Science Foundation under grant DMR-95-31009.

Literature Cited

1. deGrandpre, M.; Graziano, K.; Thompson, S. D.; Liu, H.-y.; Blum, L. Proc. SPIE **1988**, *923*, 158-171.
2. Thackeray, J. W.; Orsula, G. W.; Rajaratnam, M. M.; Sinta, R.; Herr, D.; Pavelcheck, E. Proc. SPIE **1991**, *1466*, 39-52.
3. Gamsky, C. J.; Dentinger, P. M.; Howes, G. R.; Taylor, J. W. Proc. SPIE **1995**, *2438*, 143-152.
4. Dentinger, P. M.; Nelson, C. M.; Rhyner, S. J.; Taylor, J. W.; Fedynyshyn, T. H.; Cronin, M. F. *J. Vac.Sci. Tech. B* **1996**,*14*, 4239-4245.
5. Dentinger, P. M.; Taylor, J. W. *J.Vac.Sci.Tech.B* **1997**, *15*, 2632-2638.
6. Shih, H.-Y.; Reiser, A. *Macromolecules* **1995**, *28*, 5595-5600.
7. Martin, Y.; Wickramasinghe, H. K. *Appl. Phys. Lett.* **1994**, *64*, 2498-2500.
8. Henderson, C. L.; Tsiartas, P. C.; Simpson, L. L.; Clayton, K. D.; Pancholi, S.; Pawloski, A. R.; Willson, C. G. Proc. SPIE **1996**, *2724*, 481-490.

Chapter 12

The Preparation and Investigation of Macromolecular Architectures for Microlithography by "Living" Free Radical Polymerization

G. G. Barclay[1,3], M. King[1], A. Orellana[1], P. R. L. Malenfant[1], R. Sinta[1], E. Malmstrom[2], H. Ito[2], and C. J. Hawker[2,3]

[1]Shipley Company, 455 Forest Street, Marlborough, MA 01752–3092
[2]IBM Almanden Research Center, Center for Polymeric Interfaces and Macromolecular Assemblies, 650 Harry Road, San Jose, CA 95120–6099

Copolymerization is the most important synthetic strategy available to tailor the physical, solution and mechanical properties of macromolecules. In this paper the synthesis of well defined random and block copolymers containing monomer units of very different polarity are described. Novel macromolecular architectures incorporating 4-hydroxystyrene, t-butylacrylate and styrene have been prepared to investigate the combined effects of macromolecular architecture and hydrophobicity on thin film aqueous base dissolution.

The dissolution behavior of copolymers in aqueous base is of primary importance in photoresist performance. As microlithography continues to decrease in exposure wavelength and the minimum feature size required from photoresist compositions decreases below 0.20 μm, control of macromolecular architecture will play an increasingly significant role. That is, how does the random, alternating or "blocky" nature of the matrix polymer effect its solubility in aqueous base. It has been well documented for novolac based photoresists that the molecular weight distribution profoundly affects the dissolution behavior and lithographic performance of the resist.[1,2] However, as microlithography progresses from I-line (365 nm), to DUV (248 nm) and now 193 nm, polymer architectures are changing dramatically to accommodate adequate transmittance. At DUV the principle materials consist of linear phenolic polymers and copolymers, for example poly(4-hydroxystyrene) (PHOST). In contrast, to obtain sufficient transmittance at 193 nm, hydrocarbon backbones incorporating carboxylic ácid moieties are being used. Thus, as these new exposing technologies advance, the macromolecular architecture and the hydrophobic/hydrophilic components of the matrix polymers need to be altered dramatically. Subsequently, the aqueous base dissolution characteristic of these materials are also changed significantly.

Previously we have studied the effect of molecular weight and polydispersity

[3]Corresponding authors.

on the aqueous base dissolution behavior of poly(4-hydroxystyrene) (PHOST).[3] This study suggested that the dissolution behavior of PHOST in aqueous base is primarily governed by the lower molecular weight fractions. The dissolution rate (DR) of PHOST vs M_n can be expressed as an exponential decay, and as a result the DR can be estimated if M_n is known. In contrast, there is no straightforward relationship between DR and M_w. Apart from this investigation of PHOST, little research has been carried out on the effect of macromolecular architecture on aqueous base dissolution. The role of polymer structure is becoming increasingly important considering the fact that the copolymer systems used in advanced lithography consist of monomeric units of very different polarity, for example isobornyl methacrylate and methacrylic acid. Ober et al. have investigated the lithographic behavior of block copolymers comprising monomers of very different hydrophobicities, for example silioxane monomers and t-butyl methacrylate.[4,5,6] In the present study to elucidate the effect of macromolecular architecture upon the dissolution of thin films in aqueous base, well defined random and block copolymers have been prepared containing different hydrophobic and hydrophilic components.

Recently great interest has been generated by "living" free radical polymerizations.[7] Nitroxide mediated "living" radical polymerization proceeds via the reversible capping of growing chain ends with 2,2,6,6-tetramethyl-1-piperidinyloxy (TEMPO), reducing free radical concentration and inhibits premature chain termination. This technique affords accurate control over molecular weight distribution, chain ends and also provides a simple synthetic technique for the preparation of well defined complex polymeric architectures. Copolymerization is the most versatile synthetic technique for controlling the functionality and thereby tailoring of the physical and mechanical properties of macromolecules. Variation in the nature and relative ratios of the monomer units in a random copolymer can result in a wide variety of useful materials, ranging from plastics to elastomers. However, a methodology for the accurate molecular weight control of random copolymers has long been elusive. Classical techniques for controlling polymeric architectures, such as anionic, cationic, and group transfer procedures are not well suited for the control of random copolymerizations, due to the greatly different monomer reactivities resulting from these methods. As a result the random copolymerization of disparate monomers, such as styrene and acrylates, can only be accomplished under normal free radical conditions, which leads to poorly defined and polydispersed macromolecules. In this paper, we investigate the use of "living" free radical polymerization for the preparation of well defined random copolymers of (a) 4-hydroxystyrene / styrene, and (b) styrene / t-butyl acrylate. Further, due to the stability of the nitroxide chain-end this technique is ideally suited for the preparation of block copolymers, therefore the preparation of the corresponding block copolymers was also investigated. These copolymer systems where used to investigate the combined effects of macromolecular architecture and hydrophobicity on aqueous base thin film dissolution.

Synthesis

Materials. 4-Acetoxystyrene (1) (Hoechst Celanese), styrene (2) (Aldrich) and t-butylacrylate (3) (Aldrich) were distilled under reduced pressure prior to use. 1-Phenyl-1-(2',2',6',6'-tetramethyl-1'-piperidinyloxy)ethane (4) was prepared as previously reported.[8]

Random Copolymerization. 4-Acetoxystyrene (1) (15.68 g, 0.097 mol) and styrene (2) (4.32g, 0.042 mol) were placed in a 100 mL round bottom flask and purged with N_2. 1-Phenyl-1-(2',2',6',6'-tetramethyl-1'-piperidinyloxy)ethane (4) (0.52g, 0.002 mol) was then added. After addition of the initiator, the polymerization mixture was heated to 125-130°C, under N_2, and stirred for 48 hours. During the polymerization the polymer solidified in the reaction vessel. The reaction was then cooled to room temperature. The polymer dissolved in acetone (100 mL), and isolated by precipitation into hexanes (1000 mL). The poly(4-acetoxystyrene-co-styrene) (5) was then filtered, washed with hexanes and dried in a vacuum oven overnight at 50°C. Isolated yield 92% of theory. M_n = 9230, M_w = 10330, polydispersity {PD} = 1.12 (Theoretical A.M.U = 10050).

A similar synthetic procedure was used to prepare the various random copolymers of (a) 4-acetoxystyrene and styrene; or (b) styrene and t-butyl acrylate.

Deacetylation of Poly(4-acetoxystyrene-co-styrene) (5). To a slurry of poly(4-acetoxystyrene-co-styrene) (70:30) (2) (10.0 g, 0.069 mol) in methanol at reflux (50 mL), under N_2, ammonium hydroxide (5.33 g, 0.152 mol) dissolved in water (10 mL) was added dropwise over 15 minutes. After addition, the reaction mixture was heated at reflux for 18 hours, during which time the polymer went into solution. The reaction is then cooled to room temperature, and the polymer isolated by precipitation into water (500 mL), filtered, washed well with water, and dried in a vacuum oven overnight at 50°C. Isolated yield of poly(4-hydroxystyrene-co-styrene) (6) 85% of theory. M_n = 7278, M_w = 8297, PD = 1.14.

Block Copolymers. Well defined block copolymers of 4-hydroxystyrene and styrene were prepared by firstly, the synthesis of low molecular weight blocks of styrene capped with TEMPO. The molecular weights of these styrene blocks were controlled by the ratio of unimolecular initiator (4) relative to styrene monomer (2). For example, 1-phenyl-1-(2',2',6',6'-tetramethyl-1'-piperidinyloxy)ethane (4) (1.67g, 0.0064 mol) was added to styrene (2) (20.0g, 0.192 mol) and heated to 125-130°C, under N_2, for 48 hours. The reaction was then cooled to room temperature and the polymer dissolved in tetrahydrofuran (100 mL), and isolated by precipitation into methanol (1000 mL). The TEMPO terminated polystyrene (7) was then filtered, washed with methanol and dried in a vacuum oven overnight at 50°C. Isolated yield 90% of theory. M_n = 2764, M_w = 3062, PD = 1.10 (Theoretical A.M.U = 3120).

A similar synthetic procedure was used to prepare the homopolymer of t-butyl acrylate.

These short TEMPO capped styrene blocks (7) were then used as "macroinitiators" for the further polymerization of 4-acetoxystrene (1) using the same procedure as described above. The length of the acetoxystyrene block is controlled by the ratio of the "macroinitiator" (7) to acetoxystyrene monomer. The acetoxystyrene block was then converted to 4-hydroxystyrene by quantitative base hydrolysis of the acetoxy groups as previously described, Table 1. A similar synthetic strategy was used for the preparation of poly(styrene- co-t-butyl acrylate) block copolymers.

Aqueous Base Dissolution Characteristics. Films of these copolymers were cast on 4" Si wafers and baked at 130°C for 1 minute to remove the casting solvent, ethyl lactate. The film thickness (approximately 1.0 μm) was measured on a Tencor Alphastep. The dissolution rates of these films were then measured by immersion in aqueous tetramethylammonium hydroxide (TMAH) solution (at 23°C) using a Perkin-Elmer 5900 Development Rate Monitor. The dissolution rates were then measured by immersion in a variety of TMAH solutions of increasing normality (0.14, 0.26, 0.30, 0.60 and 1.0 N).

Characterization. Compositions were determined using a GE QE300 NMR spectrometer, ^{1}H (300 MHz) and ^{13}C (75.4 MHz) NMR, in acetone d_6, with tetramethylsilane as an internal standard. For conversion studies, the samples were transferred directly from the polymerization reaction to an NMR tube and dissolved in acetone d_6. Molecular weights of the polymers were determined by gel permeation chromatography (GPC) using a Waters model 150C equipped with four Ultrastyragel columns in tetrahydrofuran at 40°C. The molecular weight values reported are relative to polystyrene standards. Differential scanning calorimetery (DSC) was performed using a Perkin-Elmer DSC-7 under N_2 (10 °C/min).

Discussion

Poly(4-hydroxystyrene-co-styrene). A series of well-defined random poly(4-hydroxystyrene-co-styrene) were prepared by the *"living"* free radical copolymerization of 4-acetoxystyrene (1) and styrene (2), using the unimolecular initiating system, 1-phenyl-1-(2',2',6',6'-tetramethyl-1'-piperidinyloxy)ethane (4) at a temperature of 130°C (Scheme 1). All of the copolymerizations were carried out neat under a nitrogen environment. The resulting well-defined poly(4-acetoxystyrene-co-styrene) polymers (5) were then quantitatively converted to poly(4-hydroxystyrene-co-styrene) (6) by deacetylation with base as previously described.[3]

Table 1. Properties of Random and Block Poly(4-hydroxystyrene-co-styrene)

Hydroxy-styrene	Styrene	Arch.	M_w	PD	DR(Å/sec)[1]
100	0	Homo	9550	1.2	2050
90	10	Random	8297	1.27	677
80	20	Random	9908	1.21	34
70	30	Random	8297	1.14	3
55	45	Random	8559	1.26	1 (9.4[2] / 114[3])
90	10	Block	10155	1.22	330
80	20	Block	8854	1.17	93
70	30	Block	6856	1.12	7
55	45	Block	10020	1.17	1 (100[2] / 305[3])

Notes: Dissolution Rate in (1) 0.26N TMAH; (2) 0.6N TMAH; (3) 1.0N TMAH

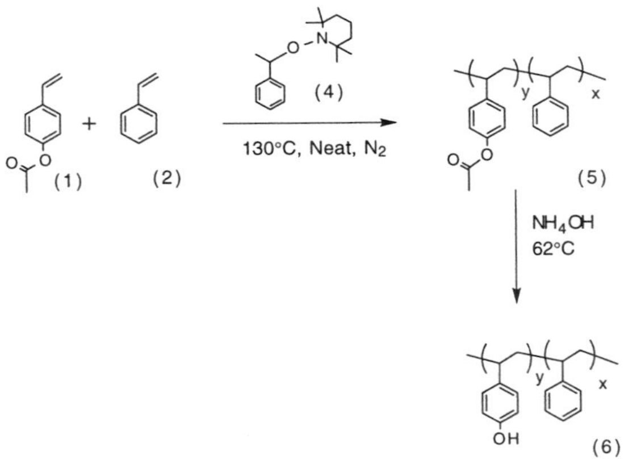

Scheme 1. Synthesis of Random Poly(4-hydroxystyrene-co-styrene)

Using this novel copolymerization technique, a range of narrow polydispersity poly(4-hydroxystyrene-co-styrene) have been synthesized (PD = 1.1 to 1.3), with a wide variety compositions, ranging from 90:10 to 55:45 (4-hydroxystyrene : styrene), Table 1.

Well-defined block copolymers of 4-hydroxystyrene and styrene (9) were prepared by the synthesis of low molecular weight blocks of styrene capped with TEMPO (7) (Scheme 2). The molecular weights of these styrene blocks were

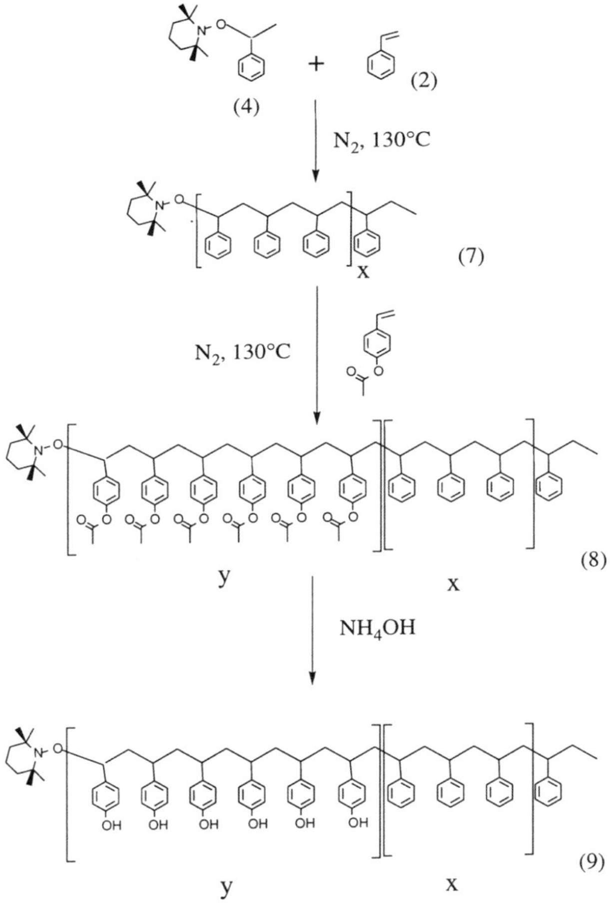

(4) + (2)

N_2, 130°C

(7)

x

N_2, 130°C

y x (8)

NH_4OH

OH OH OH OH OH OH

y x (9)

Scheme 2. Synthesis of Block Poly(4-hydroxystyrene-co-styrene)

controlled by the ratio of unimolecular initiator (4) relative to styrene monomer. These short TEMPO capped styrene blocks (7) were then used as "macroinitiators" for the further polymerization of 4-acetoxystrene (1). The length of the acetoxystyrene block was controlled by the ratio of the "macroinitiator" (7) to acetoxy monomer (1). Figure 1, illustrates the uniform increase in molecular weight and distribution of a 70:30 block copolymer of poly(4-acetoxystyrene-co-styrene) (8) (M_n 9415; PD 1.27) from the starting polystyrene macroinitiator (7) (M_n 2764; PD 1.10). The acetoxy block was then deactylated by quantitative base hydrolysis of the acetoxy groups to form a block copolymer of poly(4-hydroxystyrene-co-styrene) (9). In all cases, both random and block, the monomer feed ratio was in agreement with the composition of the isolated polymers, as determined by ^{13}C NMR. Complete deactylation of these acetoxy copolymers to form the polar hydroxyl functionality was also determined by ^{13}C NMR. These model random and block copolymers of 4-hydroxystryrene and styrene were then used to investigate the combined effects of macromolecular architecture and hydrophobicity upon thin film dissolution, Table 1.

Poly(styrene-*co*-t-butyl acrylate). One of the major issues with TEMPO mediated "living" free radical polymerizations is the very different reactivities of styrene and acrylates. It has been observed that TEMPO mediated styrene homopolymerization achieve high conversion, with low polydispersity and excellent molecular weight control. In contrast acrylate homopolymerizations exhibit considerably lower conversion with much broader polydispersities, Figure 2. However, it has been shown that "living" free radical polymerization permits the synthesis of well defined random copolymers of styrene with acrylates and methacrylates.[9] Therefore, the preparation of random copolymers of styrene (2) with t-butyl acrylate (3) were investigated under "living" free radical conditions for a variety of monomer feed ratios, Scheme 3. The polymerizations were performed neat, under N_2, at 125-130°C for 48 hours, using the unimolecular initiator (4).

(2) + (3) —(4), 125 C / N_2→ (10)

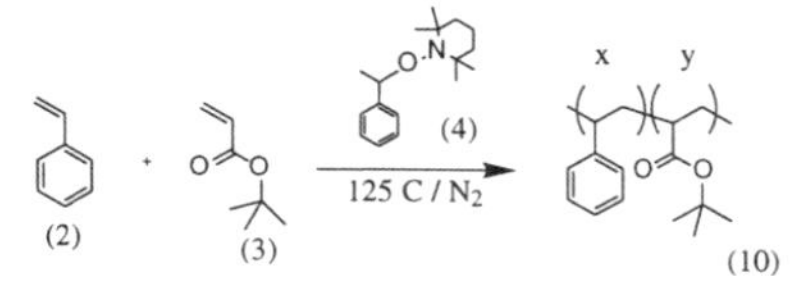

Scheme 3. Synthesis of Random Poly(styrene-co-t-butyl acrylate)

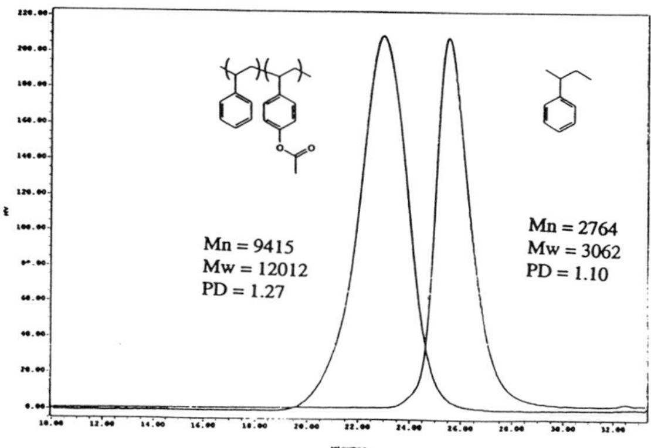

Figure 1. GPC of 70/30 Block Poly(4-acetoxystyrene-co-styrene)

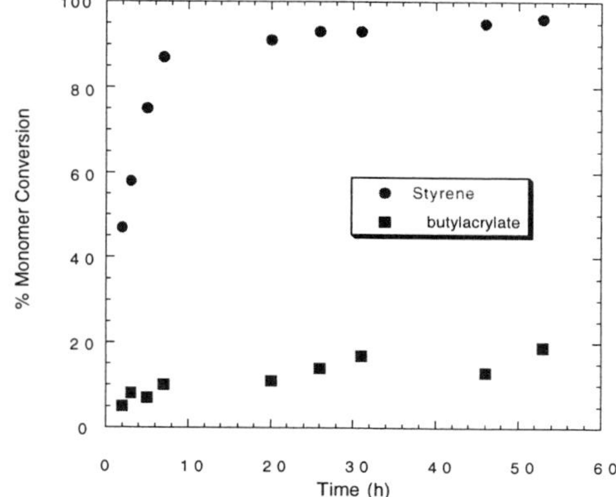

Figure 2. Monomer Conversion for the "Living" Free Radical Polymerization of Styrene vs t-Butyl acrylate

The polymerization of styrene and t-butyl acrylate (feed ratio 65/35) at 130°C gave a narrow polydipersity random copolymer (PD = 1.24) with high conversion (> 90%). Quantitative ^{13}C NMR analysis of this copolymer (10) indicated that the ratio of repeat units incorporated (68/32) was in good agreement with the theoretical feed ratio. The polymerization of styrene and t-butyl acrylate (65/35) was further investigated by following the extent of conversion, polydispersity, and molecular weight, during the polymerization. Figure 3, shows the growth of polydisperisty (PD) and molecular weight relative to total monomer conversion during this {65/35}styrene and t-butyl acrylate copolymerization. As can be seen from Figure 3, the polydispersity is controlled well below the theoretical limit (PD = 1.5) throughout this copolymerization, resulting in a polydispersity of 1.24. Further, an incremental increase in molecular weight with time was observed, indicating that the "living" nature of the TEMPO mediated radical polymerization is unaffected by the presence of the acrylate monomer for this copolymer ratio. To investigate the effect of the monomer feed ratio on molecular weight control and polydispersity, a variety of copolymers of styrene with t-butyl acrylate were prepared under "living" free radical conditions. As illustrated in Figure 4, good agreement between the observed molecular weights and the theoretical molecular weight of 9,500 a.m.u.[10] is achieved at high styrene ratios. Further, at these higher styrene ratios extremely low polydispersities obtained. For example, heating a neat mixture of styrene and t-butyl acrylate (feed ratio 80/20) at 130°C for 48 hours, resulted in an 85% isolated yield of polymer, which was shown to have a polydispersity of 1.21. As the styrene ratio decreases, it was observed that the polydispersity increases and molecular weight control is lost. Over the range of copolymer compositions investigated the polydispersity increases from a low of 1.10 for a 90/10 feed ratio (styrene / t-butyl acrylate) to high of 1.70 for a 30/60 copolymer. The reason for this increase in polydispersity and loss of molecular weight control at high acrylate ratios is most likely due to the stability of the of the acrylate TEMPO adduct chain end. It should be noted that over the composition range of interest 0-40 mol% t-butyl acrylate, the control over macromolecular structure is significantly greater than with classical free radical techniques with the observed polydispersities all well below the theoretical limit of 1.50.[9] Using this technique, a series of poly(styrene-co-t-butyl acrylate) random copolymers were prepared, Table 2.

A similar synthetic strategy was used to attempt the preparation of block copolymers of poly(styrene-co-t-butyl acrylate). In an endeavor to prepare a poly(t-butyl acrylate) macroinitiator (11), t-butyl acrylate was polymerized using 1-phenyl-1-(2',2',6',6'-tetramethyl-1'-piperidinyloxy)-ethane (4) at a temperature of 130°C (Scheme 4). However, in stark contrast to the TEMPO mediated polymerization of styrene, the polymerization of t-butyl acrylate resulted in polymers with broad polydispersities (e.g. M_n = 9415; PD = 2.2). As a result these poly(t-butyl acrylate) macroinitiators were too ill-defined for the preparation of well-defined block

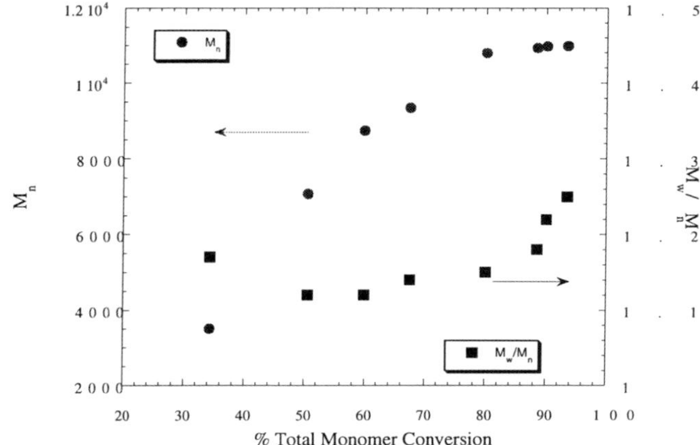

Figure 3. 65/35 Random Poly(styrene-co-t-butyl acrylate): Control of M_n and PD

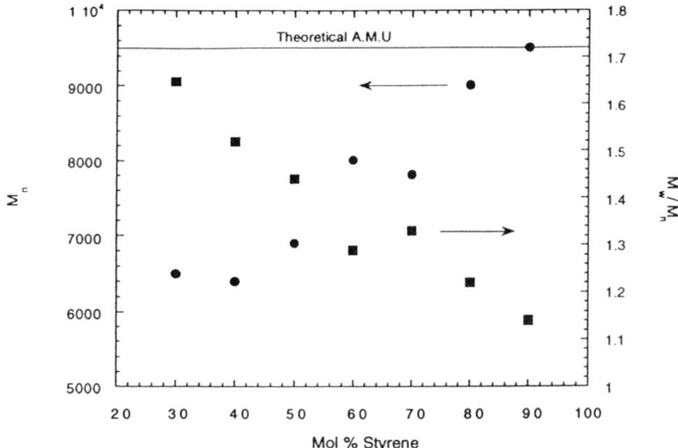

Figure 4. Effect of Monomer Feed Ratio on M_n and Polydispersity

Table 2. Random Poly(styrene-co-t-butyl acrylate)

Styrene	t-Butyl-acrylate	M_n	M_w	PD
90	10	11500	13110	1.14
80	20	11340	13720	1.21
70	30	10200	13360	1.31
65	35	10900	13520	1.24

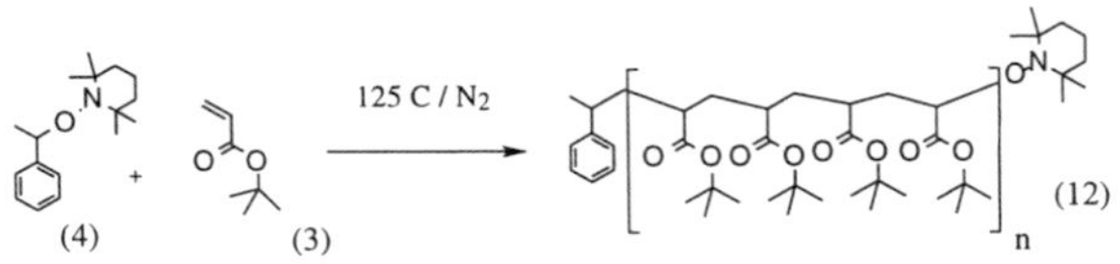

125 C / N_2

(4) (3) (12)

Scheme 4. Synthesis of Poly(t-butyl acrylate)

copolymers with styrene. Alternatively, block copolymers of poly(styrene-co-t-butyl acrylate) (12) were prepared starting from a polystyrene macroinitiator (7), Scheme 5. Although a uniform increase in molecular weight was observed for these polymerizations (Figure 5), the resulting block copolymers were too ill-defined having polydispersities ranging from 1.5 to 1.8, depending upon composition. As a result of this broad polydispersity these block copolymers did not have consistent acrylate block lengths and were therefore not suitable for this study of aqueous base dissolution behavior. The non-uniform distribution of styrene blocks is illustrated in the photodiode array spectra of the GPC traces. From Figure 6, it can be seen that the benzene π - π^* absorption (@ 260nm), characteristic of polystyrene, is non-uniformly spread throughout the molecular weight distribution. Further, as would be expected in a polydisperse system, this characteristic styrene absorption is higher in the lower molecular weight fraction of the distribution, due to simple concentration effects. At present other controlled polymerization techniques are being used to prepare block copolymers of poly(styrene-co-t-butyl acrylate).

(4) (2)

125 C / N_2

(7)

125 C / N_2

(3)

(11)

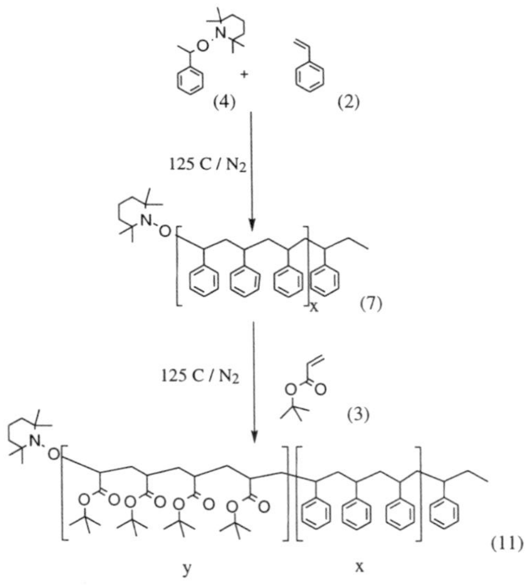

Scheme 5. Synthesis of Block Poly(styrene-co-t-butyl acrylate)

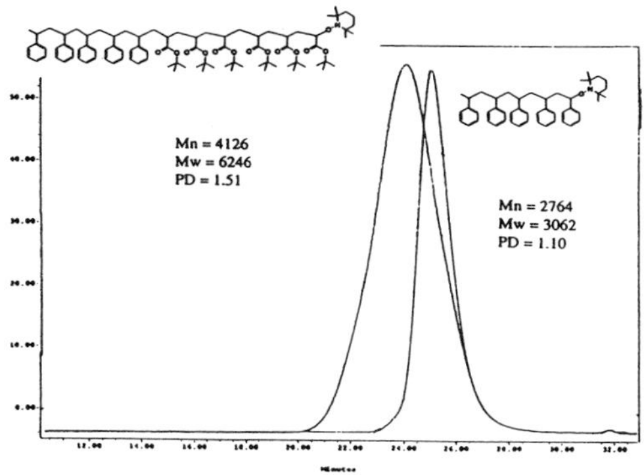

Figure 5. GPC Trace of Block Poly-(styrene-co-butyl acrylate)

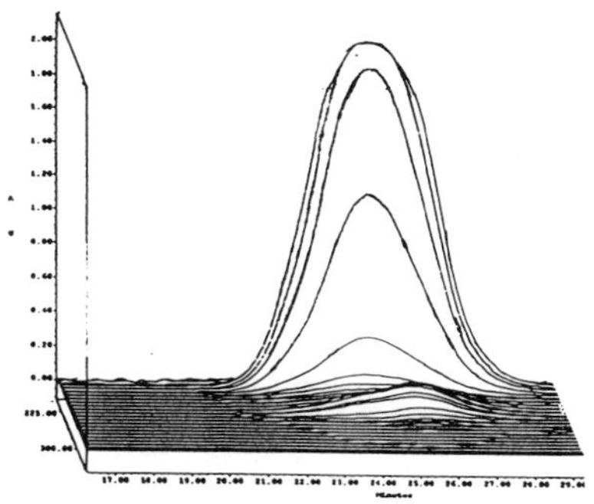

Figure 6. GPC Photodiode Array Spectra of Block Poly-(styrene-co-t-butyl acrylate)

Effect of Macromolecular Architecture Upon Aqueous Base Dissolution. To study the effect of molecular architecture upon aqueous base dissolution, as described previously, well-defined random and block copolymers of 4-hydroxystyrene and styrene were prepared with a variety of molar compositions, ranging from 90:10 mol% (hydroxystyrene:styrene) to 55:45 mol%, Table 1. The thin film aqueous base solubilization of these copolymers was then investigated.

Figure 7 shows the dissolution rates of a range of well-defined random and block copolymers of 4-hydroxystyrene and styrene in 0.26 N TMAH. As can be seen from this figure, as the styrene content is increased there is a dramatic reduction in the aqueous base solubility of both the random and block copolymers. The dissolution rate of these copolymers approaches 0 Å/sec for the 55:45 composition in 0.26 N TMAH. Interestingly, it was observed that over a composition range of approximately 20 - 30 mol% styrene the block copolymers exhibit significantly faster dissolution than the corresponding random copolymers. At compositions containing less than 10 mol% styrene the aqueous base solubility of the random and block copolymers are similar. Further, copolymers containing greater than 40 mol% styrene also exhibit no differentiation between random and block architectures in 0.26 N TMAH. However, to investigate the improved aqueous base solubility of the block copolymers compared to the random copolymers, the dissolution behavior of the composition containing 45 mol% of styrene was investigated as a function of TMAH normality, Figure 8. At normalities below 0.3 N, there is little difference in the dissolution rate between the random and block copolymers. The large hydrophobic styrene content dominates solubilization in aqueous base. However, at normalities greater than 0.3 N, there is a considerable difference in dissolution behavior between the random and block architectures. For example, the 55/45 block copolymer exhibits a 10x faster dissolution rate than the corresponding random copolymer in 0.6 N TMAH. The possible reasons for this dramatic and fundamental difference in the aqueous base dissolution behavior between random and block copolymers of 4-hydroxystyrene and styrene is under investigation. However, initial studies indicate that at these high styrene contents (> 40 mol%), microphase separation of hydrophilic phenolic and hydrophobic styrene blocks may play an important role.

A smooth dissolution, with a hard, gel-free interface between the resist and the aqueous base developer is necessary to obtain a high resolution resist. To study the solubilization of these phenolic copolymers in more detail, the development rate monitor interferograms where recorded for each composition. If a smooth dissolution interface is obtained, a perfect sinusoidal pattern results from the interference of the reflected light from the resist/developer interface and the substrate, Figure 9a. If an irregular pattern is obtained, this is characteristic of non-uniform dissolution at the developer/resist interface, which can be attributed to either multiple reflections from gel layer formation (Figure 9b) or diffuse reflection from a rough surface (Figure 9c).

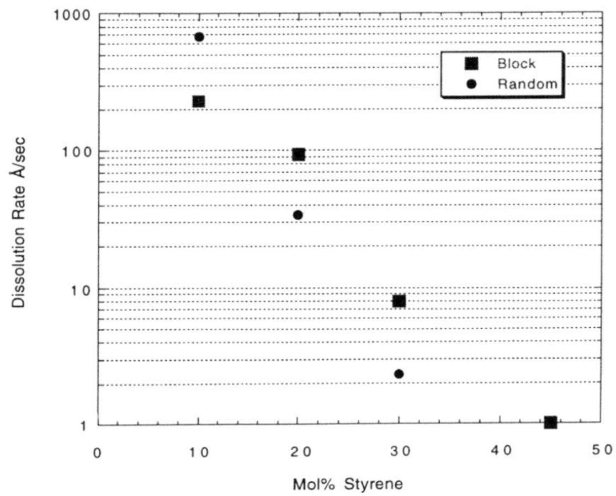

Figure 7. Dissolution Rates of Random and Block Poly(4-hydroxystyrene -co-styrene) in 0.26 N TMAH

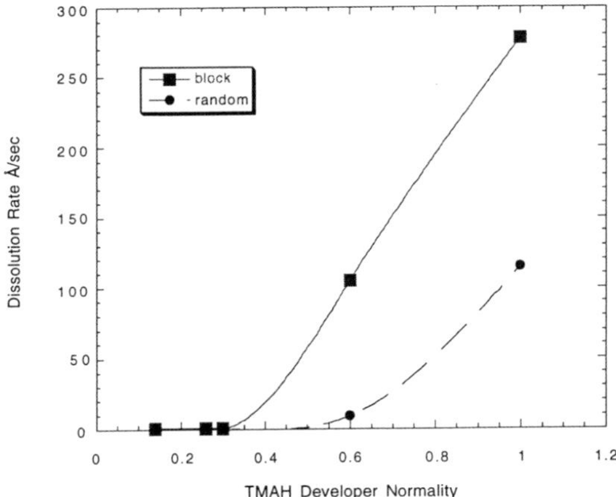

Figure 8. Dissolution Behavior of 55/45 Block Poly(4-hydroxystyrene -co-styrene) in aqueous TMAH

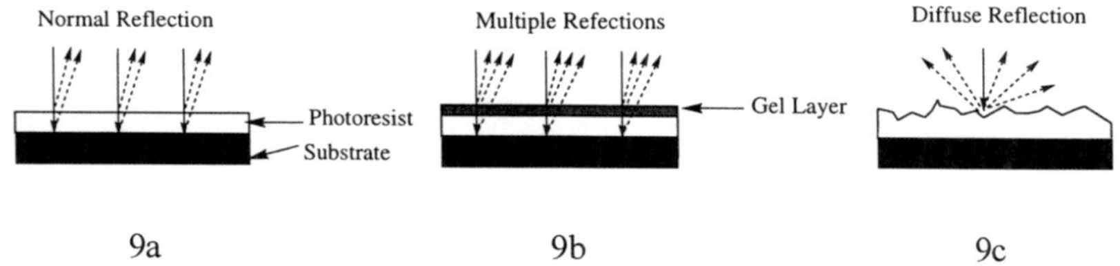

Figure 9. Modes of reflection from the resist/developer interface and the substrate.

It was observed that at high styrene contents (> 20 mol%), the block copolymers exhibit non-uniform dissolution behavior, as indicated by irregular development rate monitor (DRM) interferograms. This is in contrast to the random copolymers which exhibited normal, "smooth" dissolution behavior at all compositions investigated. Figure 10 shows the DRM interferograms of both random and block 80/20 poly(4-hydroxystyrene-co-styrene) in 0.26 N aqueous TMAH. As can be seen from this figure both the random and block copolymers exhibit a sinusoidal pattern indicative of a smooth, gel-free dissolution front. In contrast, as illustrated in Figure 11, at a composition of 55/45 poly(4-hydroxystyrene-co-styrene), while the random copolymer exhibits normal dissolution behavior, the block copolymer shows a non-uniform dissolution front in 0.6 N TMAH. The irregular aqueous base solubilization of the 55/45 block copolymer was also noted in higher normality developer (1.0N aq. TMAH). As suggested previously, one possible explanation for the abnormal dissolution behavior of the 55/45 block copolymer, is microphase separation. The small domains of differing hydrophobicity resulting in the uneven solubilization of the polymer matrix at the developer front, causing either multiply reflections due to swelling , or diffuse reflections from rough surfaces at the interface. Interestingly the same block copolymer composition (55/45) exhibited two glass transition temperatures, indicative of microphase separation, at 97 and 162°C . In contrast the random copolymer of the same composition showed only one T_g at 144°C. Small angle X-ray analysis is underway to confirm microphase separation in these higher styrene content block copolymers.

Conclusions

A series of well defined random and block copolymers of 4-hydroxystyrene and styrene have been prepared by "living" free radical polymerization. Further, to investigate the effect of the aqueous base solubilizing group a number of narrow polydispersity poly(styrene-co-t-butyl acrylate) random copolymers were prepared. However, an attempt to prepare the correponding block copolymers of styrene and t-butylacrylate was unsuccessful. The resulting block copolymers were ill-defined having polydispersities ranging from 1.5 to 1.8 depending upon composition. An investigation into the effect of molecular architecture upon the aqueous base dissolution behavior of poly(4-hydroxystyrene-co-styrene) random and block

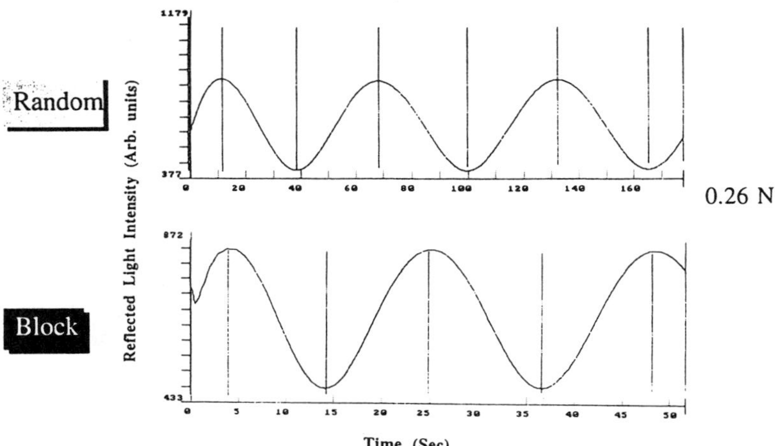

Figure 10. DRM Interferograms of 80/20 Poly(4-hydroxystrene-co-styrene)

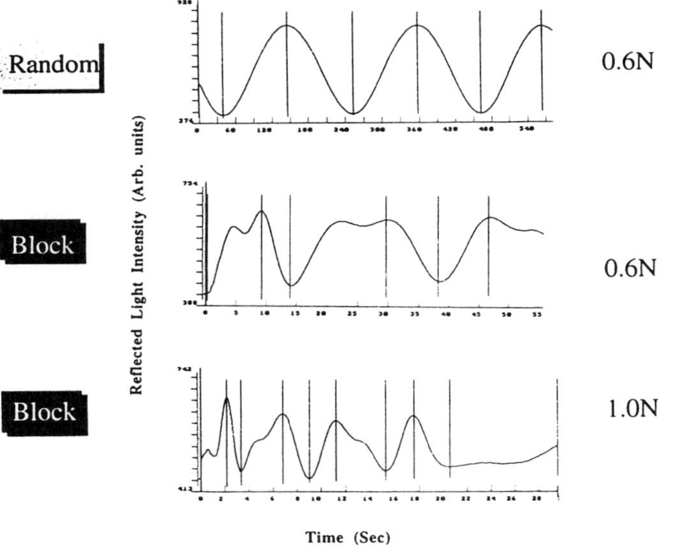

Figure 11. DRM Interferograms of 55/45 Poly(4-hydroxystrene-*co*-styrene)

copolymers was undertaken. It was observed that with increasing mol% of styrene the dissolution rate of the copolymer decreases. It was also shown that, poly(4-hydroxystyrene-co-styrene) block copolymers exhibit faster dissolution rates in aqueous base than the equivalent random copolymers. Finally, at higher styrene contents (> 20mol%) the block copolymers show non-uniform dissolution behavior. This is in contrast to the corresponding random copolymers, which exhibit smooth dissolution behavior.

References

1. Allen, R.D.; Rex Chen, K.J.; Gallagher-Wetmore, P.M. *Proc. SPIE*, **1995**, 2438, 250.
2. Tsiartas, P.C.; Simpson, L.L.; Qin, A.; Willson, C.G.; Allen, R.D.; Krukonis, V.J.; Gallagher-Wetmore, P.M. *Proc. SPIE*, **1995**, 2438, 261.
3. Barclay, G.G.; Hawker, C.J.; Ito, H.; Orellana, A.; Malenfant, P.R.L.; Sinta, R.F. *Proc. SPIE*, **1996**, 2724, 249.
4. Gabor, A.H.; Pruette, L.C.; Ober, C.K.*Chem. Mater.*, **1996**, 8, 2282.
5. Gabor, A.H.; Ober, C.K.*Chem. Mater.*, **1996**, 8, 2272.
6. Wang, J.G.; Sundararajan, N.; Ober, C.K. *Polym. Mater.: Sci. and Eng.*, **1997**, 77, 443.
7. Georges, M.K.; Veregin, R.P.N.; Kazmaier, P.M.; Hamer, G.K. *Macromolecules*, **1993**, 26, 2987. Hawker, C.J. *J. Am. Chem. Soc.*, **1994**, 116, 11314. Matyjaszewski, K.; Gaynor, S.; Wang, J.S. *Macromolecules*, **1995**, 28, 2093.
8. Hawker, C.J.; Barclay, G.G.; Orellana, A.; Dao, J.; Devonport, W. *Macromolecules*, **1996**, 29(16), 5245.
9. Hawker, C.J.; Elce, E.; Dao, J.; Volksen, W.; Russell, T.P.; Barclay, G.G. *Macromolecules*, **1996**, 29(7), 2686.
10. Theoretical value based on the relative molar ratio of initiator and monomer, and the assumption that one molecule of initiator initiates a single chain with no termination.

Chapter 13

Acid Proliferation Reactions and Their Application to Chemically Amplified Lithographic Imaging

Kunihiro Ichimura[1], Koji Arimitsu[1], Soh Noguchi[1], and Kazuaki Kudo[2]

[1]Research Laboratory of Resources Utilization, Tokyo Institute of Technology, 4259 Nagatsuta, Midori-ku, Yokohama 226, Japan
[2]Institute of Industrial Science, University of Tokyo, 7-22-1 Roppongi, Minato-ku, Tokyo 106, Japan

Three kinds of acid amplifiers liberating sulfonic acids autocatalytically were developed. They include *tert*-butyl 2-methyl-2-tosyloxymethyl-3-keto-butanoate, 1-phenyl-1-(2-benzenesulfonyloxyethyl)-1,3-dioxolane and *cis*-1-hydroxyl-1-phenyl-2-tosyloxycyclohexane. These acid-sensitive compounds undergo autocatalytic decomposition to produce acidic species in a non-linear manner so that it is reasonable to refer these reactions as acid proliferation. The addition of every acid amplifier boosts photosensitivity of conventional chemically amplified photoresist systems and improves performances including photosensitivity and pattern profiles of a photoresist for 193 nm photolithography.

A variety of polymeric materials for photoimaging based on the generation of photoacids have been developed in the framework of chemically amplified photoresists (*1*). The general process consists of the photolytic generation of an acidic species, the subsequent acidolytic reactions of a dissolution inhibitor by heating and the development with a suitable solvent. Though these systems offer improved photosensitivity owing to the involvement of catalytic processes, considerable reduction of exposure energies has required to achieve versatile photoimagings. Our basic strategy to boost the photosensitivity of polymeric materials arose from the idea to incorporate of an autocatalytic process in photopolymer systems, taking notice of the fact that silver halide photography involves the autocatalytic reduction of silver halide grains in the development procedure. We found recently that an acetoacetate derivative substituted with *p*-toluenesulfonyloxy (tosyloxy) residues generate *p*-toluenesulfonic acid (TsOH) through autocatalytic processes and proposed to call this type of compound and this kind of reaction, acid amplifier and acid proliferation, respectively, taking into account the generation of a catalytic acid in a non-linear manner through a geometrical progression (*2*). Patent literature described a novel photoimaging system which consists of the formation of carboxylic acids catalyzed by a photogenerated superacid to transform leuco-dyes into colored substances. Though acidic species are generated in this systems by an acidolysis reaction, it contains no autocatalysis (*3*). To our knowledge, there has been no distinct report on the autocatalytic acidolytic reaction except for the implication that the reversible elimination/addition reaction of cycloalkyl benzenesulfonate takes place through an autocatalytic process (*4*).

Major concern of this paper is to review the nature and the reactivity of acid amplifiers and to apply them to the enhancement of photosensitivity of chemically amplified photoresist systems.

Molecular Design and Synthesis

General Consideration. An acid amplifier should fulfill the following requirements for our purposes. First, it should liberate a strong acid to initiate an autocatalytic reaction of the acid amplifier. Second, a liberated acid should be strong enough to induce a subsequent acidolytic reaction in post-exposure bake for image formation. Thirdly, an acid amplifier should be thermally stable under conditions during the processes to achieve both of the autocatalytic reaction and the subsequent acid-catalyzed reaction of a dissolution inhibitor. We have so far developed three kinds of acid amplifiers which give a sulfonic acid as a strong acid on the basis of different organic reactions.

Initially, we designed *p*-(*tert*-butoxycarbonyloxy)benzyl *p*-toluenesulfonate (**1**) under anticipation that the acidolytic deprotection of *p*-(*tert*-butoxycarbonyloxy) (t-BOC) residue would give an acid-labile *p*-hydroxybenzyl sulfonate to liberate a sulfonic acid quite readily (Scheme 1). It was revealed that **1** does not meet the third condition stated above because of the inefficient stability of the benzyl sulfonate structure toward heat treatment so that no clear indication was obtained whether an autocatalytic process is involved (*5*). This result led us to the molecular design for acid amplifiers by introducing an alkyl sulfonate structure owing to relative thermal stability. We have so far developed the following three kinds of acid amplifiers ((a)-(c) in Scheme 2).

Acetoacetates Exhibiting Fragmentation. The first class of acid amplifiers consists of α-methylated acetoacetate derivatives having an acid-labile ester unit (**2**). This type of compounds is transformed by the acid-catalyzed deprotection into β-keto-acetic acid which is readily decarboxylated by heating into the corresponding keto-tosylate (**3**). **3** undergoes smoothly the β-elimination to afford an unsaturated ketone (**4**) and TsOH, as shown in Scheme 2(a). We prepared some acetoacetate derivatives to optimize chemical properties fulfilling the requirements state above (Scheme 3). It was confirmed that sulfonic acids including TsOH and methanesulfonic acid (MsOH) are of sufficient acidity to cause the deprotection of **2** in less polar solvents while the deprotection proceeded much more slowly with dichloroacetic acid owing to its weaker acidity (pKa = 1.25). Consequently, dichloroacetylated derivatives (**2e** and **2f**) exhibit much less ability for the acid proliferation reaction. These results imply that the introduction of a sulfonyloxymethyl residue at the α-position of the acetoacetate moiety is essentially of significance.

In respect of protected esters, we prepared both *tert*-butyl and 2-phenylpropyl-2 (PP) esters. Though PP esters are deprotected acidolytically faster than its *tert*-butyl counterpart and seems to be convenient for the present purpose, the PP tosylate (**2b** and **2d**) was rather thermally labile so that it does not fulfill the present requirements. Finally, we employed *tert*-butyl acetoacetate having a tosyloxymethyl substituent at an α-carbon (**2a**) as the acid amplifier for the present purpose. This compound of mp 52-53 °C was synthesized by the α-methylation of tert-butyl acetate, followed by α-hydroxymethylation with formalin and tosylation in a 10% overall yield (*2*).

The autocatalytic fragmentation behavior of **2a** in solution was monitored by means of NMR spectra. While the compound shows no degradation at 100 °C in a less polar mixture of diphenyl ether and toluene, the addition of TsOH results in the

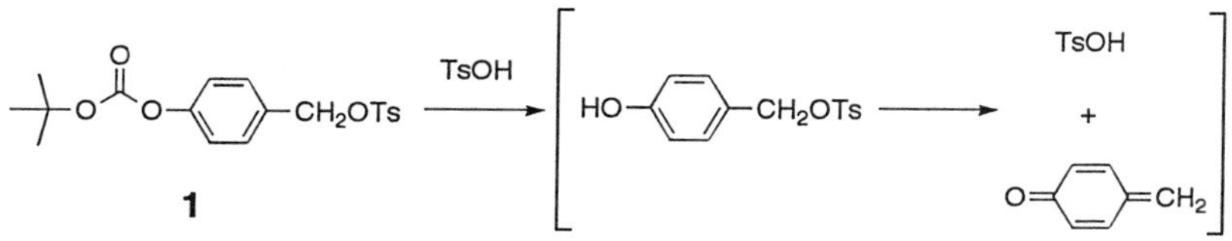

Scheme 1 An attempt to generate TsOH autocatalytically.

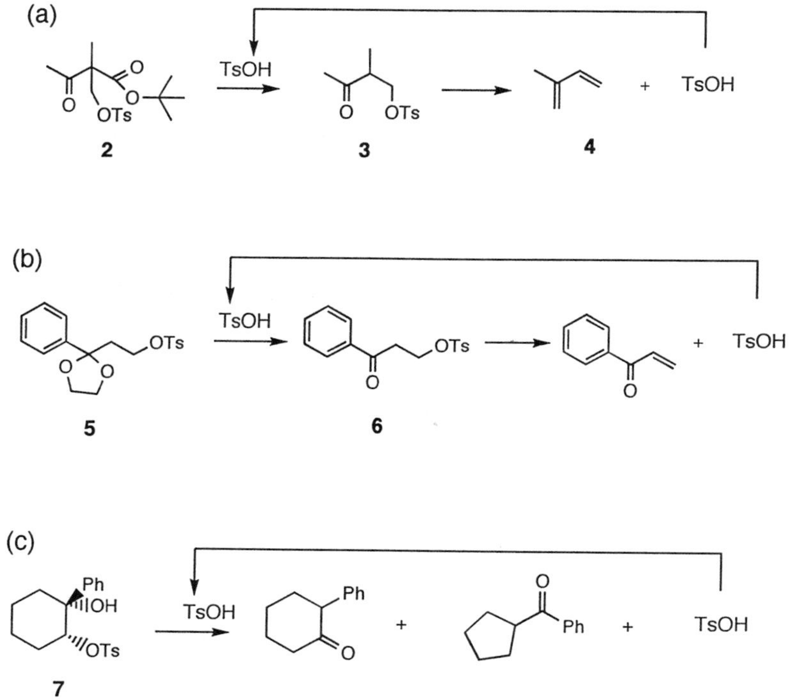

Scheme 2 Acid proliferation reactions to generate sulfonic acids.
(a) The fragmentation of *tert*-butyl acetoacetates, (b) the fragmentation of a ketal-tosylate and (c) the pinacol-type rearrangement of diol monotosylates.

disappearance of **2a** and the formation of the corresponding unsaturated ketone as a result of the β-elimination displaying a sigmoidal curve (Figure 1), suggesting the involvement of an autocatalytic process. It is worthy to note here that no acid-catalyzed reaction is essentially caused when dioxane or methanol is used as a solvent. This is obviously because of the higher basicity of these solvents (*pKa* of conjugated acids of methanol and dioxane are -2.2 and -2.92, respectively) when compared with that of an ester unit (*pKa* = *ca.* -6.0) so that the protonation on an oxygen atom of the ester of **2a** is practically suppressed in these solvents.

Ketal Tosylates. The second candidate for the acid proliferation is ketal derivatives having a tosylated residue (**5**). These were synthesized by the ketalization of benzoylacetate, followed by LAH reduction and tosylation (*5*). Compound **5** (mp 48-50 °C) had to be stored in a refrigerator since crystals of **5** were converted into black materials gradually at an ambient temperature upon prolonged storage. The ketal compound (**5**) was readily converted in the present of an acidic species into β-keto-tosylate (**6**) which is subjected to the β-elimination to give TsOH in a non-linear manner (Scheme 2(b)). NMR spectra of a solution of **5** in the presence of TsOH at 100 °C revealed that the β-keto-tosylate (**6**) as an intermediate was formed during the reaction (*6*).

Diol monotosylates. Pinacol rearrangement involves the dehydration and migration of a residue of 1,2-diol to give the corresponding ketone under acidic conditions. This suggested us the possibility that TsOH can be liberated by the pinacol rearrangement of a diol monotosylate (Scheme 2(c)). To our knowledge, the pinacol-type rearrangement of diol monosulfonates has not yet been carried out under acidic conditions.

We prepared the *cis*-isomer of 1-phenyl-1-hydroxy-2-tosyloxycyclohexane (**7**) of mp 109-110 °C starting from 1-phenylcyclohexene, which was converted into the corresponding *cis*-diol with the acid of a mixture of hydrogen peroxide and formic acid, followed by selective monotosylation. The *cis* configuration was selected here in order to minimize a stereoelectronic effect of OH group leading to the enhanced elimination of the β-tosyloxy residue.

Figure 2 shows time conversion curves of **7** in $CDCl_3$ at 100 °C. In the absence of TsOH, the disappearance of the diol monotosylate started after heating for 200 min or longer and occurred in a non-linear manner. The addition of TsOH resulted in the marked reduction of heating time for the abrupt consumption of **7**, indicating that the acidolysis reaction of **7** takes place autocatalytically. The determination of product distribution showed that 2-phenylcyclohexanone-1 and benzoylcyclopentane are obtained in 36 % and 44 % yields, respectively, while TsOH is formed almost quantitatively (Scheme 4). Considering the chemical structures of the ketonic products, it is very likely that **7** undergoes pinacol-type rearrangement.

Photosensitivity Characteristics of Photopolymers Incorporating Acid Amplifiers

p-[(*tert*-Butoxycarbonyloxy)styrene) (pBOCSt) is one of the representative acid-labile polymers giving a chemically amplified photoresist by the combination with a photoacid generator (PAG) (*7*). The effect of the acetoacetate-type acid amplifier (**2a**) on the deprotection of t-BOC of the polymer was determined by FT-IR measurements to followed the decrease in the absorption band due to the t-BOC groups of a film of the polymer containing **7** and TsOH. It was shown that the peak height of the band of a UV irradiated film decreases abruptly upon post-exposure bake after a certain

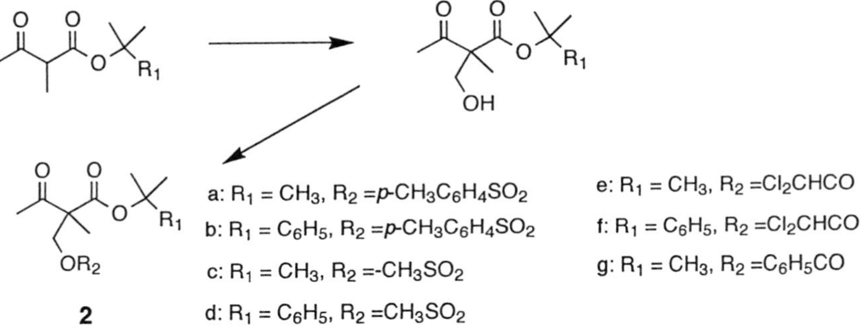

Scheme 3 Preparation of acetoacetate derivatives.

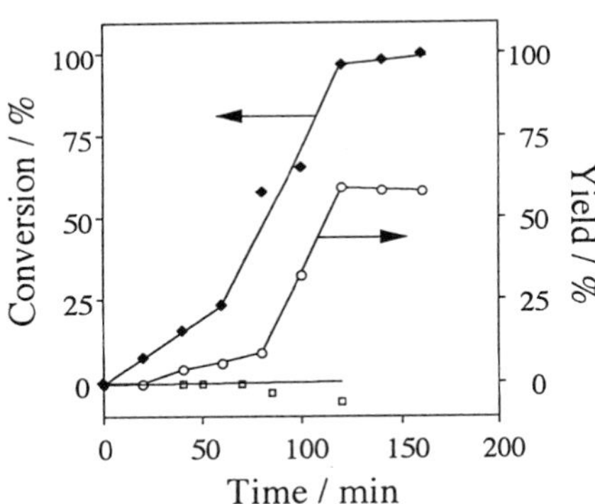

Figure 1 The disappearance of the acetate (**2a**) (- ◆ -) and the formation of 3-keto-2-butene (-○-) in a 3:1 miture of diphenyl ether and d_8-toluene at 80°C in the presence of 13 mol% of TsOH. The plots (-□-) are for **2a** in the absence of TsOH.

induction period, implying that the acidolytic deprotection takes place also in a non-linear manner in a polymer solid (*2*).

PAGs (**8**, **9**, **10**) employed in this work are summarized in Scheme 5. The effect of the addition of **7** on the photosensitivity of films of pBOCSt plus PAG was followed by conveniently by monitoring the thickness of UV-irradiated films consisting of a ternary mixture after post-exposure bake for 1 min at 100 °C. The thickness of a fully deprotected polymer film becomes about 60 % of the original one as a result of the fragmentation to release CO_2 and isobutylene. The results shown in Figure 3 are summarized as follows. First, the addition of 10 wt% of **7** induces sudden completion of the reduction of film thickness after irradiation with 313 nm light and subsequent post-exposure bake. This makes a sharp contrast to the behavior of a film containing PAG only, which displays a gradual decrease in thickness. This marked effect of **7** supports strongly that the additive plays in fact as an acid amplifier to accumulate the concentration of TsOH in a non-linear manner to lead to such drastic deprotection. Second, the effect of **7** is observed for all the systems tested irrespective of the nature of PAG. The exposure energy required for the deprotection is influenced by PAG. It is likely that the exposure energy is related closely with a quantum efficiency for the generation of a photoacid. It follows that the results shown in Figure 3 mean that the autocatalytic reaction of **7** is triggered by every strong acid which is conventionally used for chemically amplified photoresists. It should be mentioned that mesylated derivatives (**2c**) are not suitable here because of the volatility during post-exposure bake though methanesulfonic acid does work as a trigger acid. An increase in the molecular weight of this type of compound should resolve this problem.

A much more remarkable effect of the acid amplifier on photosensitivity was observed when a ternary system consisting of poly(*tert*-butyl methacrylate) (ptBM), an aromatic sulfonium salt (**8**) as PAG and the ketal-tosylate (**5**) as an acid amplifier is employed. ptBM was used in this case on account of the sensitivity of the *tert*-butyl ester group toward an acidic species though this polymer is not always favorable for practical applications. Photosensitivity determination was made by measuring thickness of films after 313 nm light irradiation, post-exposure bake and development with a dilute aqueous solution of tetramethylammonium hydroxide. As Figure 4 shows, an exposure energy required for the complete dissolution of a film was one order magnitude smaller than that of a film in the absence of **5**. Quite a similar effect was observed for a ternary system which employs the diol monotosylate (**7**) as an acid amplifier. As given in Figure 5, the photosensitivity is enhanced to a similar extent when the acid amplifier is added to a chemically amplified photoresist systems consisting of ptBM. The results suggest that the photosensitivity enhancement is leveled off at a concentration of about 10 mol%.

Application of 193 nm Photolithography

Encouraged by the results shown above, the evaluation of the additive effect of an acid amplifier has been recently achieved by Ohfuji et al., aiming at improving photolithographic characteristics to meet requirements for 193 nm photomicropatterning (*8*). The materials employed here was a base polymer incorporating 1-ethoxyethyl methacrylate (**11**) as an acid-labile unit, a tosylated benzoin (**10**) as PAG and the diol monotosylate (**7**) as an acid amplifier. Thin films were exposed to 193 nm light from an ArF laser, followed by post-exposure bake and development with a dilute alkaline solution. The results are summarized as follows. First, the addition effect of **7** was confirmed as follows. An exposure energy of about 10 mJ/cm^2 required for the novel chemically amplified photoresist was reduced

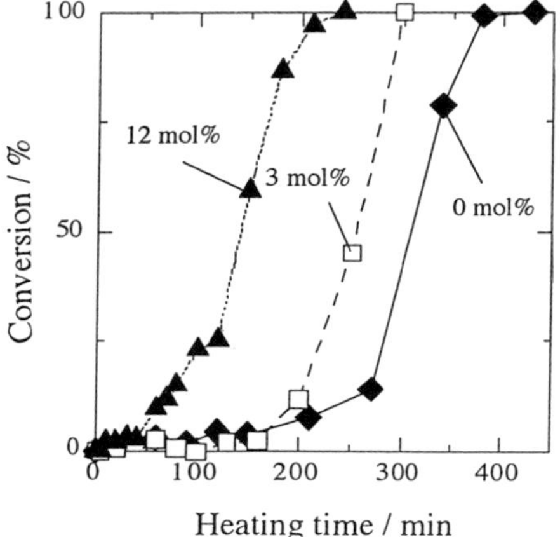

Figure 2 Autocatalytic decomposition of **7** (170 mmol dm^{-3}) in $CDCl_3$ at 100 °C in the presence of TsOH (-◆-: 0 mol%, - □-: 3 mol% and -▲ -: 12 mol%).

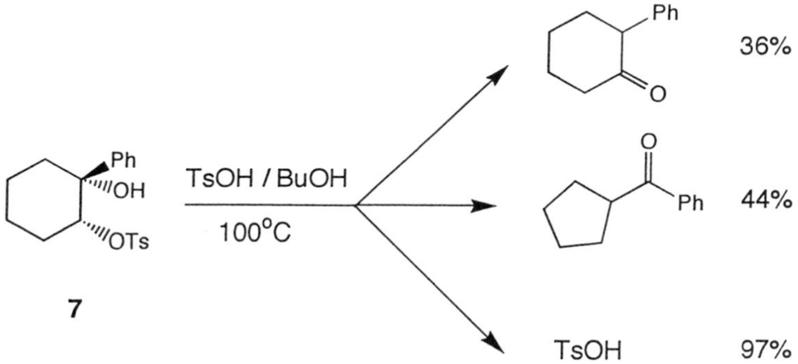

Ph
OH
OTs
7
TsOH / BuOH
100°C
Ph
O
36%
O
Ph
44%
TsOH
97%

Scheme 4 Acidolytic products of the diol tosylate (**7**).

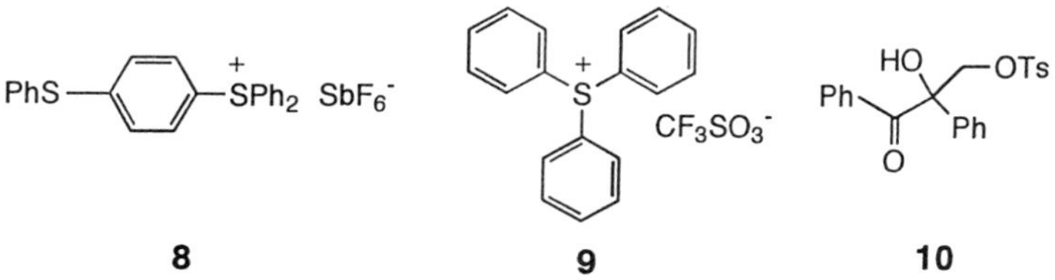

PhS
$\overset{+}{S}Ph_2$ SbF_6^-
8
$\overset{+}{S}$
$CF_3SO_3^-$
9
HO
OTs
Ph
Ph
O
10

Scheme 5 Photoacid generators used in this study.

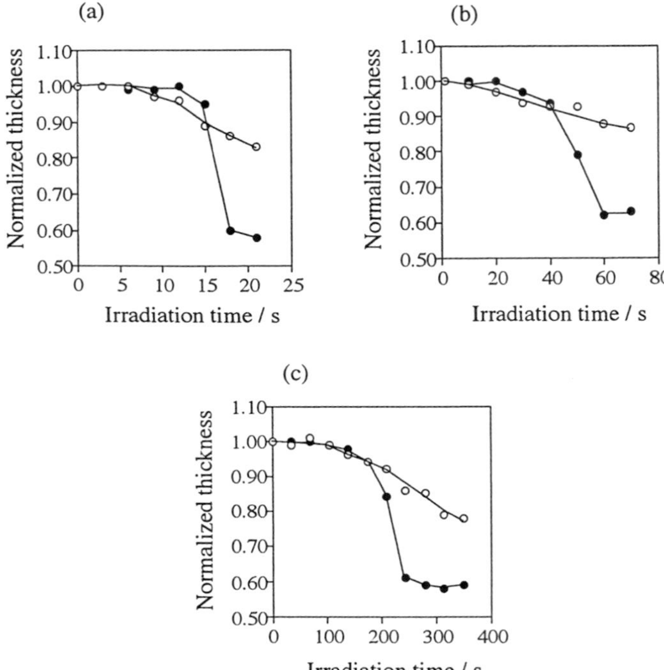

Figure 3 Photosensensitivity curves for pBOCSt films in the absence (-○-) and in the presence of (-●-) of 10 wt% of **2a** using (a) **8**, (b) **9** and (c) **10** as photoacid generators.

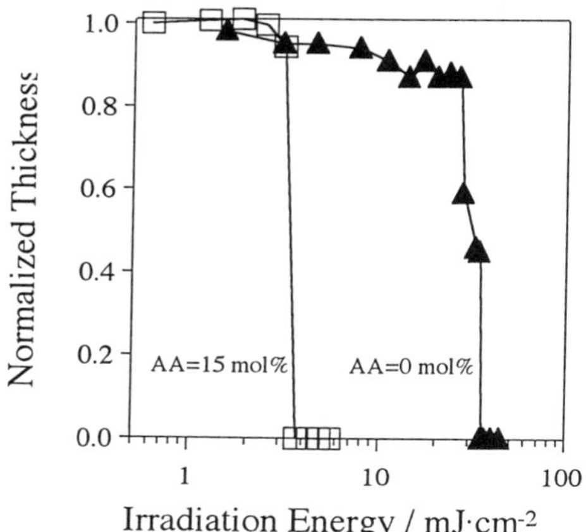

Figure 4 Photosensitivity curves for poly(*tert*-butyl methacrylate) photoactivated by 2 mol% of **8** in the absence (-▲-) and in the presence (-□-) of 15 mol% of ketal-tosylate (**5**) as an acid amplifier. Post-exposure bake; 100°C for 1 min, development; 3 wt% of Me_4N^+ OH^-.

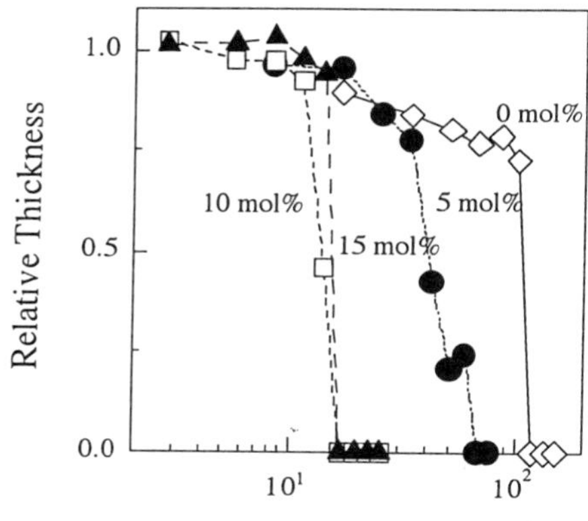

Figure 5 Photosensitivity curves for poly(*tert*-butyl methacrylate) photoactivated by 2 mol% of **8** in the absence (-◇-) and in the presence of 5 mol% (-●-), 10 mol% (-□-) and 15 mol% (-▲-) of diol monotosylate (**7**) as an acid amplifier. Post-exposure bake; 100°C for 1 min, development; 3 wt% of Me_4N^+ OH^-.

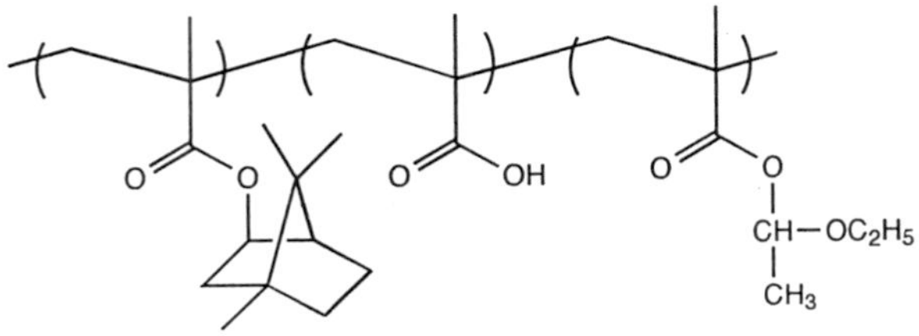

11

down to 7 mJ/cm^2 when 2 wt% of **7** is added whereas an exposure energy is reduced by half when the concentration of **7** is 5 wt%. Because **7** has the aromatic moieties which adsorb 193 nm light strongly so that the photosensitivity determination was made using thinner films of 0.3 mm in thickness. Second, although further optimization is needed from a practical viewpoint, 0.16 mm lines and spaces are clearly resolved. This is in contrast to the results for a chemically amplified photoresist without **7** which cannot resolve 0.16 mm lines and spaces at all and suffers from the generation of T-top in resolution experiments. This data means that **7** contributes to the improvement of pattern profiles to enhance the resolution power in 193 nm photolithography.

Conclusion

A novel concept of acid proliferation was proposed and applied to improve photosensitivity characteristics of chemically amplified photoresist systems. The acid proliferation involves the autocatalytic acidolytic decomposition of an acid-labile compound to produce a strong acidic species in a non-linear manner. This kind of acid-labile compound is reasonably referred to as an acid amplifier. It produces an acid in a manner of geometrical progression, leading to the enhancement of a subsequent acidolytic reaction for photoimage formation on the basis of the incorporation of the non-linear process in chemically amplified photoresist systems. An acid amplifier should liberate an acid strong enough to catalyze both the acidolysis of the acid amplifier and a subsequent deprotection to change the solubility and be thermally stable at least during post-exposure bake.

Three types of acid amplifiers have been so far developed in our laboratory. The first is a *tert*-butyl acetoacetate (**2a**) with a tosyloxymethyl residue at the a-position and confirmed to liberate TsOH as a result of the autocatalytic fragmentation to give isobutylene and CO_2. The second type is β-keto-sulfonate compound (**3**) which is readily deprotected under an acidic condition to give a b-ketoalkyl sulfonate as an intermediate, leading to the β-elimination to yield a sulfonic acid. The third one is *cis*-isomer of 1-pheny-1-hydroxy-2-tosyloxycyclohexane and produce TsOH as a result of pinacol-type rearrangement. The autocatalytic processes of these compounds were determined and confirmed both in solutions and in polymer films in the presence of a catalytic amount of TsOH.

It was found that the addition of these acid amplifiers to conventional chemically amplified photoresist systems results in marked enhancement of photosensitivity characteristics. A photoresist based on a ternary system consisting of pBOCSt, the acetoacetate-type acid amplifier (**2a**) and PAG exhibited enhanced deprotection during post-exposure bake irrespective of the nature of PAG. Marked improvement in photosensitivity was observed when the ketal-sulfonate-type acid amplifier (**3**) or the diol monotosylated-type acid amplifier (**7**) is embedded in a thin film of poly(*tert*-butyl methacrylate) as an acid-labile base polymer and PAG. The photosensitivity was about ten times higher than that without the acid amplifiers.

Practical additive effect of an acid amplifier was verified by evaluating photolithographic characteristics of a chemically amplified photoresist for 193 nm photomicropatterning; both photosensitivity and pattern profiles were improved. The optimization of the chemical structure of the acid amplifier and of its combination with a suitable dissolution inhibitor should be a promising way to give high performance photoresists for microphotopatterning.

Literature cited

1. Reichmanis, E. In *Polymers for Electronic and Photonic Applications*; Wong, C. P., Ed.; Academic Press: San Diego, 1993, pp 67-117.
2. Ichimura, K.; Arimitsu, K.; Kudo, K. *Chem. Lett.* **1995**, 551; Arimitsu, K.; Kudo, K.; Ichimura, K. *J. Am. Chem. Soc.* in press.
3. a) US 5,441,850 (Aug. 15, 1995; Polaroid Co.); b) 5,534,388 (Jul. 9, 1996; Polaroid Co.).
4. Nenitzescu, C. D.; Ioan, V.; Teodorescu, L. *Chem. Ber.* **1957**, *90*, 585.
5. Ichimura, K.; Itoh, H. unpublished results.
6. Kudo, K.; Arimitsu, K.; Ichimura, K. *Mol. Cryst. Liq. Cryst.* **1996**, *280*, 307.
7. Ito, H.; Willson, C. G.; Frechet, J. M. J. Digest of Technical Papers of 1982 Symposium on VLSI Technology, 1982, 86.
8. Ohfuji, T.; Takahashi, M.; Sasago, M.; Noguchi, S.; Ichimura, K. *J. Photopolym. Sci. Technol.* **1997**, *10,* 551.

Advanced Microlithographic Materials Chemistry and Processing

Elsa Reichmanis

In the last decade, major advances in fabricating VLSI electronic devices have placed increasing demands on microlithography, the technology used to generate today's integrated circuits. In 1976, state-of-the-art devices contained several thousand transistors with minimum features of 5 to 6 μm. Today, they have several million transistors and minimum features of less than 0.3 μm. Within the next 5 years, a new form of lithography will be required that routinely produces features of less than 0.2 μm. Short-wavelength (deep-UV) photolithography and scanning and projection electron-beam and X-ray lithography are the possible alternatives to conventional photolithography.

The consensus candidate for the next generation of lithography tools is photolithography using 193 nm light from an ArF excimer laser source. At this wavelength, the intense absorption of aromatic molecules at 193 nm severely limits the use of conventional matrix resins such as novolacs and polyvinylphenols for 193 nm lithography. This has both necessitated a paradigm shift in the approach to lithographic materials and process design and spawned the design of revolutionary resist schemes. In recognition of the strong motivation of device manufacturing engineers to retain as much of the acquired knowledge base regarding solution developed resists as possible in the design of materials for advanced lithographic applications, many of the current research efforts related to 193 nm lithographic materials involve the design of new chemistries that provide aqueous base solubility, etching resistance, resolution, photospeed and process latitude. Several novel materials solutions leading to robust, aqueous base developed 193 nm resist materials have emerged from recent research efforts. Avenues that can lead to transparent, etching resistant polymers include the incorporation of alicyclic and/or silicon bearing substituents. The former can be designed to afford polymers that effect etching performance comparable to novolacs and polyvinylphenols, while the latter affords O_2 etching resistance to allow use in traditional bilevel applications.

Approaches to the design of advanced materials for electron beam lithographic application mirror those for alternative technologies. While workhorse resists such as poly(1-butene sulfone) continue as the dominant imaging materials for photomask fabrication, issues such as dry-etching resistance and aqueous solubility must be addressed for both advanced mask and device applications. Several platforms are under consideration including single-layer and bilayer schemes. With respect to the latter approach, incorporation of silicon into the polymeric resist provides a convenient avenue to achieving an adequate differential etching with respect to substrate materials, while novel chemistries are being explored for more traditional solution developed alternatives.

For any advanced materials to be successful, materials properties must be carefully tailored to maximize lithographic performance with minimal sacrifice of other performance attributes, e.g., adhesion, solubility, RF plasma etching stability. Clearly, materials chemistry/structure plays the key role in defining performance. The challenges associated with the development of new classes of imaging materials for advanced lithographic technologies will be described in this section and the relationships between materials structure and process performance will be discussed.

Chapter 14

Deprotection Kinetics of Alicyclic Polymer Resist Systems Designed for ArF (193 nm) Lithography

Uzodinma Okoroanyanwu[1,3], Jeffrey D. Byers[2], Ti Cao[1], Stephen E. Webber[1], and C. Grant Willson[1]

[1]Department of Chemistry, University of Texas, Austin, TX 78712
[2]SEMATECH, 2706 Montopolis Drive, Austin, TX 78741

The deprotection kinetics of alicyclic polymer resist systems designed for 193 nm lithography was examined using IR and fluorescence spectroscopic techniques. A kinetic model was developed that simulates the deprotection of the resists fairly well. A new, simple, and reliable method for monitoring photoinduced acid generation in polymer films and in solutions of the kind used in 193 nm and deep-UV lithography was developed. This technique could find application in the study of diffusional processes in thin polymer films.

The need to understand and monitor the photoacid generation process in chemically amplified resists cannot be over-stated. First, with the minimum feature size expected to reach the 0.1 μm mark by 2007 (*1*), preserving the integrity of the latent image has now become a major concern in microlithography. Diffusion of the photoacid during the time between exposure and development can cause significant contrast loss and ultimately lead to loss of the latent image, especially in chemically amplified photoresists that must require a post-exposure baking step, which facilitates the diffusion of the acid due to the high temperature normally used. Thus, there is no question that the progress of microlithography within the deep-UV and vacuum-UV regimes will depend significantly on how well the acid generation and diffusion processes in photoresists are understood and controlled.

Infra-red spectroscopy and fluorescence spectroscopy provide convenient methods for studying these processes. Using FTIR, we investigated the deprotection kinetics of some of our alicyclic polymer resist systems (*2,3*) (containing triphenylsulfonium hexafluoroantimonate) that were exposed to 248 and 193 nm

[3]Current address: Advanced Micro Devices, One AMD Place, P.O. Box 3453, MS 78, Sunnyvale, CA 94088–3453.

radiations. With fluorescence spectroscopy we monitored and quantified the photogeneration of acid in some of the alicyclic polymer resist films. The photoresist polymers studied included (1) poly(methylpropyl bicyclo[2.2.1]hept-5-ene-2-carboxylate-co-bicyclo[2.2.1]hept-5-ene-2-carboxylic acid) (*trivial name*: poly(carbo-*t*-butoxynorbornene-*co*-norbornene carboxylic acid) [poly(CBN-*co*-NBCA)] and (2) poly(methylpropyl bicyclo[2.2.1]hept-5-ene-2-carboxylate-*co*-maleic anhydride) (*trivial name*: poly(carbo-t-butoxynorbornene-co-maleic anhydride) [poly(CBN-*alt*-MAH)] (See **Chart 1** below).

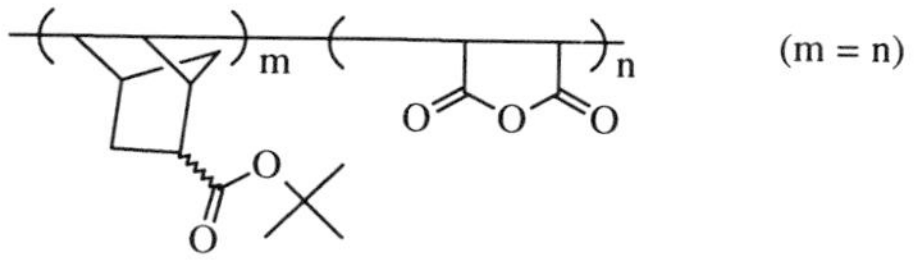

(m = n)

Poly(CBN-*alt*-MAH)

CO_2H

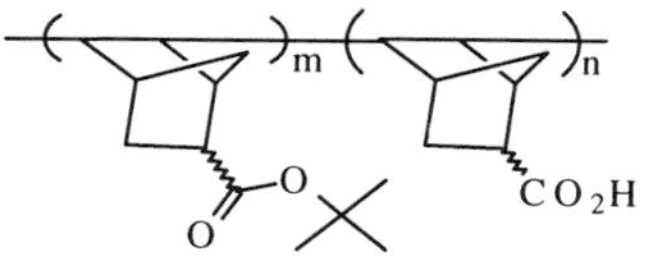

Poly(CBN-*co*-NBCA)

Chart 1. Alicyclic Copolymers

Fluorescence techniques are being used increasingly in wide range of research fields including biochemical, medical and chemical research, due primarily to the inherent sensitivity of the technique and the favorable time scale of the phenomenon of fluorescence (*4*). Some fluorophores have dramatically different emission characteristics for their protonated and unprotonated forms. Using this phenomenon, the photo-acid concentrations generated within chemically amplified photoresist systems can be monitored (*5-7*).

The new resists were formulated by dissolving the alicyclic polymer and a photoacid generator (PAG) such as triphenylsulfonium hexafluoroantimonate in an appropriate solvent such as propyleneglycol monomethyl ether acetate (PGMEA). Upon irradiation, the PAG in the resist film generates a latent image of strong acid, which upon baking, catalyzes the deprotection of the *t*-butyl ester pendant group (as illustrated for a typical photoresist in **Scheme 1**), leading to the formation of isobutylene and a polymer with a norbornene carboxylic acid unit which has a higher solubility in basic developing solvent than its masked precursor. Development of the exposed and baked film in aqueous base generates a positive tone image of the mask.

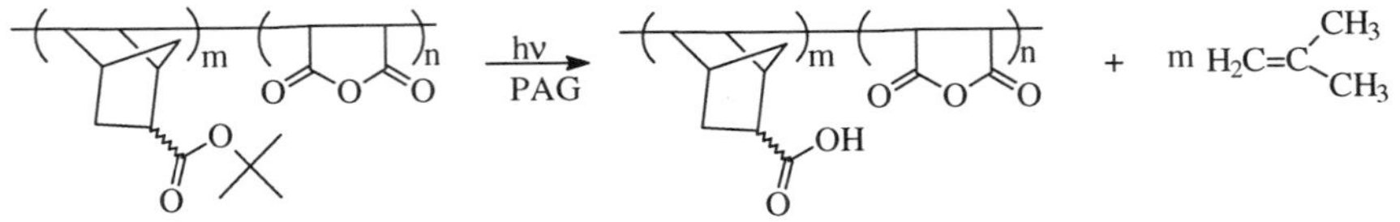

Scheme 1

Experimental Section

Materials. Alicyclic polymers such as poly(CBN-*alt*-MAH) and poly(CBN-*co*-NBCA) were synthesized by free radical and Pd(II)-catalyzed addition polymerization techniques, respectively, the details of which are given in References (*2,3*) and in (Okoroanyanwu, U.; Byers, J.; Shimokawa, T.; Willson, C.G. *Chem. Mater.*, 1998, in press) (Okoroanyanwu, U.; Shimokawa, T.; Byers, J.; Willson, C.G. *Chem. Mater.*, 1998, in press). Triphenylsulfonium hexafluoroantimonate (**TPSHFA**) was prepared according to the literature procedure (*8*). CD11 anti-reflective coating was obtained from Brewer Scientific Company. BARL antireflective coating was obtained from IBM Corporation. PD-523AD developer, 0.21N aqueous tetramethylammonium hydroxide solution with surfactant was obtained from Japan Synthetic Rubber Company. LDD-26W developer, 0.26N aqueous tetramethylammonium hydroxide, was obtained from Shipley Company.

N-(9-acridinyl)acetamide (**ACRAM**) was prepared from 9-aminoacridine hydrochloride hydrate with acetic anhydride in pyridine, using the literature procedures of (*9,10*). The crude product was purified by washing with ice cold chloroform before being dried in vacuum. THF was distilled from activated charcoal and KH_2PO_4.

IR Measurements on Radiation-Exposed Alicyclic Polymer Photoresist Films. Infrared spectra were measured using a Nicolet Magna-IR FTIR/550 spectrometer. The resist solutions were spin coated at 2500 RPM to produce ~0.7µm films on 8 inch double polished silicon wafers and heated at 150°C for 60s (unless otherwise stated) in hard contact with the bake plate. Following exposure to known doses ranging from 0-50 mJ/cm^2 on a GCA XLS KrF (248 nm) or an ISI MicroStep (193 nm) exciplex laser stepper, the wafer was baked with minimal delay (<5 min) for 60s at 150 °C. This experiment was repeated for post-exposure bake temperatures of 140, 130 and 120°C.

Optical Measurements on Alicylic Polymer Films. A Hewlett-Packard 8450A diode array UV-visible spectrophotometer was used to obtain ultraviolet spectra as follows: Resist polymer was dissolved in PGMEA at 16% by weight relative to solvent and spin coated at 2500 RPM on 1 inch quartz wafers and then heated at 150 °C for 60 s, after which the thickness of the film and the UV absorbance were recorded. Thickness was determined by an average of at least three values obtained at different positions along a scratch using the Alpha Step profilometer.

A Woolam Variable Angle Scanning Ellipsometer (VASE) was used to measure the Cauchy coefficients of the photoresist compositions. The photoresist solution comprising 15% by weight of polymer relative to solvent and 3wt % of triphenylsulfonium hexafluoroantimonate (relative to polymer weight) in PGMEA was filtered through a 0.45 μm Teflon filter. The resist was spin coated at 2500 RPM to ~0.4μm thickness on silicon wafers. The post-apply-bake process was 150°C for 60s in hard contact with the bake plate. Post-exposure bake of the wafer was conducted with minimal delay (<5 min) for 60s at 140°C, after which the wafer was placed in the VASE and scanned to determine its Cauchy coefficients. These Cauchy coefficients were used to determine the film thickness of the resist films by means of a Prometrix SM-200 film thickness probe.

Fluorescence Titration Experiments to Determine the Amount of Photoacid Generated in Thin Photoresist Films

General techniques. Absorption spectra were monitored using HP 8453 and HP 8451A spectrophotometers. Fluorescence spectroscopy was carried out using a SPEX Fluorescence Spectrophotometer. All emission spectra were scanned from 410-600 nm at right angle in S/R mode. A bandwidth of 1.72 nm was used for both excitation and emission monochromators. The integral intensity from 413 to 600 nm was considered as the fluorescence intensity of the sample. **ACRAM**/THF solution (2 mL) with a concentration ranging from 4.39×10^{-6} to 1.53×10^{-5} M was prepared and used for the titration of trifluoroacetic acid (TFA). A solution of TFA in THF with concentration of 1.89×10^{-3} M was made and used as the standard acid solution. A mixed solution of TFA with polyNBCA (TFA/polyNBCA = 1:9 M/M) was prepared and used for evaluating the influence of polyNBCA (the deprotected form of the t-butyl ester-protected acid cleavable group of the resist polymers) on titration.

Experiments in Solution. A solution of **TPSHFA** (2.82×10^{-2} M) was irradiated with 193 nm radiation in a quartz curvette equipped with a cap in dose ranges of 14, 500, 1000, and 5000 mJ. Following exposure, the irradiated solutions were titrated against 2mL of **ACRAM** and the fluorescence spectra acquired.

Experiments in Photoresist Films. Quartz disk wafers (1 inch in diameter) were cleaned, using a protocal that involved soaking them in 30 % H_2O_2/conc. H_2SO_4 for 15 minutes, followed by a 2 minute soak in 10 % HNO_3, followed by a 2 minute soak in CH_3OH/conc. HCl, followed by a 2 minute soak in conc. H_2SO4. Finally, the wafers were rinsed in de-ionized water for 5 minutes (3×) and then dried in an oven at 120 °C overnight.

Photoresist formulations comprising 20 wt % of polymer, 3–10 wt % of PAG (relative to total polymer weight) in PGMEA were spin coated at 2000 RPM on 1 inch quartz disks. The resulting films were baked at 150 °C for 60 seconds. Typical film thickness was 1 μm. The films were then exposed to 193 nm radiation using an ISI ArF Stepper. Following this, the exposed films were dissolved off the disc with THF and titrated against 2 mL of a solution of **ACRAM**. The photoresist polymers studied included poly(CBN-*alt*-MAH) and poly(CBN-*co*-NBCA).

Results and Discussions

IR Study of the Deprotection Kinetics of Radiation-Exposed Resist Films. Upon exposure to UV-radiation, the PAG decomposes with a rate constant k_C to produce the photoacid (designated as Acid), as illustrated in equation 1. Here hν represents photon energy, where h is the Planck's constant and ν is the frequency of the radiation.

$$PAG + h\nu \xrightarrow{k_C} Acid \quad (1)$$

The generally accepted mechanism for the generation of acid from irradiation of the triphenylsulfonium salt was reported by Dektar and Hacker (*11*), Knapzyck and McEwen (*12*), and Crivello and Lam (*13*). The excited state of the sulfonium salt is believed to undergo homolytic cleavage of the carbon-sulfur bond to give an intermediate sulfur-centered radical cation along with phenyl radical. The photoacid (a Brönsted acid) is believed to arise from hydrogen atom abstraction by the radical cation followed by dissociation (*11*). Some investigators have observed phenylthiobiphenyl rearrangement products, which suggests that the acid may arise by photorearrangement followed by dissociation (*11*). Also, evidence has been presented which suggest that phenyl cation is produced by heterolytic cleavage of the excited state of the sulfonium salt (*14*).

The rate of decomposition of the PAG or the rate of the formation of the acid is therefore given by equation 2. Integrating equation 2 gives the acid concentration as a function of dose, as shown in equation 3.

$$d[PAG]/dt = k_C\,[Acid] \quad (2)$$

$$[Acid]_{Dose} = [PAG]_0\,.(1 - e^{-kC.Dose}) \quad (3)$$

The value of k_C for the resists exposed at 248 nm was taken from Reference (*15*) as the product of the PAG absorption and photoacid quantum yield at 254 nm. Here it is assumed that the quantum yield does not change with minor changes in wavelength and that there are no charge transfer or other indirect mechanisms for acid generation in our resist system. The quantum yield for the resists exposed at 193 nm was determined experimentally.

In the presence of acid quenchers like plasticizing additives and dissolution inhibitors, which may consume some of the generated acids (as shown in equation 4), the rate of acid depletion due to the quencher (Q) is given by equation 5:

$$Acid + Q \xrightarrow{k_{loss}} \quad (4)$$

$$d[Acid]/dt = -k_{loss}.[Q].[Acid] + D\nabla^2[Acid] \quad (5)$$

where ∇^2 is the Laplacian operator and D is the diffusion coefficient. In the presence of antireflection coatings, the second term of equation 5 can be neglected since the acid concentration gradient is negligibly small due to the lack of standing waves. If we assume quenching to be instantaneous and irreversible, then the concentration of the photoacid at any time is given by equation 6

$$[\text{Acid}]_t = [\text{Acid}]_{\text{Dose}} - [\text{Q}] \tag{6}$$

$$\text{If } [\text{Q}] > [\text{Acid}]_{\text{Dose}},\ \text{then } [\text{Acid}]_t = 0 \tag{7}$$

If [M] is the concentration of the *tert*-butyl ester unit of the polymer, the reaction of the photoacid-catalyzed thermolysis (deprotection reaction) of the acid cleavable group of the polymer is given by equation 8, and the rate of deprotection of the acid cleavable group of the polymer (amplification rate) is given by equation 9. Here k_a is the amplification rate constant, M is the *tert*-butyl ester unit and X is the deprotected carboxylic acid unit.

$$\text{M} + \text{Acid} \xrightarrow{k_a} \text{X} + \text{Acid} \tag{8}$$

$$d[\text{M}]/dt = -k_a.[\text{Acid}].[\text{M}] \tag{9}$$

The amplification rate constant is given by equation 10, where E_a is the activation energy of the deprotection reaction, R is the universal gas constant, A is the Arrhenius pre-exponential factor and T represents the bake temperature.

$$K_a = Ae^{-Ea/RT} \tag{10}$$

Equation 10 can be linearized by logarithmic transformation, as shown in equation 11.

$$\ln k_a = \ln A - (E_a/RT) \tag{11}$$

A plot of ln K_a versus 1/T gives a straight line, the slope of which is related to the activation energy, and the vertical intercept of which is related to the Arrhenius pre-exponential factor. Using numerical simulation of equations 3, 5, and 9, the deprotection kinetics can be modeled for various experimental conditions and values for k_a obtained.

When the photoacid generator, triphenylsulfonium hexafluoroantimonate, is exposed to radiation, it decomposes to release the super acid, hexafluoroantimonic acid, in the resist film. While this photochemical reaction can occur at room temperature, the acid-catalyzed deprotection of the pendant *t*-butyl group of the resist polymer occurs at reasonable rates only at elevated temperature. It is therefore necessary to heat the resist film to an appropriate temperature (postexposure bake) to provide the energy that is required for the acid-catalyzed deprotection of the *t*-butyl group of the ester, which in turn, affords the base-soluble norbornene carboxylic acid unit; isobutylene volatilizes. The extent of deprotection at constant temperature is dependent on the dose of applied radiation. By monitoring the carboxylic acid OH stretch 3000-3600 cm^{-1} and the ester carbonyl (C=O) around 1735 cm^{-1}, acid carbonyl (C=O) around 1705 cm^{-1}, and ester (C-O-C) stretch around 1150 cm^{-1}, it is possible to determine the extent of dose-dependent deprotection, as well as the influence of baking temperature on the extent of deprotection for each resist system by means of IR spectroscopy. Doses ranging from 0-50 mJ/cm^2 were applied to each resist system, after which they were baked at 120, 130, 140, and 150 °C for 60s and analyzed by FTIR.

Figure 1 shows a typical family of IR spectra of a resist formulated with triphenylsulfonium hexafluoroantimonate and poly(CBN-co-NBCA), samples of which were exposed to 248 nm radiation doses of 0-50 mJ/cm^2 and baked at 130°C after exposure. Similar spectra were collected at 120°C, 140°C and 150°C and also under 193 nm exposure for poly(CBN-*alt*-MAH).

Figure 2 shows the dose-dependent absorbance profiles of the carboxylic OH stretch (3100-3500 cm^{-1})and the ester C-O-C stretch (around 1150 cm^{-1}) of a poly(CBN-co-NBCA) exposed to 248 nm radiation and baked afterwards at 120°C, 130°C, 140°C and 150°C. The carboxylic acid OH stretch and the carboxylic acid carbonyl (C=O) stretch (1695-1705 cm^{-1}) both increase, while the ester carbonyl (C=O) stretch (1730-1735 cm^{-1}) decreases with increasing dose of exposure (0 to 50 mJ/cm^2), which allows us to follow the deprotection of the *t*-butyl ester group and the consequent conversion to a carboxylic acid group. The C-O-C (1150 cm^{-1}) stretch of the ester also decreases with dose of exposure, indicating the loss of the isobutylene group from the resist polymer.

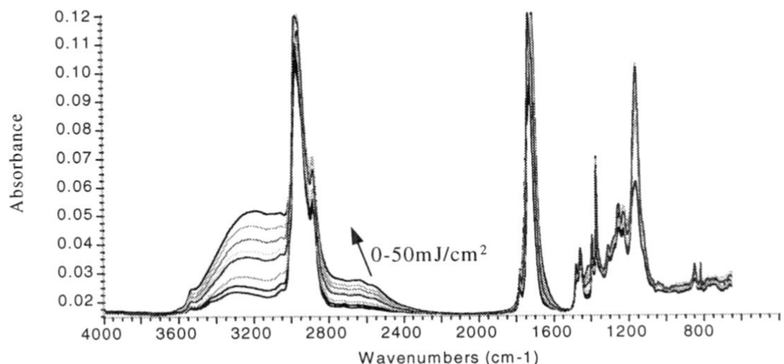

Figure 1. Change in FTIR Spectra of a typical resist copolymer (poly(CBN-*co*-NBCA) as a function of dose. Post-exposure bake temperature = 130°C, time = 60 seconds. Resist was exposed at 248 nm.

Using the model, it was possible to simulate the deprotection kinetics of these resists. **Figures 2 through 4** are plots of the deprotection under 248 and 193 nm exposures. It shows the normalized absorbance profile of the carboxylic acid (monitored with the carboxylic OH stretch in 3000-3600 cm^{-1} band) formed during the baking process and the normalized absorbance profile of the *t*-butyl ester protecting group as monitored with the C-O-C stretch around 1150 cm^{-1}), as a function of exposure dose. The baking time was 60 seconds. The circles (o) in this figure represent experimental values obtained at 150°C, while the lines represent the values predicted by the model at baking temperatures of 120°C, 130°C, 140°C and 150°C. The experimental values obtained at 120°C, 130°C and 140°C are not shown in this figure for the sake of clarity, but they too were consistent with the values predicted by the model. In all the cases, the modeling results are reasonably consistent with the experimental values. Thus, the model is a fairly accurate model for describing the deprotection kinetics of these resist systems within limits of experimental error and within the temperature and dose ranges investigated in this study.

Figure 5 is an Arrhenius-type plot of the deprotection kinetics of resists formulated with poly(CBN-co-NBCA). The activation energy for the deprotection of the pendant-*t*-butyl group of resists formulated with poly(CBN-*co*-NBCA) made by Pd(II)-catalyzed addition and free radical polymerization techniques was determined to be 6.7 and 9.4 Kcal/mol, respectively, over the temperature range of 120-150°C. **Figure 6** is an Arrhenius-type plot of the deprotection kinetics of resist formulated with poly(CBN-co-MAH). The activation energy for the deprotection of the pendant-*t*-butyl group of resist polymer was determined to be 18.3 Kcal/mol over the temperature range of 120-150°C. This Arrhenius relationship is not considered valid outside of the region experimentally covered.

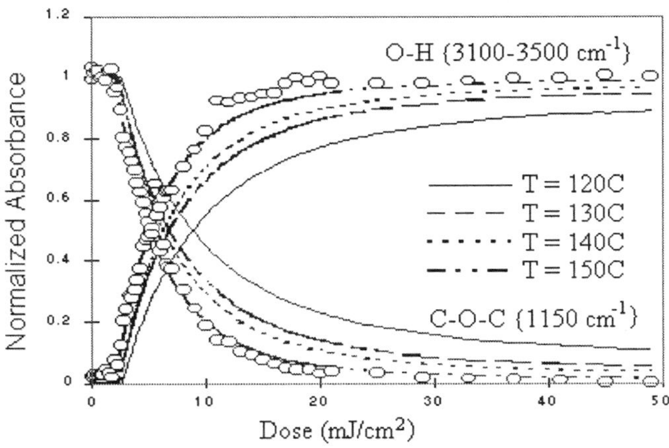

Figure 2. The carboxylic OH and ester C-O-C stretch profile upon 248 nm irradiation of resist.

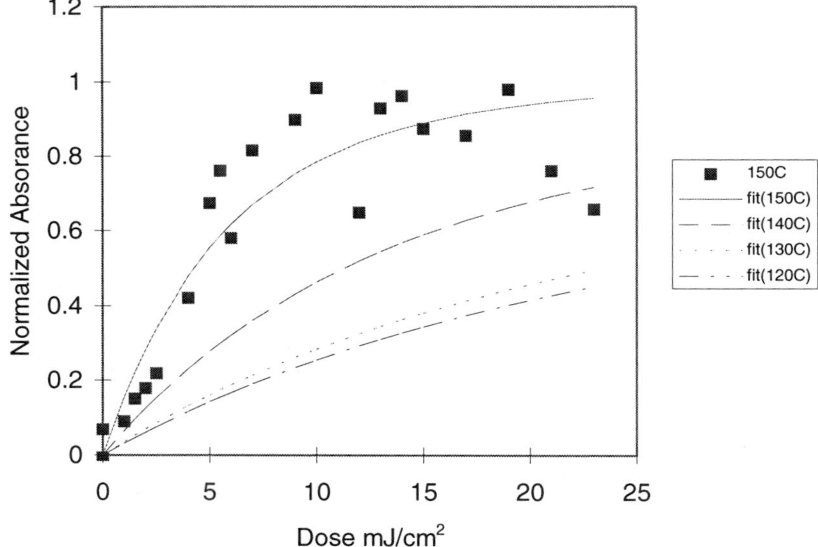

Figure 3. The carboxylic OH stretch profile upon 193 nm irradiation of resist formulated with poly(CBN-alt-MAH) and 3 wt% of TPSHFA.

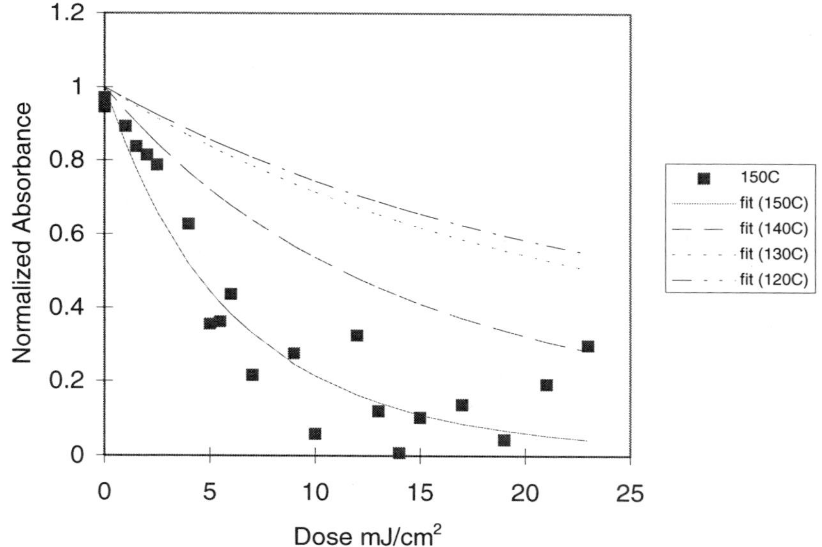

Figure 4. The ester C-O-C stretch profile upon 193 nm irradiation of resist formulated with poly(CBN-alt-MAH) and 3 wt% of TPSHFA

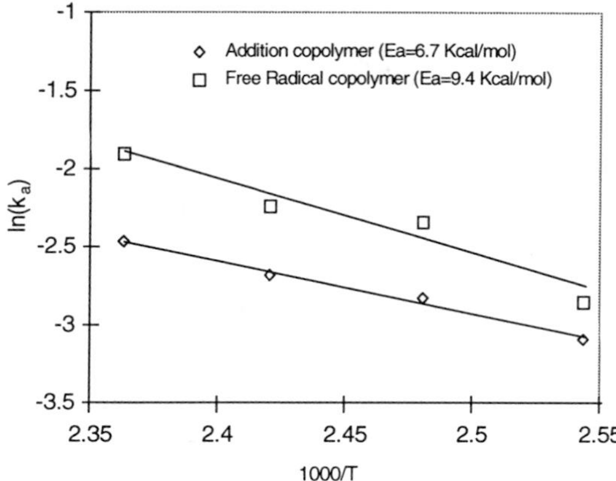

Figure 5. Arrhenius plot for resist formulated with poly(CBN-co-NBCA) and 3 wt% of TPSHFA, and exposed under 248 nm radiation.

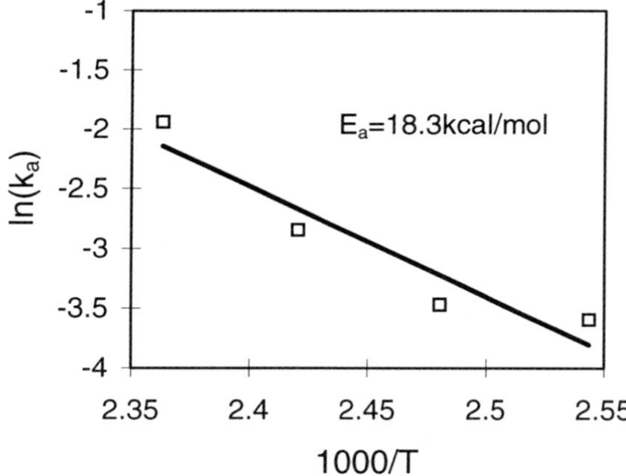

Figure 6. Arrhenius plot for resist formulated with poly(CBN-co-MAH) and exposed under193 nm radiation.

Monitoring Photoacid Generation in Thin Photoresist Films By Means of Fluorescence. Many molecules with pH-dependent fluorescence properties are known and used only in aqueous media, which essentially precludes them from resist applications. To be of use in resist applications, fluorescent molecules must be soluble in organic solvents of moderate polarity and also be compatible with polymer films

(*16,17*). Aromatic monazines like acridine have been known to be essentially nonfluorescent in non-hydrogen-bonding solvents, whereas their protonated forms are highly fluorescent in aqueous solutions, which makes this class of materials interesting candidates for potential acid sensors (5).

Commercially available acridine is usually provided as the HCl salt; if this acid is not completely removed, it will undoubtedly mask and even become a source of interference when the acridine salt is titrated against the photo-generated acid. We therefore employed N-(9-acridinyl)acetamide (**ACRAM**) in our study. **ACRAM** was readily obtained from acridine amine with acetic anhydride in pyridine, as shown in **Scheme 2**

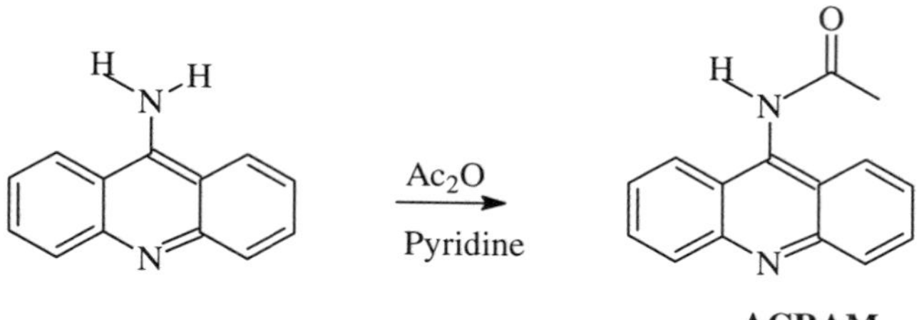

Ac_2O

Pyridine

ACRAM

Scheme 2

We investigated the effect of acid on the absorption and fluorescence properties of **ACRAM** in THF, employing both direct addition of TFA, polyNBCA and photoacid generation using **TPSHFA**. We also investigated the fluorescence properties of these PAGs in resist polymers like poly(CBN-alt-MAH) and poly(CBN-co-NBCA). These polymers have UV absorbance ca. 0.3 - 0.58/μm at 193 nm.

Figure 7 shows the change of fluorescence spectrum of 27.1 nM ACRAM in THF with addition of TFA/THF solution. The fluorescence spectra in the figure show both a hypsochromic shift of the peaks around λ_1: 440 → 430 and λ_2: 470→460 nm and an increase in the intensity of these peaks relative to the spectrum of the sample with no TFA. Since similar fluorescence spectral changes have been observed in other monoazines upon protonation in aqueous solution (*6,7*) , we conclude that the changes observed are due to protonation of **ACRAM** by TFA as shown in the **Scheme 3**

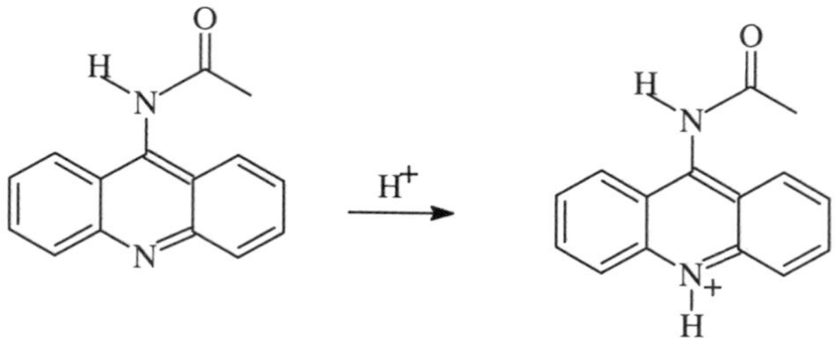

H^+

Scheme 3

Figure 8 shows fluorescence intensity (normalized to the pure **ACRAM** spectrum) of TFA/THF solution. The two straight lines on the graph are linear fits to

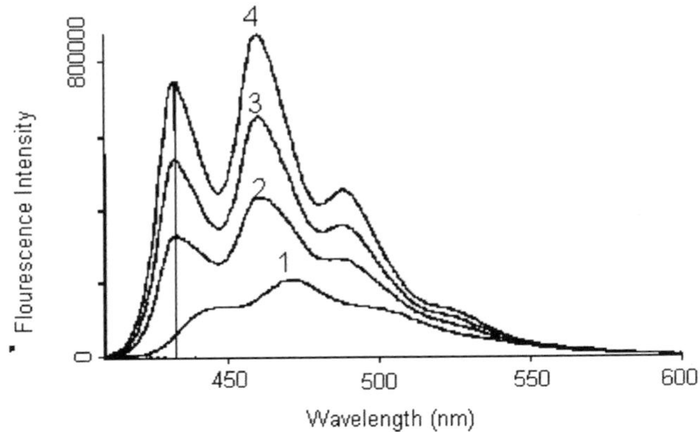

Figure 7. Fluorescence spectra of 27.1 nM **ACRAM** in THF with and without TFA/THF solution. (1) + 0 uL of TFA/THF solution (2) + 6 μL of TFA solution. (3) + 10 μL of TFA/THF solution. (4) + 40 μL of TFA/THF solution The excitation wavelength was $\lambda_{ex} = 406$ nm.

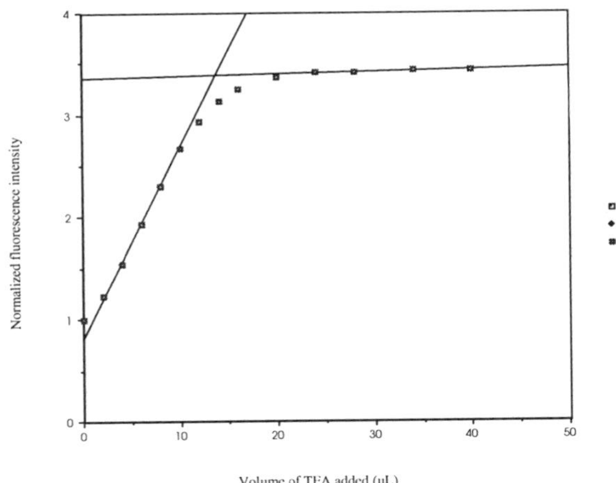

Figure 8. Fluorescence intensity (normalized to the pure **ACRAM** spectrum around λ_1: 440 → 430) of TFA/THF solution. The excitation wavelength was $\lambda_{ex} = 406$ nm.

the initial and final parts of the intensity profile. The intersection of these two straight lines occurs at the point of maximum protonation limit (MPL), where the protonation site of the acid sensor (probe) is fully protonated. At MPL, the number of moles of acid is exactly equal to the number of moles of **ACRAM** in solution. Beyond the MPL, any further addition of the titrant does not significantly change the observed spectral properties of the probe. Multiplying the volume of the added acid corresponding to the MPL (14.2 μL) by the concentration of the stock solution of TFA (1.89×10^{-3}M) afforded the number of moles of TFA (needed to fully protonate **ACRAM**. This procedure was employed in the determination of the MPPs for all the other titration cases investigated in this study.

Figure 9 shows fluorescence intensity (normalized to the pure **ACRAM** spectrum) of TFA/THF and TFA/polyNBCA/THF solutions. The addition of polyCBN has an insignificant effect on observed fluorescence spectra of **ACRAM** (**Figure 9 and Table I**).

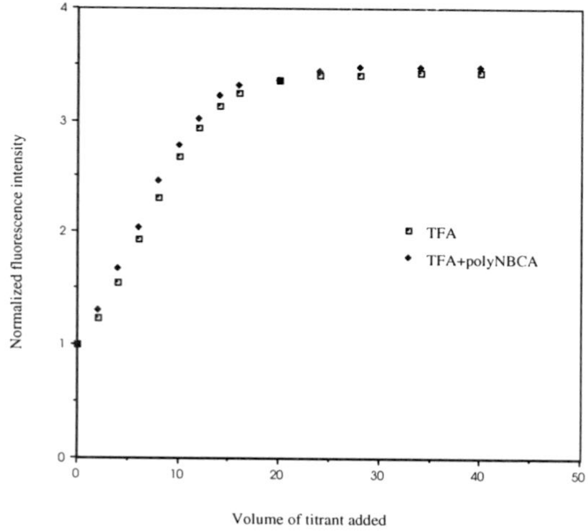

Figure 9. Fluorescence intensity (normalized to the pure **ACRAM** spectrum) of TFA/THF and TFA/polyNBCA/THF solutions around λ_l: 440 → 430). The excitation wavelength was λ_{ex} = 406 nm.

From **Figures 8** and **9**, it can be seen that beyond the MPL, any further addition of the titrant does not significantly change the observed spectral properties of the probe. This implies that most of the **ACRAM** is fully protonated beyond this point, which occurs essentially at a 1:1 molar ratio between the two reactants. This observation is consistent with the fact that **ACRAM** has only one protonation site, namely at the nitrogen in the ring. The amide carbonyl oxygen or even the amide nitrogen does not get protonated under the conditions we studied. Even with the addition of polyNBCA, the molar ratio between the titrant and the titrand remained essentially the same, indicating that the titrant (TFA) is by far a stronger acid than polyNBCA (**Figure 9**).

Table I. Summary of the Results of the Titration of TFA/THF and TFA/PolyNBCA/THF Solutions against **ACRAM**

Titrant	Moles **ACRAM** (nM) at MPE	Moles TFA (nM) at MPE	TFA/**ACRAM**
TFA	27.1	26.8	0.99
TFA /polyNBCA (1 : 9 M/M)	28.3	26.6	0.94

Therefore, in principle, it is possible to monitor the photogeneration of acid in photoresist films by fluorimetric titration of solutions of the exposed resist with **ACRAM**. To check whether this inference is correct, we irradiated solutions of **TPSHFA** at 193 nm with dose ranges of 500 to 2000 mJ. **Figure 10** shows typical fluorescence spectra obtained for a solution of 27.1 nM **ACRAM** in THF. Again, there is a clear hypsochromic shift in the peaks around λ_1: 440 $\rightarrow$ 430 and λ_2: 470$\rightarrow$460 nm and an increase in the intensity of these peaks relative to **ACRAM** samples without added solutions of the irradiated **TPSHFA**. This increase in fluorescence intensity is consistent with photogeneration of acid and subsequent protonation of **ACRAM**.

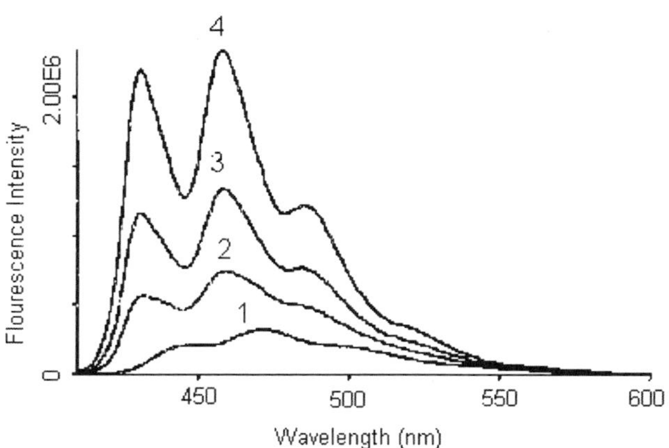

Figure 10. Fluorescence spectra of 27.1 nM **ACRAM** in THF with and without **TPSHFA** solution (2.82 $\times 10^{-2}$ M) in THF. (1) + 0 μL of **TPSHFA** solution. (2) + 75 μL of **TPSHFA** /THF solution. (3) + 125 μL of **TPSHFA**/THF solution (4) + 275 μL of **TPSHFA**/THF solution.

Figure 11 shows the amount of photoacid generated from solutions of **TPSHFA** in THF at different irradiation doses. It is noteworthy to mention that the high doses shown in this Figure that were employed in this experiment is a reflection of the difficulty of finding suitable solvents that have enough transparency to 193 nm radiation to permit this type of experiment. Our choice of THF as the solvent for the experiment stems from the fact that of all the organic solvents that are able to dissolve the resist polymers, THF has the least optical density, *ca.* 1.5 for a 2 mL solution in a curvette with 1 cm optical path length, which implies that only 3.2 % of the radiation gets to the PAG, with most of the radiation being actually absorbed by THF. By irradiating with such high doses, we sought to increase the amount of radiation that the PAG molecules get. At doses much lower than 500 mJ, no significant photoacid was produced.

Experiments in Films. Given that lithography is done on polymer films, not solutions, it was necessary to determine whether the results obtained in solution could be reproduced in polymer films. The effect of photogenerated acid on the fluorescence spectra of **ACRAM** was investigated by dissolving the 193 nm-exposed resist films in THF and titrating the resulting solution against the former. **Figure 12** shows the change of fluorescence intensity (normalized to the pure **ACRAM** spectrum) with the addition of photoresist solution formulated from poly(CBN-*alt*-MAH) and **TPSHFA** (4.41×10^{-7} M). These spectra show hypsochromic spectral shifts of the peaks around

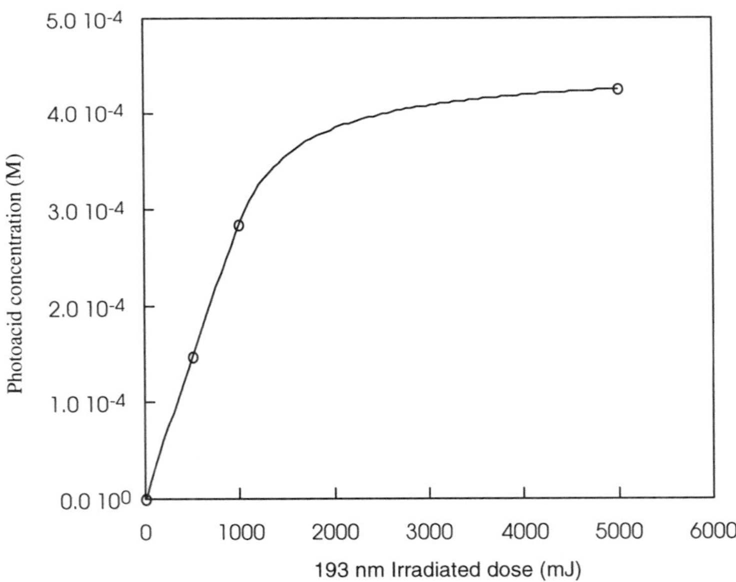

Figure 11. Photoacid generation from solutions of **TPSHFA** in THF as a function of irradiation dose at 193 nm. Concentration of **TPSHFA** = 2.82×10^{-2} M

λ_1: 440 → 430 and λ_2: 470→460 nm, as well as an increase in the intensity of these peaks, which are identical to those observed in the solution experiments. This indicates the protonation of **ACRAM**/THF solution by the photogenerated acid.

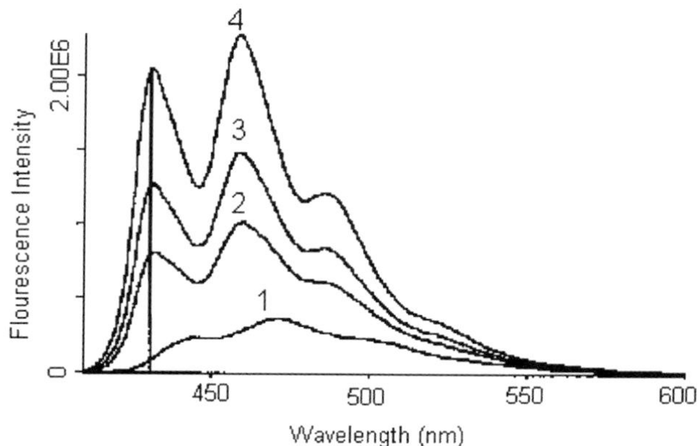

Figure 12. Fluorescence spectra of 27.1 nM **ACRAM** in THF with and without photoresist solution formulated from poly(CBN-*alt*-MAH) and **TPSHFA** (4.41×10^{-7} M). (1) + 0 μL of photoresist/THF. (2) + 35 μLphotoresist/THF. (3) + 60 μL of photoresist/THF (4) + 120 μL of photoresist/THF.

Table II. Photoacid generation at 193 nm from photoresist formulated with poly(CBN-*alt*-MAH) and different TPSHFA concentrations. The exposure dose was with 50 mJ/cm^2

PAG	Dose (mJ/cm^2)	Conc. of **TPSHFA** (M)	Conc. of photoacid(M)
TPSHFA	50	4.41×10^{-7}	1.04×10^{-7}
TPSHFA	50	6.61×10^{-8}	4.13×10^{-8}

Table II summarizes the amount of photoacid generated at 193 nm from photoresists from different PAGs with 50 mJ/cm^2. Using the values of Table II, the quantum yield for acid production was determined from the number of acid molecules produced per photon absorbed by the film. In this way, a quantum yield of 1.1×10^{-4} was determined for the PAG (**TPSHFA**) in poly(CBN-alt-MAH) resist system that was exposed to 193 nm radiation. Such low quantum yield is indicative of the fact that most of the 193 nm radiation that is absorbed by the resist is not utilized in the production of photoacid from the PAG (**TPSHFA**). The value of k_C used in the modeling was determined as the product of the quantum yield and the resist absorption coefficient.

Summary. The deprotection kinetics of alicylic polymer resist systems designed for 193 nm lithography was examined. A kinetic model was developed that simulates the deprotection of the resists fairly well. The activation energies of 6.7 and 9.4 Kcal/mol were determined for resists formulated with poly(CBN-co-NBCA) made by Pd(II)-

catalyzed addition and free radical polymerization techniques, and exposed to 248 nm radiation, while an activation energy of 18.3 Kcal/mol was determined for resist formulated with poly(CBN-alt-MAH) that was exposed to 193 nm radiation. A quantum yield of 1.1×10^{-4} at 193 nm was determined for the PAG (**TPSHFA**) in the resist system formulated with poly(CBN-alt-MAH).

A new simple and reliable method for monitoring photoinduced acid generation in polymer films and in solutions of the kind used in 193 nm and deep-UV lithography was developed. By using N-(9-acridinyl)acetamide, a fluorescent acid-sensitive sensor, we have been able to study the effects of trifluoroacetic acid and photoacids generated from triphenylsulfonium hexafluoroantimonate on the spectral properties of the acid sensor in THF solution and in alicyclic polymer resist films exposed at 193 nm. In both cases a hypsochromic spectral shift and an increase in fluorescence intensity were observed upon protonation. This technique could find application in the study of diffusional processes in thin polymer films.

Acknowledgments: Financial support of this study by SEMATECH and Semiconductor Research Corporation is gratefully acknowledged.

Literature Cited

1. SIA *The National Technology Roadmap for Semiconductors*, Semiconductor Industry Association (SIA), San Jose, CA, **1994**, pp. 81–93.
2. Okoroanyanwu, U.; Shimokawa, T.; Medeiros, D.; Willson, C.G.; Byers, J.; Allen, R.D. *Proc. SPIE.* **1997**, 3049**,** 92.
3. Okoroanyanwu, U. *Ph.D. Thesis*, University of Texas at Austin, **1997.**
4. Lackowicz, J.R. *Principles of Fluorescence Spectroscopy.* Plenum Press: New York, **1986**, Chps. 1 & 2.
5. Pohlers, G.; Virdee, S.; Sciano, J.C.; Sinta, R. *Chem. Mater.* **1996**, 8, 2654.
6. Weller, A.Z. *Elektrochem.* **1957**, 61, 956.
7. Kasama, K.; Kikuchi, K; Yamamoto, S.; Uji-ie, K.; Nishida, Y.; Kokubun, H. *J Phys. Chem.* **1981**, 85, 1291.
8. Crivello, J.V.; Lam, J.H.W. *J. Org. Chem.* **1978**, 43**,** 3055.
9. Shields, C.J.; Falvey, D.E.; Schuster, G.B.; Burchardt, O.; Nielsen, P.E. *J Org. Chem.* **1987**, 53, 3501.
10. Tomoska, H.; Omata, S.; Anzai, K. *Biosci. Biotech. Biochem.* **1994**, 58, 1420.
11. Dectar, J.L.; Hacker, N.P. *J. Chem. Soc., Chem. Commun.* **1987**, 1591.
12. Knapzyck, J.W.; McEwen, W.E. *J. Org. Chem.*, **1970**, 35, 2539.
13 Crivello, J.V. ; Lam, J.H.W. *J. Polym. Sci. Polym. Chem. Ed.*, **1979**, 17, 977.
14. Davidson, R.S.; Goodin, J.W. *Eur. Polym. J.* **1982**, 18**,** 487.
15. McKean, D.R. Schaedeli, U.; MacDonald, S.A. In *Polymers in Microlithography Materials and Processes*, Reichmanis, E.; Iwayanagi, T., Eds., ACS Symposium Series 412, American Chemical Society, Washington, DC, **1990**, pp. 26-38.
16. McKean, D.R. Schaedeli, U.; MacDonald, S.A. *J. Polym. Sci., Part A: Polym. Chem.* **1989**, 27, 3927.
17. Carey, W.P.; Jorgensen, B.S. Appl. Spectrosc., **1991**, 45, 834.

Chapter 15

193 nm Single Layer Resist Based on Poly(norbornene-alt-maleic anhydride) Derivatives

The Interplay of the Chemical Structure of Components and Lithographic Properties

F. M. Houlihan, A. Timko, R. Hutton, R. Cirelli, J. M. Kometani, Elsa Reichmanis, and O. Nalamasu

Bell Laboratories, Lucent Technologies, 600 Mountain Avenue, Murray Hill, NJ 07974

A retrospective of our work on novel cycloolefin-maleic anhydride copolymers for 193 nm imaging applications is presented. Free radical induced copolymerization of cycloolefins with maleic anhydride afford hydrolytically robust alternating copolymers. Aqueous base solubility in these copolymers is induced by incorporation of acrylic acid and/or acrylate esters that can be acidolytically cleaved. The amount of acrylate incorporated in the resin is a linear function of the starting monomer ratio. These ter- or quartenary copolymers are thermally stable, hydrolytically robust and possess good plasma etching resistance. Formulation of these resins with appropriate dissolution inhibitors (DI) and photoacid generators (PAG) provide resist formulations that exhibit very good lithographic properties at 193 nm. We have found that a blend of a monomeric polar cholate with an oligomeric cholate material exhibits optimum lithographic performance when used in conjunction with a PAG capable of generating photoacids of lower volatility such as nonafluorobutanesulfonic acid. These formulations are compatible with 0.262 N tetramethyammonium hydroxide (TMAH) and have been shown to resolve at least 0.15 micron line/space pairs.

Photolithography with 193 nm light (ArF excimer laser) is a prime candidate for future microelectronics manufacturing of ≤0.15 μm design rules. However, the design of 193 nm single-layer photoresists that are functionally similar to novolac and poly(hydroxystyrene) based resists poses several interrelated challenges: the transparency, adhesion, and plasma-etching stability which have been, for the most part, inherent characteristics of novolac, poly(hydroxystyrene)-based resists must all be intentionally engineered into materials for 193 nm lithography. To date, these efforts have focused primarily on acrylates and methacrylate copolymers derivatized with pendant alicyclic groups(*1*). Although these materials can be made plasma etching resistant with high carbon/alicyclic content(*1e*), the resulting polymers tend to be brittle and display poor adhesion and poor aqueous base solubility. These effects can be remedied by lowering the alicyclic content, but plasma etching resistance is compromised. Some approaches to address the challenge of designing resists that are both aqueous base soluble and plasma etching resistant include: i) careful tailoring of polymer properties to balance lithographic performance and plasma etching resistance, and ii) development of three component systems in which high carbon content dissolution inhibitors (DI) improve both imaging and plasma etching characteristics.

An obvious approach to impart greater plasma etching resistance to a given resin is to incorporate alicyclic moieties directly into the polymer backbone. Several synthetic routes can be employed to synthesize polymers of this type including ring-opening metathesis polymerization (ROMP), metal catalyzed vinyl polymerization, radical promoted vinyl polymerization and alternating polymerization of cycloolefins with maleic anhydride. We have chosen the latter approach as it is most suitable to synthesize metal-ion free polymers(*2*). In addition to our own work on cycloolefin-maleic anhydride copolymers(*2*) there is an early reference(*3*) on the potential of poly(norbornene-alt-maleic anhydride) as a plasma etching resistant polymer matrix for use at 248 and 193 nm. More recently, there has been an account on the use of poly(tert-butyl bicyclo[2.2.1]hept-5-ene-2-carboxylate) in an e-beam resist scheme(*4*) and reference to the potential application of substituted polynorbornenes obtained via addition polymerization for 193 nm lithography has recently appeared followed by several other schemes that employ cycloolefins in the polymer backbone for lithographic applications (*5*).

Figure 1 outlines the three resist approaches that we have examined employing resins based upon alternating copolymers of maleic anhydride and cycloolefins. Two of these, the three component and hybrid approaches depend upon the use of a DI to improve resolution and further enhance plasma etching resistance. In this paper we will detail our work on the design, synthesis and use of cycloolefin-maleic anhydride based resins, photoacid generators and novel multifunctional DIs for 193 nm lithography.

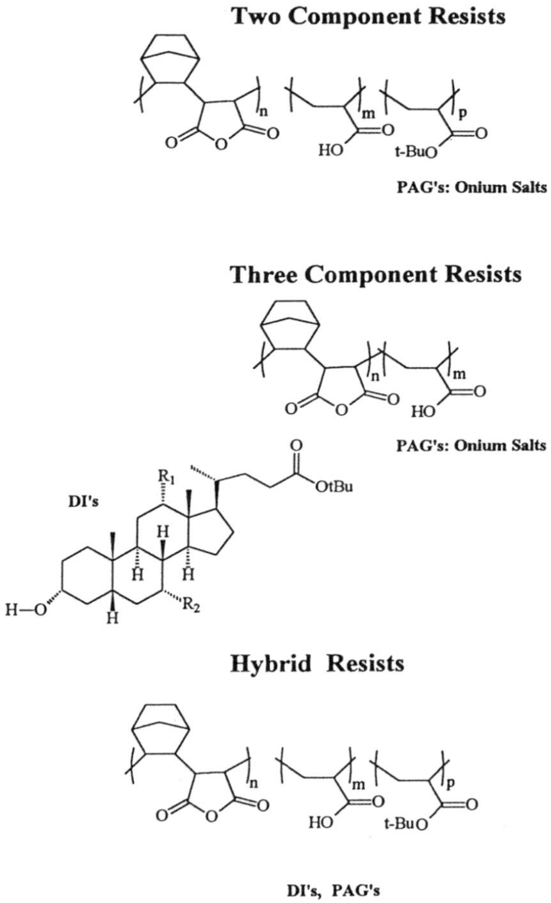

Two Component Resists

PAG's: Onium Salts

Three Component Resists

PAG's: Onium Salts

DI's

Hybrid Resists

DI's, PAG's

Figure 1. 193 nm resist systems based on a combination of P(NB/MA) polymer, monomeric and oligomeric DI derivatives (R=H, OH in three component schemes and aliphatic diester moiety in hybrid resists), and PAGs.

Alternating Copolymer Matrix Resins

Synthesis and Materials Properties.

Alternating Copolymerizations. Cycloolefin-maleic anhydride alternating copolymers are an attractive alternative to methacrylate-based matrix resins. Compelling features of these copolymers include: i) facile synthesis *via* standard radical polymerization; ii) a potentially large pool of cycloolefin feed stocks; and iii) a generic structural motif that incorporates alicyclic structures directly in the polymer backbone and provides a latent water-solubilizing group that may also be useful for further structural elaboration. A large number of cycloolefins are known to copolymerize with maleic anhydride(*6*). As a rule, they yield high-T_g copolymers with a 1:1 alternating structure. While we have screened a number of cycloolefins as components, our efforts to date have largely concentrated on norbornene, which copolymerizes cleanly and yields materials with promising properties.

Poly(norbornene-alt-maleic anhydride) [P(NB/MA)] is a colorless powder with T_g > 300°C and an onset of decomposition (under argon) of 370°C. Experiments reveal that P(NB/MA) is hydrolytically robust: polymer dissolved into cyclohexanone may be stored for at least a month without significant changes in characteristics. Unfortunately, this hydrolytic stability also means that films of P(NB/MA) do not dissolve at useful rates in conventional aqueous base developer (0.262 N TMAH) solution. A partial hydrolysis of P(NB/MA) yields more readily soluble materials, but this approach suffers from drawbacks outlined below.

Norbornene-maleic anhydride copolymerizations were first described in a patent that gave two key insights (*7*); *i*) copolymerization provides a material with a 1:1 composition regardless of monomer feed ratio, and *ii*) incorporation of small percentages (<10% is claimed) of other vinyl monomers without disruption of the essentially alternating nature is tolerated. We have found (*2*) that materials useful for microlithography may be obtained by copolymerization with acrylate monomers such as t-butyl acrylate(TBA) and acrylic acid(AA). Copolymerization with AA and/or TBA provides a controllable method of synthesizing aqueous base soluble resins. Specifically, we have found that there is no significant deviation from a 1:1 ratio of the NB:MA repeat units as the relative amount of acrylate repeat unit is increased. Furthermore, there is a simple linear relationship between the feed of acrylate monomers and their incorporation into the polymer(*2d*). Gel permeation chromatographic analysis of these materials gives clean monomodal peaks with a polydispersity of < 2.5 regardless of acrylate loading. The M_w ranges(*2d*) from 4,000 to 8,000. The tertiary and quaternary polymers possess thermal and optical properties, and solubility in organic solvents similar to the parent resin P(NB/MA), but adhere significantly better to semiconductor substrates than the P(NB/MA). Thin films of tertiary and quartenary polymers show desirable absorption properties at 248 and 193 nm [e.g. Absorbance per micron of poly(norbornene-alt-maleic anhydride-co acrylic acid 15 %), P(NB/MA/AA15)(*8*), with 20 % (wt) t-butyllithocholate is 0.05 and 0.27 AU/μm, respectively].

Alternative Monomers. We have tested several other cycloolefins for copolymerization with MA(*2d*). Of these we have found that the following can be successfully polymerized. 5,6-Dihydrodicyclopentadiene (DDPD) copolymerizes with maleic anhydride in low yield (17%) under conditions analogous to the norbornene copolymerization. This copolymer may be of interest as a higher-carbon content analog of P(NB/MA). 1,5-Cyclooctadiene (COD) and 1,5-dimethyl-1,5-cyclooctadiene (DMCOD) produced fully soluble copolymers; however, 1,2-copolymerization competed significantly with cyclopolymerization. As for the NB/MA resins these materials had polydispersities lower than 2.5 with a M_w which ranged between 4,000 and 8,000.

Because of their high alicyclic/carbon content these materials give better etching selectivity and are attractive potential matrices for 193 nm lithography. For instance, in a chlorine based plasma P(COD/MA) and P(DDPD/MA) give relative rates of 1.15 and 1.11, respectively, compared to a Shipley HB1811, a typical Novolak based resist.

Chemical Structure And Dissolution Behavior.

As mentioned previously, the anhydride moieties in P(NB/MA) are remarkably hydrolytically stable and do not appreciably facilitate resin dissolution in aqueous TMAH (0.262 N TMAH). Poly(norbornene-alt-maleic anhydride-co-acrylic acid) P(NB/MA/AA) dissolution in aqueous base is similarly not dependent on anhydride hydrolysis; rather, it varies as a function of acrylic acid content. These findings suggest that the anhydrides remain predominantly latent (i.e., non-hydrolyzed) during the development under the conditions employed, or conversely, that the developer, while basic enough to solubilize acidic functionalities such as carboxylic acids, lacks the reactivity to efficiently promote anhydride hydrolysis in the hydrophobic and sterically congested polymer. We have explored three approaches towards modifying the dissolution behavior of P(NB/MA) polymer matrices: i)chemical modification of the MA moiety to promote base dissolution, ii)enhancement of reactivity of MA moieties towards base hydrolysis by the addition of nucleophilic additives to the developer, and iii) use of cholate based dissolution inhibitors in P(NB/MA) resins that contain acrylate repeating units.

P(NB/MA) Functional Group Transformations. Functional group transformation is one approach which is possible for the alteration of the hydrolytic latency of P(NB/MA) copolymers. Of the potentially useful post-polymerization functionalizations of P(NB/MA), hydrolysis, alcoholysis (to yield half-esters), imidization, amidization, and reduction were explored briefly(*2d*).

P(NB/MA) proved to be surprisingly resistant to hydrolysis, both in solution and as a thin film(*2d*). Moreover, the diacid prepared by hydrolysis in concentrated TMAH reverts to the anhydride upon heating(*2d*). Similarly, alcoholysis with methanol is difficult and the methyl half-ester reverts to the anhydride on heating(*2d*). Reactions with amines to yield poly(amic acids) and poly(imides) proceed more readily. However polyimides are too strongly absorbing at 193 nm to make suitable resist matrices. Also, because primary amines are potent imidization catalysts for amic acids(*9*), amidization of P(NB/MA) with these also causes strong absorbance at 193 nm. Selective amidization can be achieved with bulky secondary amines which yield stable poly(amic acid) with aqueous base solubility and acceptable transparency (absorbance = *ca.* 0.5 AU/μm at 193 nm). These materials which contain basic amide moieties however heavily level the pH of potential photoacids. Finally, P(NB/MA) can be reduced with $NaBH_4$ to a poly(norborne-alt-cis-4-hydroxy-2-butenoic acid). This polymer can be isolated without lactonization, is soluble in aqueous base, displays excellent film-forming properties, and is thermally stable to at least 150°C in the absence of acid. Lactonization and insolubilization occur readily in films at 100-120°C using triphenylsulfonium triflate as a photoacid generator(*2d*). These materials are negative imaging resists upon development with 0.1 N TMAH and have sensitivities in the range of 20 mJ/cm^2.

Dissolution Accelerators. An alternative approach to chemically modifying MA is to add nucleophilic additives capable of catalyzing anhydride hydrolysis into developer. This has several advantages: i) it may be applicable not just to 193 nm resins, but also to 248 nm resins containing similar latent functionalities such as acetoxystyrene moieties; ii) it may present an alternative to the common, but environmentally unacceptable practice of using isopropanol/water developer mixtures; and iii) it may influence development characteristics of resists in lithographically useful ways.

Acylation catalysts, typically nucleophilic amines and imines that activate carboxylates by forming highly nucleophile-susceptible acyliminium intermediates, promote a wide variety of synthetically useful carboxylate transformations. Many are sufficiently water-soluble to evaluate as anhydride hydrolysis catalysts. Scheme 1 depicts a representative example of this class of reaction, the hydrolysis of a P(NB/MA) polymer matrix in aqueous base (TMAH) catalyzed by 4-(dimethylamino)pyridine (DMAP).

P(NB/MA/AA15) was used to evaluate developer accelerators(*2d*). Screening of imidazole, histidine, 1,2,4-triazole, DMAP and guanidine showed that imidazole and DMAP are potent dissolution-rate accelerators. While DMAP appears to be slightly better dissolution accelerators than imidazole, imidazole is more attractive from toxicity and cost considerations. Additionally, P(NB/MA/AA15) dissolution rates can be continuously varied by nearly an order of magnitude over a range of imidazole concentrations from 0 to 0.262 M (*2d*).

Dissolution Inhibitors. One approach towards effectively modifying the dissolution behavior of P(NB/MA/AA) and P(NB/MA/AA/TBA) based resists is to use the matrix resin in conjunction with a dissolution inhibitor. Design criteria for an effective 193 nm DI include low absorbance at 193 nm, solubility in spin casting solvents, loading capacity in resin matrix, contrast enhancement of resin matrix and dark erosion suppression. One candidate for such applications is the cholate based esters.

The first example of a cholate based dissolution inhibitor was 2-nitrobenzylcholate which was used in conjunction with copolymers of methyl methacrylate and methacrylic acid for 248 nm applications (*10, 11*). Upon photolysis of the o-nitrobenzyl moiety, this ester, which is initially insoluble in aqueous base, is converted to the very base soluble cholic acid. The high contrasts obtained with such systems are made possible by the large molar volume of resist that can be made hydrophilic by a single bond scission. Similarly, t-butyl cholate has been described for use in various phenolic matrices(*12,13*) for long-UV and 248

Scheme 1

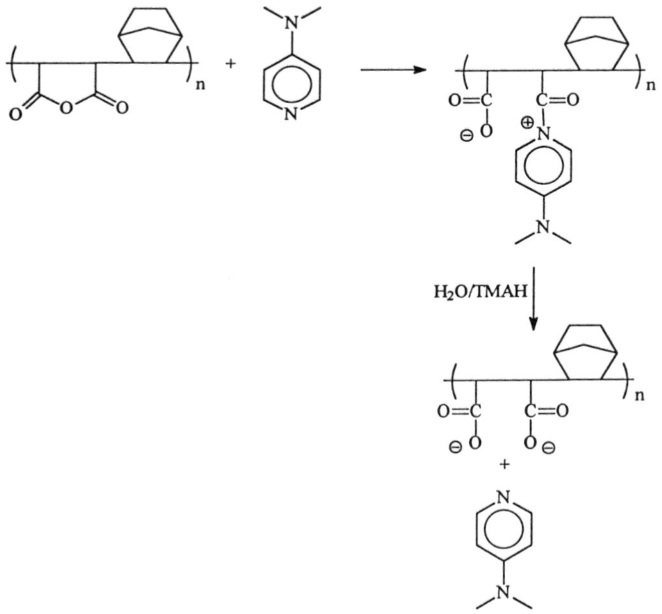

H_2O/TMAH

nm deep UV applications in a "chemically amplified" acidolysis reaction initiated by photolysis of a photoacid generator (PAG). These workers and others have also discussed the utility of multifunctional chemically amplified phenolic based dissolution inhibitors for long-UV and 248 nm applications(*12, 13, 14*).

For 193 nm applications, evaluation of a series (*1a*) of cholate based dissolution inhibitors suggested that the dissolution inhibition of methacrylate-based resins by these derivatives is largely a function of the hydrophobicity of the cholates employed. The observed relative order of hydrophobicity and dissolution inhibition was: lithocholate (1 pendant hydroxyl) > deoxycholate = ursocholate (2 pendant hydroxyls) > cholate (3 pendant hydroxyls). Experiments using monomeric dissolution inhibitors such as t-butyl cholate (**1a**), t-butyl deoxycholate (**1b**), t-butyl lithocholate (**1c**), t-butyl lithocholate acetate (**2**) with the P(NB/MA) acrylate resins afforded resist systems that exhibited low contrast, poor adhesion, dark erosion (unexposed resist film loss) and were incompatible with industry standard 0.26 N TMAH developers.

Drawing from the vast body of research related to diazonaphthoquinone (DNQ) design for conventional resist applications, the design principles for 193 nm DI's can be identified. The use of DNQ dissolution inhibitors in resists predates their application to microelectronics and starts their use in making printing plates (*15,16a*). For G- and I-line lithography, multifunctional DNQs have been shown to dramatically improve contrast and side-wall profiles relative to monofunctional analogs in Novolak resists (*16b*). For instance, one of the more commonly used multifunctional DNQ's, is a trihydroxybenzophenone functionalized with multiple DNQ moieties. In this approach, the fully converted tris-indenecarboxylic acid is formed in appreciable amounts only above a certain threshold energy, thereby greatly increasing the non-linear relationship between the exposure and dissolution rate of the resist. Using the principles behind the design of multifunctional DNQ DI's, a series of dimeric (**3**) and oligomeric (**4**) cholates derivatives where synthesized (Scheme 2, 3). Resist properties such as dark erosion resistance, adhesion and contrast were used to determine the efficacy of the approach. Also considered in our evaluation were certain physical properties of DI's such as their solubility in safe spin casting solvents such as 1-methoxy-2-propanol acetate and compatibility with the matrix.

Chemical Structure And Resist Properties

A successful resist design must be robust enough to withstand standard processing conditions which are employed in the industry for manufacture of microelectronic components. In particular, the resist must have sufficient thermal stability and resistance to flow to be able to withstand typical pre-exposure and post-exposure bake temperatures (a range of 100 to 150°C is typical), the resist must have sufficient etching resistance towards plasmas employed during pattern transfer processes and finally, the resist must possess the desired thin film properties. Therefore, an important area of investigation is to ascertain how both cholate based DIs and PAG structure affects resist properties in P(NB-alt-MA)/acrylate based systems.

Scheme 2

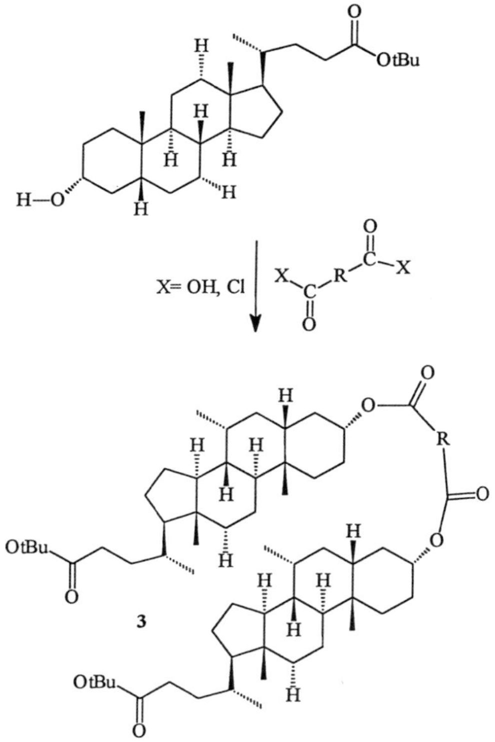

O
H
OtBu
H
H
H
H
H—O
H
H
X= OH, Cl
O
X
C
R
C
X
O
O
H
O
R
H
H
O
H
H
O
OtBu
H
H
O
H
3
H
H
OtBu
O
H

Scheme 3

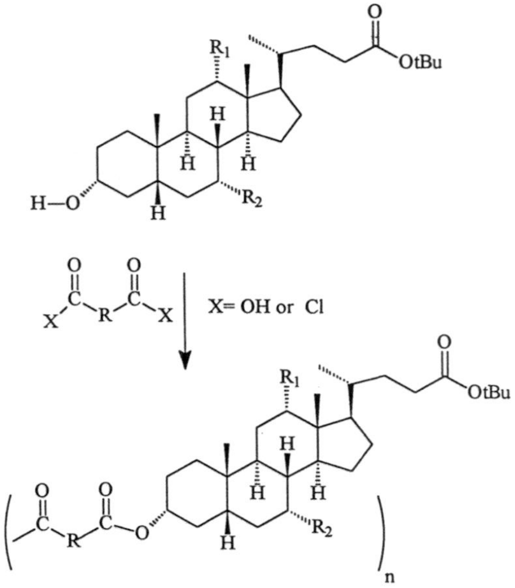

4a, b

Compound	R_1	R_2
4a	O—	H
4b	O—	O—

PAG Considerations. The t-butyl protecting group was employed in both the DI's and the polymer components of the hybrid and three component resist systems. Since high activation energies are required for the cleavage of this group, formulation of the resists with a PAG capable of generating a super acid such as triflic, nonafluorobutanesulfonic or hexafluoroarsenic acid is necessary for efficient cleavage at typical post-exposure bake temperatures of ≤150 °C. Consequently, we have concentrated our research on PAG's such as diphenyliodonium triflate(DPI-Tf); diphenyliodonium nonaflate (DPI-NF), triphenylsulfonium triflate (TPS-Tf), bis(4-t-butylphenyliodonium nonaflate (DTBP-NF) and in some instances triphenylsulfonium hexafluorarsenate (TPS-AsF_6).

As expected, PAG loading affects the sensitivity,. with increased loading tending to increase sensitivity. However, at high PAG loadings (eg. 2.8% (w/w TPS-Tf), bimodal behavior is seen, with low doses producing the expected positive-tone behavior while high doses result in insoluble residues apparently due to cross-linking of resist. As a result, only a narrow dose range exists for clearing of the resist film (Figure 2). At lower PAG loadings (1.4%), this effect was not observed. The probable cause is radical initiated cross-linking, stemming from radicals or radical cations generated by the onium salt. A similar onium salt loading related cross-linking effect (*17*) has been described for a resist formulated with poly(4-hydroxystyrene-co-styrene) and t-butyl cholate.

We have found that the nonaflate materials are preferred over the triflate analogs because of decreased volatility of the super acid, and thus fewer complications emanating from the surface depletion of acid are observed. Similar findings have been reported by other workers recently (*18*).

Resist Properties and DI Structure.

Effect of Polymer Matrix on DI Loading Effects. Two types of polymer matrices were examined in our studies, P(NB/MA/AA) and P(NB/MA/TBA/AA) (Figure 1). The nature of the resist polymer matrix plays a role in governing DI loading effects. As evident from Figure 3, while both resin types show an increase in contrast with loading of DI, the hybrid system gives consistently higher contrast. This is due to the pendant, hydrophobic t-butyl ester groups in the hybrid formulation.

Effect of DI on Thermal Flow Resistance. P(NB/MA) based resins with acrylate derived monomer content below 30% have T_g's of above 300°C. Even resins with much higher acrylate content still show T_g's higher than 150°C. For instance, P(NB/MA/TBA/AA) (45% TBA, 25% AA) has a T_g of 190°C. The addition of most cholate based DI additives at loadings up to 35% only slightly plasticizes these resins. For instance, the high acrylate content resin discussed above, gives a T_g of 185°C upon addition of 35 wt % of a 50/50 blend of a cholate based oligomer and a polar cholate derivative. Thus high thermal resist processing temperatures are possible for a wide range of P(NB-alt-MA) acrylate resin formulations without significant thermal flow occurring.

Effect of DI on Plasma Etching Resistance. Relative etching rates in a chlorine based plasma (compared to Shipley HB1811) of ~ 1.3 are observed for P(NB/MA/TBA/AA) or P(NB/MA/AA) formulated resists in which the total acrylate content is in the range of 20 mole %. In comparison, the highest acrylate

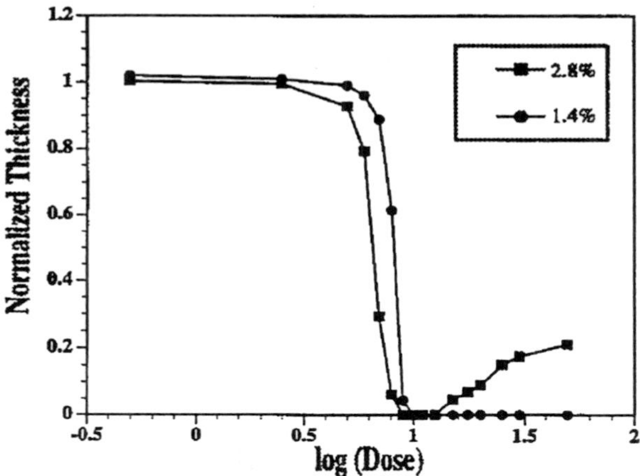

Figure 2. Development Behavior of P(NB/MA/AA15)/ 20% 2 Resist at High and Low PAG Loadings.

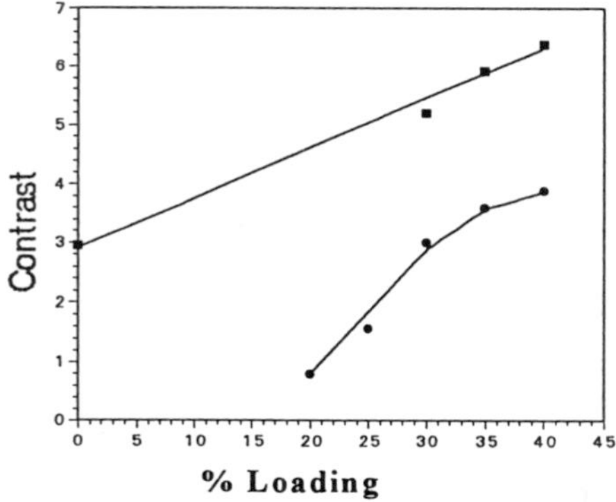

Figure 3. Comparison between loading effect of a dissolution inhibitor (t-butyl deoxycholate, 1b) on the contrast of 3-component (●) and hybrid resists (■).

loading P(NB/MA) resin that we have examined P(NB/MA/TBA/AA (45%TBA, 25%AA) gave a relative etching rate of 1.68 which is still appreciably better than poly(methyl methacrylate) under the same conditions (2.25). Suitable DI additives can appreciably lower the relative etching rate because of their high carbon/alicyclic content. For instance, P(NB/MA/TBA/AA) (45%TBA, 25%AA), showed a relative etching resistance of 1.37 when 35-40% of an oligomeric cholate based blend was added (**4b**). Similar improvements in etching selectivity are seen for resins with lower acrylate loadings. Not all DI's, however, give improvements in etching selectivity; t-butyl lithocholate for instance does not impart any improvement. This may be because of the higher volatility of this non-polar DI derivative compared to the polar monomeric or oligomeric DIs. Another possibility for the poor etching performace of t-butyl lithocholate systems is phase separation of the non-polar DI from the matrix leading to a heterogeneous mixture that is more susceptible to etching.

These results compare favorably with the needs of 193 nm lithographic patterning and show the general expected trend towards increased reactive ion etching resistance of materials with increased carbon/oxygen ratio and alicyclic content (*1a*).

Effect of DI Structure on Resist Film Properties. The effect of changing the nature of the DI on select resist properties for three component and hybrid resists systems is shown in Table I. Development selectivity trends in P(NB/MA/AA) and P(NB/MA/TBA/AA) resins do not appear to mirror previous investigation in methacrylate based resins (*1a*) where contrast was linked mainly to the hydrophobicity of the cholate-based DI.

Several chemical structural features appear to govern the ultimate contrast achievable by a cholate DI. For instance **1a** which has a high hydroxyl content tends to give better adhesion and higher contrasts. The dark erosion resistance of three component resists imparted by the monomeric DI **1a** and its loading capacity in both hybrid and three component resist are reduced. In comparison, low hydroxy content, hydrophobic dissolution inhibitors such as t-butyl lithocholate (**1c**) and t-butyl lithocholate acetate (**2**) impart good resistance to dark erosion but they exhibit poor solubility in typical spinning solvent, poor adhesion and tend to give lower achievable contrasts. The intermediate hydroxyl content of monomeric DI **1b** imparts good solubility in spinning solvents, good loading capacity in resist matrices which both combine to give higher ultimately achievable contrasts. Resists formulated with **1b** also have good adhesion. Very high contrasts are achieved with the dimeric lithocholates (**3**) but these formulations have poor adhesion to device substrates and poor solubility in spinning solvents. Alternately, oligomerized cholates or deoxycholates (**4a,b**) impart poor adhesion but afford very high contrasts and improve most other resist material properties.

The explanation of the difference in contrast between t-butyl lithocholate (**1c**) and the oligomeric cholates **4a,b** may arise because the latter are multifunctional dissolution inhibitor similar in design to the multifunctional DNQ's employed in commercial I line resists. Consequently, the need to deprotect multiple t-butyl groups on a single oligomer molecule may be boosting the non-linear relationship between exposure and dissolution rate leading to enhanced contrast compared to a monofunctional DI such as **1c**. Moreover, the ability of oligomer molecules to form multiple interaction through either carbonyl groups or hydroxyl moities may be playing a role. In contrast the monomeric DI, **1c,** has but a single carbonyl and hydroxyl per molecule. In this manner the oligomeric DI's **4a,b** may act as better "mortar" to reinforce the resist matrix against attack by aqueous base much in the same manner as for diazonaphthoquinone-Novolaks resists where interaction between the DI and the resin play a key role in promoting effective dissolution inhibition.

Effect of Blending DI's on Resist Properties. As discussed above, resists formulated with the oligomeric DI's tended to exhibit poor adhesion to substrates. In an effort to combine the high contrast imparted by oligomeric DI's and the good adhesion imparted by of monomeric polar DI's, a high hydroxyl content monomeric DI was added to the formulation. Indeed, formulations employing this blend approach with either P(NB/MA/AA) or P(NB/MA/TBA/AA) afforded materials with high contrast and good adhesion (Table I). Notably, the contrast obtained for the hybrid resist formulation with P(NB/MA/TBA/AA) is higher (contrast = 9) than that found for three component resist formulated with P(NB/MA/AA) (contrast =5) because of the aforementioned effect of polymer bound t-butyl esters.

Unfortunately, resists formulated with similar blends of the high contrast inducing dimeric DI **3** and the polar DI's did not show a similar synergy giving poor adhesion. This is possibly due to loading constraints in spinning solvents imposed by the solubility of the dimeric derivatives examined so far. Another explanation for the difference in behavior of blends of dimers and oligomers with polar monomeric cholates may reside in the multifunctional nature of **4a** and **4b** which would allow for multiple interactions with both the resist matrix and the polar monomer. A tighter hydrogen bonded resist matrix may be better able to withstand dark erosion by base, giving rise to better adhesion. In contrast, **3,** the dimeric lithocholate DI, has no free OH's moieties and only four ester functionalities per molecule thus showing less potential for interaction with other resist components.

Conclusion

Figure 4 show an SEM image obtained with a hybrid resist formulation containing P(NB/MA/TBA/AA) formulated with a blend of DI's and DTBP-NF which shows high contrast, good adhesion no dark erosion and is able to resolve 0.15 μm lines and spaces upon 193 nm exposure. This shows how judicious choice of matrix polymer, PAG and DI chemistries may be employed to obtain desirable imaging properties.

Acknowledgments

We would like to thank T. Wallow for his contributions to this work during his tenure at Bell Laboratories.

Table I. A Comparison of Dissolution Inhibitor Properties in P(NB/MA/AA) or P(NB/MA/TBA/AA) Resins

DI	Adhesion to Device substrate	Contrast enhancement	Loading capacity	Solubility	Dark Erosion Resistance
1a	++	+	-	++	--
1b	+	-	++	++	-
1c	-	-	nd	-	+
2	-	-	nd	-	+
3	-	++	nd	-	++
4a,b	-	++	++	++	++
1a,b:4a,b Blends	++	++	++	++	++

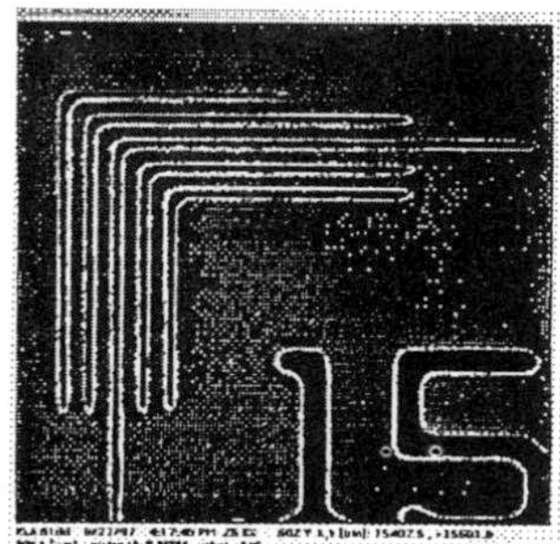

Figure 4. SEM image of 0.15 μm L/S derived from P(NB-alt-MA-co-AA-co-TBA) formulated with a blend of DI's and DTBP-NF obtained upon 193 nm exposure (20 mJ/cm^2).

References

1. For some leading references on 193 nm resists, see: (a) Allen R. D.; Wan I. Y.; Wallraff G. M.; DiPietro R. A.; Hofer D. C.; Kunz R. R. *J. Photopolym. Sci. Technol.* **1995**, 8, 623. (b) Allen R. D.; Wan I. Y.; Wallraff G. M.; DiPietro R. A.; Hofer D. C.; Kunz R. R. in *Microelectronics Technology, Polymers for Advanced Imaging and Packaging,* Reichmanis E.; Ober C. K.; MacDonald S. A.; Iwayanagi T.; Nishikubo, T., Eds.; *ACS Symposium Series 614,* American Chemical Society: Washington, DC, **1995**; p 255-270. (c) Nakano K.; Maeda K.; Iwasa S.; Ohfuji T.; Hasegawa E. *Proc. SPIE* **1995**, 2438, 433-440. (d) Nozaki N.; Kaimoto Y.; Takahashi M.; Takeshi S.; Abe N. *Chem. Mater.* **1994**, 6, 1492-1498 (e) Kunz R.R.; Palmateer S.C.; Forte A.R.; Allen R.D.; Wallraff G.M.; DiPietro R.A.; Hofer D.C. Proc SPIE 1996, 2724, 365.
2. (a) Wallow T. I.; Houlihan F.M.; Nalamasu O.; Chandross E.A.; Neenan T.X.; Reichmanis E *Proc SPIE* **1996,** 2724, 355 (b) Houlihan F.M.; Wallow T.I.; Timko A.; Neria E.; Hutton R.; Cirelli R.; Nalamasu O.; Reichmanis E. *Proc SPIE* **1997**, 3049, 84 (c) Houlihan F.M.; Wallow T.I.; Timko A.; Neria E.; Hutton R.; Cirelli R.; Nalamasu O.; Reichmanis E. *J. Photopolym. Sci. Technol.,* **1997,** 10, 511 (d) Houlihan F.M.; Wallow T.I.; Nalamasu O.; Reichmansis E. *Macromolecules,* **1997**, 30, 6517.
3. Takechi S.; Kaimoto Y.; Nozaki K.; Abe N. *J. Photopolym. Sci. Technol.* **1992**, Vol 5, No 3, 439.
4. Crivello J.C.; Shim S.Y., *Chem Mater* **1996**, 8, 376.
5. (a) Allen, R.D.; Sooriyakumaran R.; Opitz J.; Wallraff G. M.; DiPietro R. A.; Breyta G.; Hofer D.; Kunz R.; Jayaraman S.; Shick R.; Goodall B.; Okoroanyanyu U.; Willson C.G. *Proc SPIE,* **1996**, 2724, 341 (b) Niu Q. J.; Meagley R.P.; Frechet J.M.J.; Okoroanyanwu U.; Willson C.G. *Proc SPIE* **1997,** 3049, 113 (c) Okoroanyanwu U.; Shumikawa T.; Medeiros D. Willson C.G.; Niu J. Q.; Frechet J.M.J.; Byers J.; Allen R. *Proc SPIE,* **1997**, 3049, 92.
6. See Cowie, J. M. G. in *Comprehensive Polymer Science*; Allen, G.; Bevington, J. C.; Eastmond, G. C.; Ledwith, A.; Russo, S.; Sigwalt, P. Eds; Pergamon: Oxford, **1989**; Vol. 4, Chapter 22, and references therein.
7. Potter, G. H.; Zutty, N. L. U.S. Patent 3 280 080, **1966**.
8. The P(NB/MA/AAXX) nomenclature refers to feed ratio rather than experimentally determined composition as outlined in the experimental section of reference 2d.
9. Volksen, W.; Pascal, T.; Labadie J. W.; Sanchez M. I. in *Polymers for Microelectronics*; Thompson L. F.; Willson C. G.; Tagawa S. Eds; ACS Symposium Series 537; American Chemical Society: Washington DC, **1994**; p. 403.
10. Reichmanis E.; Wilkins Jr C.W.; Chandross A.E.; *J. Vac Sci Technol,* **1981**, 19(4) 1338.
11. Wilkins Jr C.W.; Reichmanis E.; Chandross A.E.; *J. Electrochemical Soc.,* **1982**, 129, No11, 2552.
12. O'Brian MJ *Polym Eng Sci*, **1989,** 29, 846.
13. Crivello JV *Chem Mater Sci*, **1994,** 6 2167.

14. Aoai T.; Yamanaka T; Kokubo T. *SPIE Proc.*, **1994,** 2195, 111
15. Reiser A. In PhotoReactive Polymers The Science and Technology of Resists; Wiley Interscience New York, New York 1989, 1
16. a) Dammel R. In *Diazonapthoquinone-based Resist*; D. Shea Ed.; SPIE Optical Engineering Press, Bellingham Washington, 1993, p 1. b) ibid p 86. c) ibid p 70
17. Crivello, J.V; Shim, S-Y.; Smith B.W.; *Chem Mater*, **1994**, 6, 2167.
18. Allen, R.D.; Opitz J.; Larson C.E.; DiPietro R. A.; Breyta G.; Hofer D. *Proc SPIE*, **1997**, 3049, 44

Chapter 16

Synthesis and Evaluation of Alicyclic Backbone Polymers for 193 nm Lithography

Hiroshi Ito[1], Norbert Seehof[1], Rikiya Sato[1,2], Tomonari Nakayama[1,2], and Mitsuru Ueda[2]

[1]IBM Almaden Research Center, 650 Harry Road, San Jose, CA 95120
[2]Yamagata University, Yonezawa, Yamagata 992, Japan

In attempts to prepare new polymers containing a bicyclic structure in the backbone for use in 193 nm lithography, we selected two approaches; 1) radical alternating copolymerization of substituted norbornenes with sulfur dioxide and 2) anionic ring-opening polymerization of sultams (cyclic sulfonamides), especially camphorsultam. The second synthetic scheme was supposed to incorporate novel base-soluble sulfonamide functionality in the backbone. In addition to the sulfonamide, we selected fluoroalcohol as another base-soluble functionality, which was incorporated into the norbornene structure, for replacement of carboxylic acid which has been the primary base-solubilizing group employed in 193 nm resists while 248 nm and i-line resists utilize a phenolic functionality for this purpose. Whereas ethane- and propanesultams undergo ring-opening polymerization, camphorsultam failed to polymerize. The radical co- and terpolymerizations of substituted norbornenes with sulfur dioxide proceeded readily to high conversions in a few hours. The synthesis and preliminary lithographic evaluation of substituted poly(norbornene sulfones) are described.

The quest for higher resolution continues in the microelectronics industry, which is currently shifting from i-line (365 nm) to deep UV (248 nm, KrF excimer laser) lithography for manufacture of 256 megabit memory and related logic devices requiring 0.25 μm resolution. This transition has been made possible by the revolutionary resist systems known as chemical amplification resists (*1*). This concept based on photochemically-induced acid-catalyzed imaging mechanisms has provided a breakthrough in sensitivity enhancement and material requirements. In order to achieve even higher resolution for a next generation device program, further shift to 193 nm (ArF excimer laser) has become a major thrust recently.

The shift from 365 to 248 nm demanded a drastic change in the imaging mechanism and material selection (*1,2*). Although the imaging mechanism of choice is most likely to be acid-catalyzed deprotection for positive imaging in ArF lithography as is the case with KrF, polymer backbone structures require a total departure from the current phenolic systems. A novolac resin employed in i-line resists has been replaced with poly(4-hydroxystyrene) in chemically-amplified deep UV lithography for its high transmission at 248 nm. These phenolic resins provide both aqueous base developability and dry etch resistance. Unfortunately, however, aromatic polymers cannot be used at the ArF excimer laser wavelength due to their excessive absorption. Thus, polymethacrylates have been the dominant polymer backbone structure with a pendant bialicyclic group providing dry etch resistance (*3-5*).

Excellent dry etch resistance of main chain alicyclic polymers such as polynorbornene has recently prompted efforts in development of chemically amplified resists based on such back bone structures (*6,7*). Since radical polymerization of nonconjugative vinyl monomers is sluggish in general, although radical polymerization of norbornene has been reported (*8*), two major approaches have been pursued (*7*), coordination cationic polymerization (*9*) and ring-opening metathesis polymerization (ROMP) (*10*) followed by hydrogenation of main chain double bonds, using transition metal catalysts. Since radical polymerization is still the most practical method of polymer preparation, we have decided to investigate alternating radical copolymerization of substituted norbornenes (**1**) with sulfur dioxide (Scheme I) (*11*). Similar attempts have been made using maleic anhydride (MA) as a comonomer (*7,12,13*). MA has been also utilized in radical copolymerization with other cycloolefins (*14, 15*).

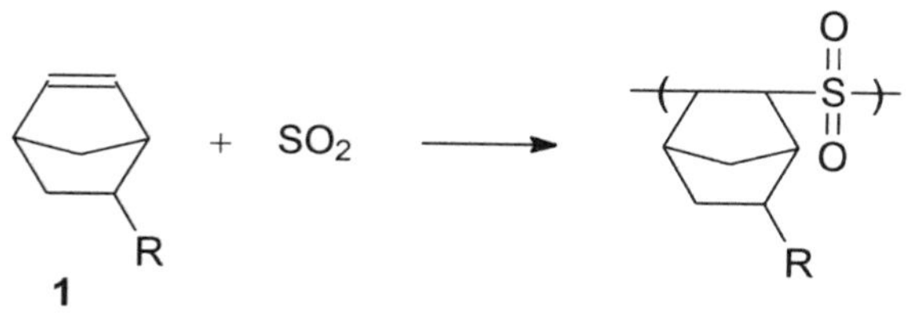

R
1
+ SO_2
O
S
O
R

Scheme I

In addition to the alternating radical polymerization of substituted norbornenes with sulfur dioxide, we also attempted anionic ring-opening polymerization of sultams (cyclic sulfonamides), especially, camphorsultam (**2**, Scheme II). This synthetic scheme was supposed to incorporate a novel base-soluble sulfonamide functionality in the backbone along with a bicyclic structure in the case of camphorsultam.

While the phenolic OH group has been the primary base-solubilizing functional group employed in the 248 nm resist design, the 193 nm resists have been exclusively based on carboxylic acid, which provides extremely fast dissolution rates in aqueous base, necessitating use of a non-standard weak aqueous base developer.

Thus, another goal we set forth was to identify a new base-solubilizing group which could be incorporated in 193 nm resists. In addition to the sulfonamide structure, we selected fluoroalcohol as another novel base-soluble functionality in the case of the norbornene-SO_2 co- and terpolymerizations.

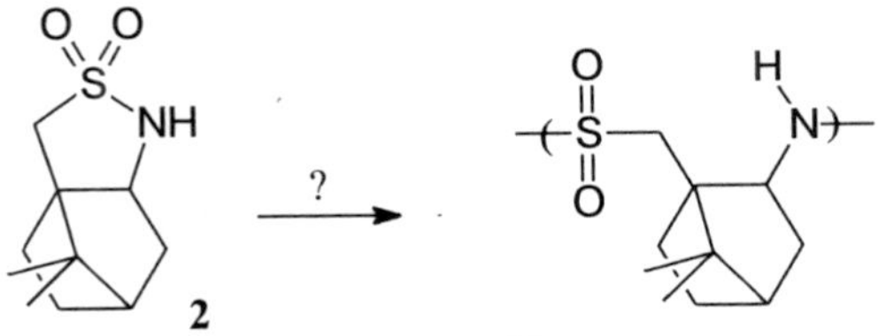

Scheme II

Experimental

Monomer Syntheses.

Camphorsultam. 10,10-Dimethyl-3-thia-4-aza-tricyclo[5.2.1.0^{1,5}]decane 3,3-dioxide (camphorsultam, **2**, Scheme II) was prepared according to the literature (*16*), using camphorsulfonic acid as a starting compound.

Substituted Norbornenes. The norbornene monomers were synthesized by the classical Diels-Alder reaction using commercially available (meth)acrylate derivatives. However, the Diels-Alder reaction between cyclopentadiene and 1,1,1-trifluoro-2-(trifluoromethyl)pent-4-en-2-ol (**3**) (tetrahydrofuran complex), which had been prepared by a literature procedure (*17*), to synthesize bicyclo[2.2.1]hept-5-ene-2-(1,1,1-trifluoro-2-trifluoromethylpropan-2-ol) (**1a**) required a high temperature (170 £C,#in an autoclave (Scheme III).

In a 500 mL steel autoclave were placed cyclopentadiene (22.00 g, 0.33 mol), **3** (109.00 g, 0.389 mol), and hydroquinone (450 mg) under a slight argon pressure. The reactor was heated at 170 °C for 17 hrs. The unreacted fluoro compound was distilled out and the desired product **1a** was isolated and purified by distillation under a reduced pressure. Yield of **1a** (*exo*/*endo*=1/4): 28.60 g (31 %). bp: 105-106 °C/55 torr. Anal. Calcd. for $C_{11}H_{12}F_6O$: C, 48.18; H, 4.41; F, 41.50. Found: C, 48.11; H, 4.53; F, 41.47. ^{13}C NMR (62.9 MHz, CD_2Cl_2): 32.9 (*exo* CH), 33.0 (*endo* CH), 34.5 (*endo* CH_2), 35.1 (*exo* CH_2), 35.4 (*endo* CH_2), 37.1 (*exo* CH_2), 77.48 (*exo* COH, J_{CF}=28.5 Hz), 77.56 (*endo* COH, J_{CF}=28.5 Hz), 124.3 (CF_3, J_{CF}=286.9 Hz), 132.6 (*endo* CH=), 137.0 (*exo* CH=), 137.5 (*exo* CH=), 138.9 (*endo* CH=).

The OH functionality of **1a** was protected with a *t*-butoxycarbonyl (tBOC) group. To a suspension of 0.80 g (0.033 mol) sodium hydride (60 % in mineral oil) in 40 mL anhydrous tetrahydrofuran (THF) cooled with an ice bath was slowly added a solution of **1a** (9.00 g, 0.033 mol) in 10 mL of THF. A vigorous hydrogen evolution occurred. After the addition was completed and the gas ceased to evolve, the cooling bath was removed and a solution of 7.40 g (0.033 mol) of di-*t*-butyl dicarbonate in 10 mL of THF was added dropwise to the alcoholate solution. Stirring was continued for 18 hrs. The reaction mixture was washed with water until the aqueous layer became neutral and then twice with a brine solution. The organic layer was dried over magnesium sulfate, the solvent removed, and the colorless oil distilled *in vacuo* to give norbornene-5-(1,1,1-trifluoro-2-trifluoromethyl-2-*t*-butoxycarbonyloxypropan) **1b**. Yield: 7.8 g (63 %). bp: 63-64 °C/4 torr. Anal. Calcd. for $C_{16}H_{20}F_6O_3$: C, 51.34; H, 5.39; F, 30.45. Found: C, 51.76; H, 5.46; F, 34.41. ^{13}C NMR (62.9 MHz, CD_2Cl_2): 27.7 (CH_3), 32.2 (*endo* CH_2), 32.7 (*endo* CH), 33.2 (*exo* CH_2), 32.85 (*exo* CH), 33.2 (*exo* CH_2), 34.9 (*endo* CH_2), 35.3 (*exo* CH_2), 42.4 (*exo* CH), 43.3 (*endo* CH), 45.9 (*exo* CH_2), 47.5 (*endo* CH), 48.25 (*exo* CH), 50.0 (*endo* CH_2), 83.67 (*exo* $\underline{C}(CF_3)_2$, J_{CF}=28.5 Hz), 83.79 (*endo* $\underline{C}(CF_3)_2$, J_{CF}=28.5 Hz), 85.3 ($\underline{C}(CH_3)_3$), 122.6 (CF_3, J_{CF}=289.9 Hz), 132.3 (*endo* CH=), 136.7 (*exo* CH=), 137.5 (*endo* CH=), 139.1 (*endo* CH=), 149.47 (*endo* C=O), 149.52 (*exo* C=O).

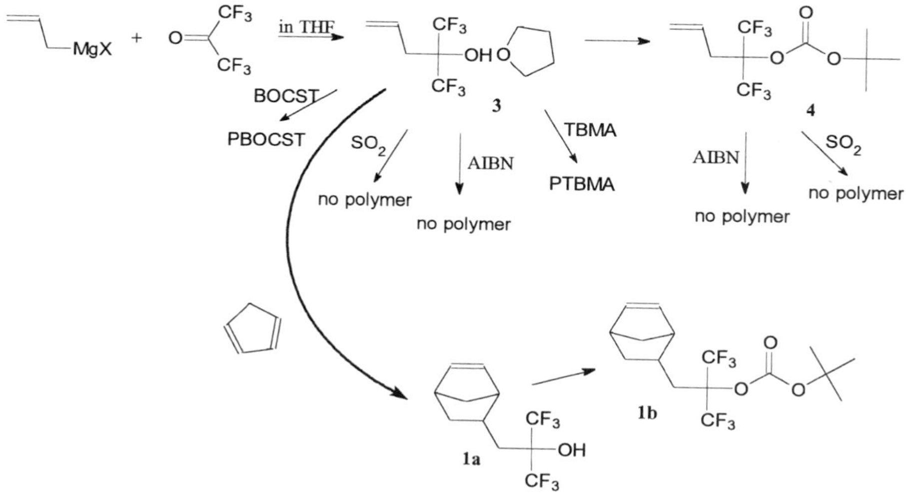

Scheme III

Attempted Anionic Ring-Opening Polymerization of Sultam 2. The procedure for ring-opening polymerization of propiosultam (*18*) was employed in our attempts to polymerize another five-membered sulfonamide **2**. *N-p*-Toluenesulfonyl camphorsultam (0.0924 g, 0.25 mmol), which had been prepared by reacting **2** with *p*-toluenesulfonyl chloride in the presence of base, and **2** (1.0765 g, 5 mmol) were dissolved in diglyme. After addition of 0.01 g (0.25 mmol) of sodium hydride (60 % in mineral oil) to the solution, the mixture was heated at 180 °C for 48 hrs under a nitrogen atmosphere with no increase in viscosity. Pouring the mixture into stirred methanol produced no precipitates. Replacement of NaH with a preformed sodium salt of **2** did not induce polymerization either. Bulk polymerization also failed.

Radical Co- and Terpolymerization of Norbornenes with SO_2. A typical polymerization procedure is described below. The fluoroalcohol monomer **1a** (2.74 g, 0.01 mol) was placed in 25 mL of liquid sulfur dioxide at -60 °C, to which were added 8 mL of anhydrous THF and 1.94 g (0.01 mol) of 5-carbo-*t*-butoxynorbornene **1c**. The mixture was stirred until a homogeneous solution was obtained. The polymerization was initiated with 0.2 mmol of *t*-butyl hydrogenperoxide (0.04 mL of a 5 M solution in decane, 1 mol% to the monomer) at -60 °C. The polymerization was carried out at -45±5 °C. After termination of polymerization by adding a solution of 80 mg of *p*-methoxyphenol in 5 mL THF and evaporation of SO_2, the remaining viscous polymer solution was diluted with 20 mL of THF. The polymer was isolated by precipitation in a mixture of 850 mL pentane, 250 mL isopropanol, and 15 mL triethylamine (to neutralize acid present in SO_2). The white polymer powder was dried at 80 °C for 6 hrs *in vacuo*. The yield was 95 %.

Other terpolymerizations and copolymerizations were performed in a similar fashion. In some cases dichloromethane was used as a diluent and in other cases neat SO_2 as a polymerization solvent.

Measurements. NMR spectra were obtained on an IBM Instruments NR-250/AF or Brucker AM300 spectrometer. DEPT was employed as an aid to assign the carbon resonances of the monomers. ^{13}C NMR spectra of the polymers were run in the inverse gate mode for better quantification. IR spectra were measured on a Nicolet Magna-IR 550 spectrometer using ~1-μm-thick polymer films spin-cast on NaCl plates. UV spectra were recorded on a Hewlett-Packard Model 8452A diode array spectrophotometer using thin films cast on quartz discs. Molecular weight determination was made by gel permeation chromatography (GPC) using a Waters Model 150 chromatograph equipped with four ultrastyragel columns at 40 °C in THF. Thermal analyses were performed on a Perkin Elmer TGS-2 at a heating rate of 5 °C/min for thermogravimetric analysis (TGA) and on a Du Pont 910 at 10 °C/min for differential scanning calorimetry (DSC) under N_2. Film thickness was measured on a Tencor alpha-step 200 or Nanometrics Nanospec/AFT 4150 Film Thickness Analyzer. A quartz crystal microbalance (QCM) (Maxtek TPS-550 sensor probe and PI-70 driver, Phillips PM6654 programmable high resolution frequency counter) was used to study the dissolution behavior of the polymer films in aqueous tetramethylammonium hydroxide (TMAH).

Results and Discussion

Polysulfonamides by Ring-Opening Polymerization. The sulfonamide functionality has a pKa in the range of 10-13 (*19*), which is similar to the value for phenol (11.8) (*20*) while the pKa value of carboxylic acid is 6.7 (*20*). Thus, we were interested in using the sulfonamide functionality as a novel base soluble group in 193 nm resist polymers for replacement of phenol and carboxylic acid. Furthermore, in contrast to the phenol or carboxylic acid group, the sulfonamide structure can be placed in a polymer backbone without losing its acidity. We decided to employ ring-opening polymerization of sultams in our attempts to prepare new polymers for 193 nm lithography, considering the facile polymerization of ethanesultam (*21*) and propanesultam (*18*). We were particularly interested in anionic ring-opening polymerization of camphorsultam (**2**, Scheme II), which is a substituted propanesultam. We repeated the literature procedure to polymerize ethanesultam (*22*) and also 4,4-dimethyl-1,2-thiazethizine-3-one 1,1-dioxide (*23*) with ease (Scheme IV). Unfortunately, however, these polymers were not soluble in common organic solvents. Since the sulfonamide NH group undergoes a replacement reaction (*18*), we prepared polysulfonamides partially protected with tBOC in an attempt to increase the solubility, which did not bear a fruit.

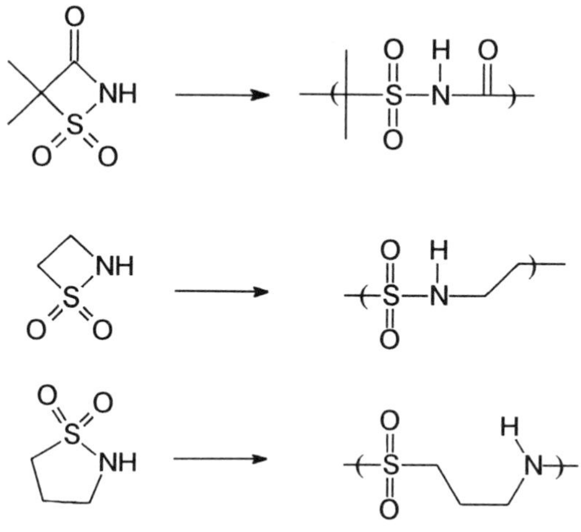

Scheme IV

Because the five-membered propanesultam undergoes facile anionic ring-opening polymerization, though at a relatively high temperature of 130 °C, camphorsultam **2**, which is a propanesultam fused with a bicyclic structure, was expected to polymerize under similar conditions to provide a new base-soluble, dry etch resistant polymer for potential use at 193 nm. However, our attempts to polymerize **2** failed even at a higher temperature of 180 °C.

Since the critical step in the anionic ring-opening polymerization of propanesultam is the rapture of the ring of the co-catalyst, *N*- benzenesulfonylsultam (activated sultam), induced by the attack of the cyclic sulufonamide anion, we treated *N*-toluenesulfonyl camphorsultam with sodium methoxide, potassium *t*-butoxide, and a preformed camphorsultam Na salt and analyzed the products by ^{1}H and ^{13}C NMR. In all the cases the NMR analysis suggested that the sulfonamide ring had been opened by the base. It is not clear at the moment why camphorsultam is reluctant to polymerize.

Trifluoromethylcarbinol as a Base-Soluble Functionality. According to the literature (*20*), pKa of trifluoromethylcarbinol derivatives varies from 6.7

{$(CF_3)_3COH$} (similar to 6.3 of acetic acid) to ~11 (11.2 for {$CF_3)_2CHOH$ and 11.5 for $(CF_3)_2CH_3COH$} (similar to 11.8 of phenol). Thus, we became interested in utilizing the $(CF_3)_2COH$ moiety as a base-soluble functionality potentially compatible, like phenol, with the industry-standard strong base developer (0.26 N TMAH) whereas the carboxylic acid group predominantly employed in most 193 nm resists provides too fast a dissolution rate in such a strong developer. The $(CF_3)_2COH$ group has been attached directly to polystyrene to produce a base-soluble polymer for 248 nm lithography (*24*).

Considering the ease of synthesis, we reacted anhydrous hexafluoroacetone with allylmagnesium chloride (and also with vinylmagnesium chloride) to produce a vinyl monomer with a pendant fluoroalcohol group (Scheme III) (*17*). As expected, this α-olefin isolated as a THF complex does not undergo radical homo- or copolymerization under standard conditions (Scheme III). Homopolymerization with 2,2-azobis(isobutyronitrile) (AIBN) produced no polymer. Copolymerization with 4-*t*-butoxycarbonyloxystyrene (BOCST) or *t*-butyl methacrylate (TBMA) resulted in formation of only poly(4-*t*-butoxycarbonyloxystyrene) (PBOCST) or poly(*t*-butyl methacrylate) (PTBMA). Radical copolymerization with SO_2 did not work at all, which was unexpected. As the OH group can interfere with radical polymerization, we prepared a vinyl monomer fully protected with tBOC and subjected this allylic monomer to radical homopolymerization and also to copolymerization with SO_2. No polymer was obtained.

Thus, we decided to incorporate the fluoroalcohol structure into norbornene by the Diels-Alder reaction as this cycloolefin has been known to undergo alternating radical copolymerization with SO_2 readily (*11*).

Poly(norbornene Sulfones) by Radical Polymerization. Cycloolefins such as norbornene copolymerize with SO_2 in an alternating fashion by radical initiation at room temperature or even at 50 °C, reflecting their high ceiling temperatures (*11*). In order to avoid the use of a pressure bottle, however, we carried out the polymerization of norbornenes with SO_2 at cryogenic temperatures. A partial list of poly(norbornene sulfones) we prepared is presented in Scheme V. All the polymers were prepared readily in high conversions but some of the polymers were not very soluble (pendant carboxylic acid, for example). This paper primarily discusses about the poly(norbornene sulfones) with a pendant *t*-butyl ester and fluoroalcohol. The 50/50 incorporation of the norbornene and SO_2 has been confirmed by elemental analysis for P(**1a**-*co*-SO_2) (Anal. Calcd. for $C_{11}H_{12}F_6O_3S$: C, 39.06; H, 3.58; F, 33.70; S, 9.48. Found: C, 39.28; H, 3.96; F, 32.41; S, 9.61).

Because we were interested in making resist polymers in high yields for an economical reason, we compared SO_2 with the most commonly employed comonomer, MA, in terms of the yield in radical copolymerization with norbornenes as shown in Table I. Table I clearly indicates that while the SO_2 copolymerization reaches a high yield of >70 % in 2-4 hrs, the MA systems are all sluggish, providing only a 33 % yield after 20 hrs in the best case with **1c**. The norbornene **1c** with a pendant *t*-butyl ester produced poly(norbornene sulfone) with high M_n of ~20,000 and a rather narrow molecular weight distribution of 1.5-2.0. Although the norbornene **1a** bearing a pendant fluoroalcohol copolymerized with SO_2 very rapidly providing almost quantitative yields in ~4 hrs, the polymers obtained were of low molecular weight, suggesting that the fluoroalcohol group functions as a chain transfer agent. It should be noted that **1a** yielded no copolymer with MA. Protection of the OH group of **1a** with tBOC to form **1b** improved the molecular weight of the resulting polysulfone dramatically (M_n=63,500) and the yield of the MA copolymer slightly to 11 %. It is apparent that SO_2 is a better comonomer than MA in alternating radical copolymerization with norbornene.

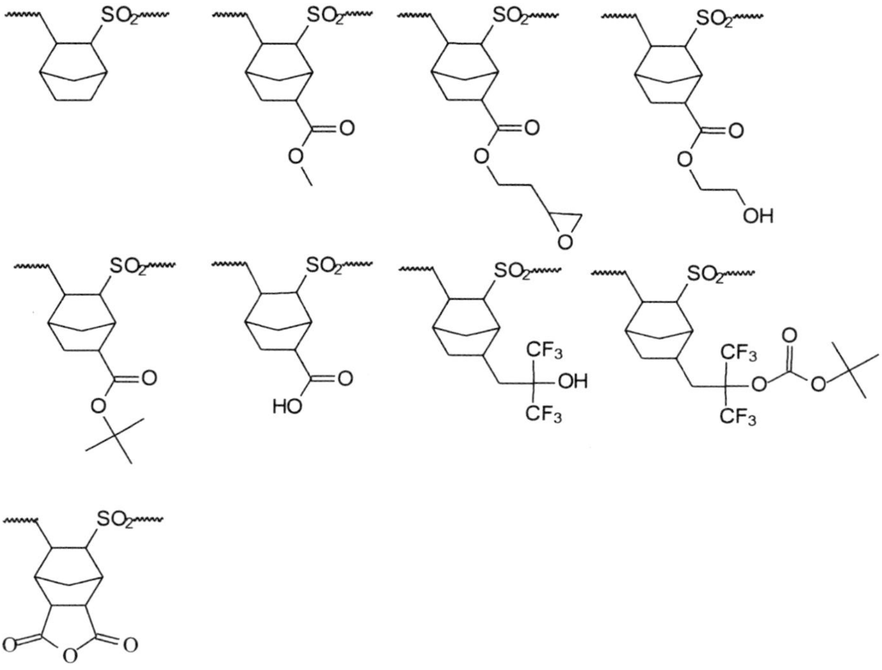

SO2 SO2 SO2 SO2
O O O
O O O
OH
O
SO2 SO2 SO2 SO2
CF3 CF3 O
O O OH O O
O HO CF3 CF3
SO2
O O O

Scheme V

Terpolymerization of **1a**, **1c**, and SO_2 were carried out to prepare resist polymers by varying the **1a**/**1c** ratio in the feed. The results are summarized in Table II. The terpolymerization proceeded smoothly, providing the polymer in >90 % yields in 3-4 hrs. Since the terpolymerizations were carried to near completion, the **1a**/**1c** ratios in the terpolymers are similar to the feed ratios. We suspect that the reactivity ratios of **1a** and **1c** are not much different. What is noteworthy is that M_n and M_w become exponentially smaller as the concentration of the fluoroalcohol unit increases in the polymer, pointing to the chain transfer involving the OH group.

The composition of the terpolymer was determined by the inverse gate ^{13}C NMR technique. ^{13}C NMR spectra of the **1a**-SO_2 and **1c**-SO_2 copolymers in acetone-d_6 are presented in Figure 1. The quaternary carbons in the two copolymers absorb at ~80 ppm with good separation. Thus, integration of these two well-separated sharp resonances in the terpolymer spectra allowed us to determine the terpolymer compositions using ^{13}C NMR as reported in Table II.

Table I. Alternating Radical Copolymerization of Norbornenes with SO_2 and MA

norbornene	comonomer	initiator (mol%)	temperature (°C)	time (hr)	yield (%)	M_n	M_w
1c	SO_2	tBuOOH (2.8)	-45±5	4	79	18,900	38,000
1c	SO_2	tBuOOH (2.0)	-45±5	4	74	23,200	36,600
1c	SO_2	tBuOOH (2.0)	-45±5	2	70	20,600	31,000
1c	MA	AIBN (3.0)	60	19.5	33		
1a	SO_2	tBuOOH (1.25)	-45±5	3.5	97	3,800	6,500
1a	SO_2	tBuOOH (1.50)	-45±5	4.0	99	5,000	9,300
1a	MA	AIBN (2.5)	60	18	0		
1b	SO_2	tBuOOH (2.0)	-45±5	4.0	84	63,500	220,700
1b	MA	AIBN (5.0)	60	21	11		

1c + MA: 30 wt/wt% in THF, **1a** + MA: 30 wt/wt% in THF, **1b** + MA: 54 wt/wt% in THF

Table II. Radical Terpolymerization of Norbornenes 1a and 1c with SO_2

#	[**1a**]/[**1c**] in feed	yield (%)	**1a**-SO_2 in polymer (mol%)	M_n	M_w
P46	100/0	97	100	3,800	6,500
P51	81/19	99	80.3	5,650	11,340
P53	70/30	96	69.3	8,470	13,300
P50	61/39	98	62.8	9,700	19,450
P48	60/40	97	61.8	9,770	18,870
P52	50/50	95	50.8	10,520	20,340
P45	40/60	91	44.5	19,000	39,000
P41	40/60	90	36.6	23,000	49,000
P43	40/60	88	34.5	22,000	47,000

1.3-1.5 mol% tBuOOH, -45±5 °C, 3-4 hrs.
cosolvent: THF, CH_2Cl_2 (P43), none (P41

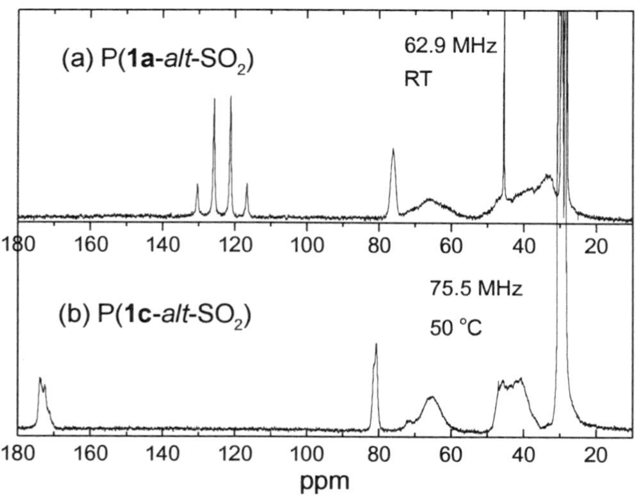

Figure 1 Inverse Gate ^{13}C NMR Spectra of Alternating Copolymers of SO_2 with (a) 1a (62.9 MHz, RT) and (b) 1c (75.5 MHz, 50 °C)

TGA curves of the poly(norbornene sulfones) bearing a pendant *t*-butyl ester and carbonate group are shown in Figure 2. The *t*-butyl ester group of P(**1c**-*alt*-SO_2) is stable thermally to ca. 170 °C and loses isobutene in the temperature range of 170-225 °C. Thus, the thermal deprotection temperature of this polymer is significantly lower than that of PTBMA (*25*). The *t*-butyl carbonate group in P(**1b**-*alt*-SO_2) begins to decompose at about 125 °C, more than 50 °C below the decomposition temperature of the *t*-butyl ester in P(**1c**-*alt*-SO_2) and ~65 °C below that of PBOCST (*26*). The glass transition was not observed in these co- and terpolymers below the thermal deprotection temperatures according to DSC analysis.

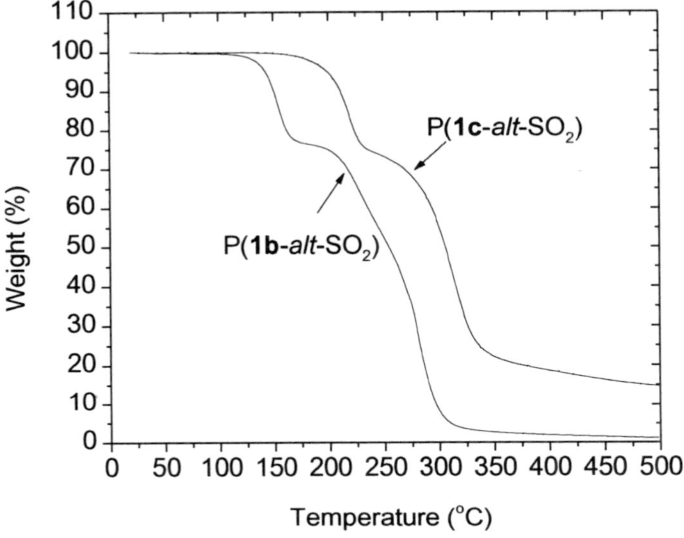

Figure 2 TGA Curves of Poly(norbornene Sulfones) with a Pendant *t*-Butyl Ester and Carbonate (heating rate: 5 °C/min)

UV spectra of 1-μm-thick films of poly(norbornene sulfones) are presented in Figure 3 in comparison with that of an alternating copolymer of **1c** with MA. The polysulfones are transparent at 193 nm with an optical density (OD) of 0.25/μm for P(**1a**-*alt*-SO_2) and 0.33/μm for P(**1c**-*alt*-SO_2). While the MA copolymer has a slightly higher but similar OD of 0.35/μm at 193 nm, it exhibits a higher and broader absorption above 200 nm. The peak at 220 nm of P(**1a**-*alt*-SO_2) completely disappears when the OH group is protected with tBOC while the absorption below 200 nm is not much affected by the masking.

The dissolution behavior of P(**1a**-*alt*-SO_2) was studied on QCM in a 0.21 N TMAH solution using thin films cast from ethyl lactate and baked at 130 °C for 60 sec. Figure 4 presents the dissolution kinetics curve of the polysulfone film. The 450-nm-thick film dissolved away in 0.15 sec, with an extremely fast dissolution rate of 30,000 Å/sec, which was nicely observed by our QCM setup. The dissolution rate of poly(4-hydroxystyrene) (PHOST) in the same 0.21 N developer is in the range of 3000-200 Å/sec depending on its molecular weight (*27*). Thus, the fluoroalcohol polymer dissolves at least one order of magnitudes faster than PHOST, which was unexpected considering the similar pKa values of the fluoroalcohol and phenol. In addition to pKa, the dissolution rate in aqueous base is very much affected by the degree of hydrogen bonding, polymer backbone structure, free volume in the film, etc. The rigid backbone structure of the poly(norbornene sulfone) may be responsible at least in part for the fast dissolution rate.

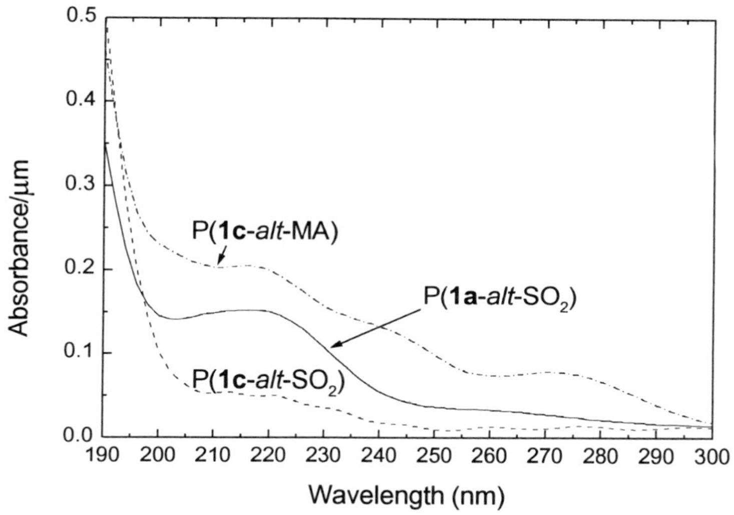

Figure 3 UV Spectra of 1-μm-thick Films of Alternating Norbornene Copolymers

The dissolution rates of the **1a**-**1c**-SO_2 terpolymer films were also measured under the same conditions (bake: 130 °C for 60 sec, developer: 0.21 N TMAH) and plotted in Figure 5 as a function of the terpolymer composition. As the norbornene **1c** bearing a *t*-butyl ester group is introduced into the polymer, the dissolution rate becomes exponentially smaller. The dissolution rate for **1a**/**1c**=1/1 is ca. 170 Å/sec and the terpolymers with more than 55 mol% *t*-butyl ester do not dissolve at all in this developer. It thus appears that the concentration of the fluoroalcohol unit in the terpolymer must be below 45 % for an unexposed resist film to provide minimum thinning in the industry-standard 0.26 N TMAH solution.

P(**1c**-*alt*-SO_2) and 5.2 wt% of triphenylsulfonium trifluoromethanesulfonate were dissolved in propylene glycol methyl ether acetate (PM Acetate) to formulate a resist. A film spin-cast on a NaCl plate was baked at 100 °C for 60 sec, an IR spectrum of which is presented in Figure 6a. The film was exposed to 20 mJ/cm^2 of 254 nm radiation from an Oriel illuminator and baked at 100 °C for 90 sec (Figure 6b). The strong tBOC carbonyl absorption at 1755 cm^{-1} almost disappeared with concurrent appearance of a broad OH absorption at 3500-3000 cm^{-1}, indicating acid-catalyzed deprotection of the tBOC group to generate free fluoroalcohol.

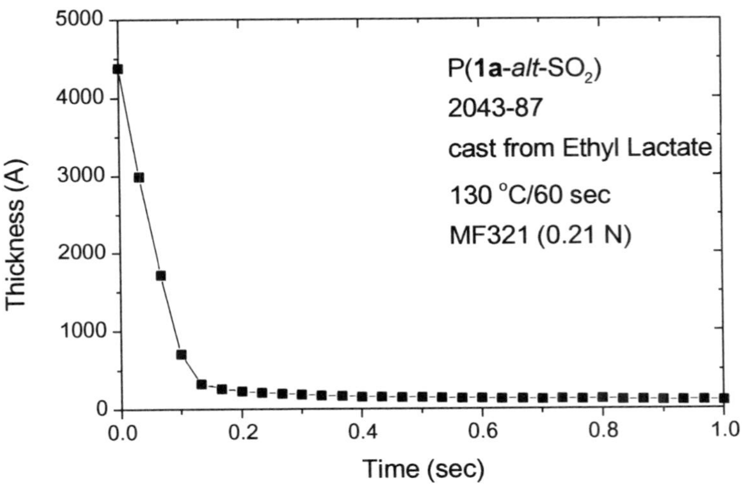

Figure 4 Dissolution Kinetics of P(1a-*alt*-SO_2) in 0.21 N TMAH studied by QCM

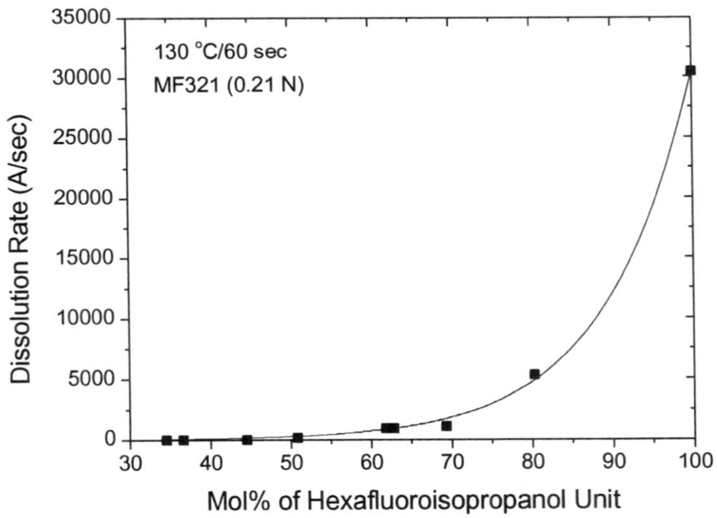

Figure 5 Dissolution Rate of Terpolymers of 1a, 1c, and SO_2 in 0.21 N TMAH as a function of Composition

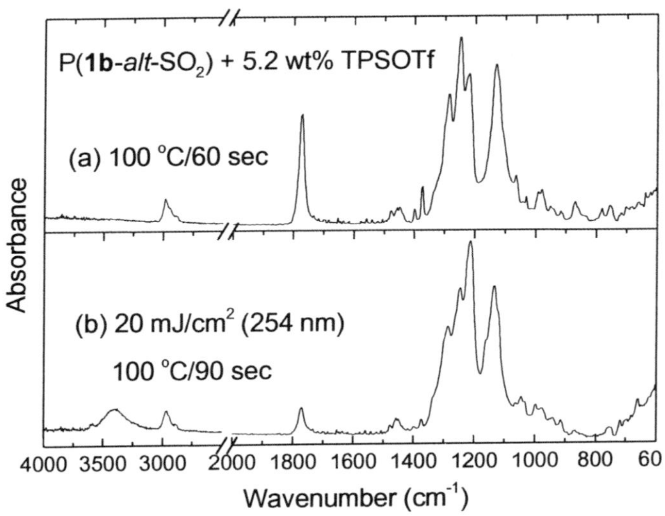

Figure 6 IR Spectra of P(1c-*alt*-SO_2) Film Containing 5.2 wt% Triphenylsulfonium Triflate Prebaked at 100 °C for 60 sec (a) and Exposed to 20 mJ/cm^2 of 254 nm Radiation Followed by Baking at 100 °C for 90 sec

The dry etch resistance of the poly(norbornene sulfones) were compared with a novolac resin and the IBM's methacrylate terpolymer 193 nm resist (*28*) using 100 sccm of pure Cl_2 at 2 mTorr (2500 W source, 75 W chuck) and the etch rates relative to novolac are tabulated in Table III. The poly(norbornene sulfones) we had prepared exhibited surprisingly poor plasma resistance! The unsubstituted poly(norbornene sulfone) was only slightly better than the methacrylate terpolymer resist (IBM V1.0B). It should be noted that introduction of an ester functionality into the ring resulted in an significant increase in the etch rate. Thus, both methyl and *t*-butyl ester polymers etch twice faster than novolac in Cl_2 plasma. Although poly(olefin sulfones) have been known to degrade very efficiently by electron beam and γ-irradiation and therefore in plasmas (*29*), we expected decent etch resistance with poly(norbornene sulfones) because they have rather high ceiling temperatures like poly(styrene sulfones) and because poly(styrene sulfones) were promoted as ion-millable positive electron beam resists (*30*). Furthermore, poly(BOCST sulfone) was developed into commercial deep UV chemically amplified positive resists named CAMP (*31*). Thus, we expected poly(norbornene sulfones) to be as dry etch resistant as poly(styrene sulfones), which turned out to be a false expectation.

Table III. Relative Etch Rates of Poly(norbornene Sulfones) in Cl_2 Plasma

polymer	etch rate relative to novolac
novolac	1.0
IBM V1.0B methacrylate resist	1.89
Poly(norbornene sulfone)	1.73
Poly(carbomethoxynorbornene sulfone)	2.15
Poly(carbo-*t*-butoxynorbornene sulfone)	2.14

Summary

An attempt to incorporate a novel base-solublizing sulfonamide functionality and a bialicyclic structure into a polymer backbone through anionic ring-opening polymerization of camphorsultam failed.

Radical copolymerization of substituted norbornenes with sulfur dioxide produced alternating copolymers in excellent yields in a few hours. The copolymers had an OD of 0.25-0.33/μm at 193 nm. The acidic di(trifluoromethyl)carbinol structure incorporated into the poly(norbornene sulfone) provided an extremely fast dissolution rate in aqueous base. Terpolymerization with carbo-*t*-butoxynorbornene resulted in an exponential decay of the dissolution rate. In contrast to poly(styrene sulfones), however, poly(norbornene sulfones) exhibited unacceptably fast etching in Cl_2 plasma.

Acknowledgments

The authors thank Rod Kunz for his Cl_2 plasma etch studies, Debra Fenzel-Alexander for her QCM work, Mark Sherwood for obtaining a 75 MHz ^{13}C NMR spectrum, and Melissa Bower and Albert Owen for their thermal analysis and GPC support.

Literature Cited

1. Ito, H. In *Desk Reference of Functional Polymers: Syntheses and Applications;* Arshady, R., Ed.; American Chemical Society: Washington, D. C., 1997, Chapter 2.4; pp 341-369.
2. Ito, H. *Solid State Technol.* **1996,** *36(7)*, 164.
3. Kaimoto, K.; Nozaki, K.; Takechi, S.; Abe, N. *Proc. SPIE,* **1992**, *1672,* 66.
4. Allen, R. D.; Wallraff, G. M.; DiPietro, R. A.; Hofer, D.; Kunz, R. R. *J. Photopolym. Sci. Technol.* **1994**, *7*, 507.
5. Nakano, K.; Maeda, K.; Iwasa, S.; Yano, J.; Ogura, Y.; Hasegawa, E. *Proc. SPIE* **1994**, *2195*,194.
6. Allen, R. D.; Sooriyakumaran, R.; Opitz, J.; Wallraff, G.; DiPietro, R.; Breyta, G.; Hofer, D.; Kunz, R.; Jayaraman, S.; Schick, R.; Goodall, B.; Okoroanyanwu, U.; Willson, C. G. *Proc. SPIE* **1996**, *2724*, 334.

7. Okoroanyanwu, U.; Shimokawa, T.; Byers, J.; Medeiros, D.; Willson, C. G.; Quingshang, J. N.; Frechet, J. M. J.; Allen, R. *Proc. SPIE* **1997**, *3049*, 92.
8. Gaylord, N. G.; Mandal, B. M.; Martan, M. *J. Polym. Sci., Polym. Lett. Ed.* **1976**, *14*, 555.
9. Risse, W.; Breunig, S. *Makromol. Chem.* **1992**, *193*, 2915.
10. Sorenson, W. R.; Campbell, T. W. *Preparative Methods of Polymer Chemistry;* Interscience Publishers: New York, NY,1968; pp 427-428.
11. Tokura, N. In *Encyclopedia of Polymer Science and Technology;* Mark, H. F., Ed.; John Wiley & Sons: New York, NY, 1968, Vol. 9; pp 460-485.
12. Wallow, T. I.; Houlihan, F. M.; Nalamasu, O.; Chandross, E. A.; Neenan, T.; Reichmanis, E. *Proc. SPIE* **1996,** *2724,* 355.
13. Choi, S.-J.; Kang, Y.; Jung, D.-W.; Park, C.-G.; Moon, J.-T.; Lee, M.-Y. *J. Photopolym. Sci. Technol.* **1997**, *10*, 521.
14. Niu, Q. J.; Fréchet, J. M. J.; Okoroanyanwu, U.; Byers, J. D.; Willson, C. G. *Proc. SPIE* **1997**, *3049,* 113.
15. Hattori,T.; Tsuchiya, Y.; Yamanaka, R.; Hattori, K.; Shiraishi, H. *J. Photopolym. Sci. Technol.* **1997**, *10*, 535.
16. Franklin, A.D.; Towson, J. C.; Weismiller, M. C.; Lal, S.; Carroll, P. J. *J. Amer. Chem. Soc.* **1988**, *110*, 8477.
17. Okamoto, Y.; Yeh, T. F.; Skotheim, T. A. *J. Polym. Sci., Part A, Polym. Chem.* **1993,** *31,* 2573.
18. Sorenson, W. R.; Campbell, T. W. *Preparative Methods of Polymer Chemistry;* Interscience Publishers: New York, NY,1968; p 351.
19. Dauphin, G.; Kergomard, A.; Veschambre, H. *Bull. Soc. Chim. France* **1967,** 3395, 3404.
20. Gandler, J. R..; Jencks, W. P. *J. Amer. Chem. Soc.* **1982**, *104*, 1937.
21. Imai, Y.; Hirukawa, H. *J. Polym. Sci., Polym. Lett. Ed.* **1973**, *11*, 271.
22. Imai, Y.; Hirukawa, H.; Ueda, M. *Kobunshi Ronbunshu* **1974**, *31*, 755.
23. Imai, Y.; Ueda, M.; Okuyama, K. *Polym. J.* **1979**, *11*, 613.
24. Przybilla, K. J.; Röschert, H.; Pawlowski, G. *Proc. SPIE* **1992**, *1672,* 500.
25. Ito, H.; Ueda, M. *Macromolecules* **1988**, *21*, 1475.
26. Fréchet, J. M. J.; Eichler, E.; Ito, H.; Willson, C. G. *Polymer* **1983**, *24*, 995.
27. Barclay, G. G.; Hawker, C. J.; Ito, H.; Orellana, A.; Malenfant, P. R. L.; Sinta, R. F. *Proc. SPIE* **1996**, *2724,* 249.
28. Kunz, R. R.; Allen, R. D.; Hinsberg, W. D.; Wallraff, G. M. *Proc. SPIE* **1993**, *1925*, 167.
29. Bowden M. J. In *Materials for Microlithography;* Thompson, L. F.; Willson, C. G.; Fréchet, J. M. J., Eds.; ACS Symposium Series No. 266; American Chemical Society: Washington, D. C., 1984, Chapter 3; pp 39-117.
30. Bowden M. J.; Thompson, L. F. *J. Electrochem. Soc.* **1974**, *121*, 1620.
31. Tarascon, R. G.; Reichmanis, E.; Houlihan, F. M.; Shugard, A.; Thompson, L. F. *Polym. Eng. Sci.* **1989**, *29*, 850.

Chapter 17

Progress in 193-nm Single Layer Resists: The role of Photoacid Generator Structure on the Performance of Positive Resists

Robert D. Allen, Juliann Opitz, Carl E. Larson, Thomas I. Wallow, Richard A. DiPietro, Gregory Breyta, Ratnam Sooriyakumaran, and Donald C. Hofer

IBM Almaden Research Center, 650 Harry Road, San Jose, CA 95120

The impact of photoacid generator (PAG) structure has been largely ignored in the early evolution of 193-nm positive resists. Most published work to date has involved the use of triflic- or metallic (antimonic or arsenic) acid producing compounds. With the relatively low reactivity (high-bake) protecting groups that are employed in our resist design, greater control of photoacid diffusion and evaporation is required than would be needed in low bake (high reactivity) deprotection chemistry. In this paper, we document the negative consequences of triflic acid on the 193-nm resist performance, including the impact of acid volatility. Model resists based on relatively hydrophobic acrylic polymers were observed to amplify the impact of triflic acid volatility on the degradation of performance of 193-nm positive resists. A family of acid generators which combine thermal stability, high reactivity (when formulated in 193-nm resists), low photoacid volatility and improved resolution will be described. These compounds are structurally described in Figure 1. Most importantly, iodonium sulfonates based on haloalkyl and activated aromatic sulfonic acids efficiently deprotect t-butyl esters and other high activation protecting groups when formulated in 193-nm resist materials and exposed to 193-nm light.

Photolithography using 193-nm light has evolved far and fast enough to be considered the most likely manufacturing technology for preparing tomorrow's semiconductor devices. It seems likely that the combination of Deep-UV (248-nm) lithography and advanced single layer chemically amplified resists (SLR's) will be pushed to its limit, and may in fact be used for the manufacturing of 0.18 micron devices (and possibly smaller!). When the limit of this single layer resist-based DUV technology is reached,

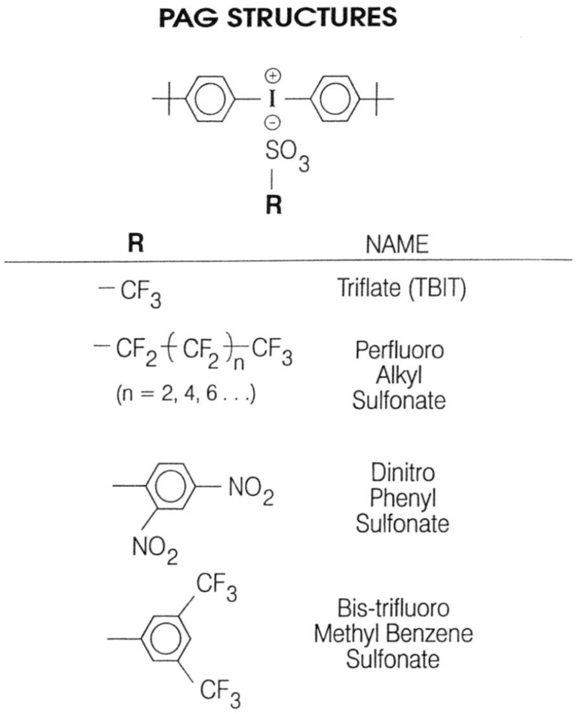

PAG STRUCTURES

R	NAME
$-CF_3$	Triflate (TBIT)
$-CF_2(CF_2)_nCF_3$ (n = 2, 4, 6 . . .)	Perfluoro Alkyl Sulfonate
NO_2, NO_2	Dinitro Phenyl Sulfonate
CF_3, CF_3	Bis-trifluoro Methyl Benzene Sulfonate

1. Chemical structures of the Iodonium sulfonates discussed in this paper.

using high NA (0.6-0.7) photolithography tools with some level of optical enhancements, and extremely refined, ultra-high performance chemically amplified 248-nm single layer resists, the most attractive solution will again be SLR-based, this time using 193-nm light and a completely new generation of chemically amplified resists (1).

The successful introduction of DUV lithography into critical level manufacturing of devices ranging from 0.35 to 0.25 micron dimensions and R&D "pilot" manufacturing of the 0.18 micron device generation, followed a decade long evolution of technological development of the total DUV lithographic package (chemically amplified photoresists, DUV photolithography tools, the KrF excimer lasers, and special environmental enclosures for (ppb) basic airborne contaminant removal)(2). The introduction of 193-nm lithography will most likely occur in the 0.15 micron critical dimension device generation. This presents a real problem for the semiconductor industry, because advanced R&D on this generation will begin in 1998. At present, the 193-nm resists and photolithography tools (steppers and scanners) are not yet ready, although substantial progress in both areas has occurred over the last year (1996-7). This unprecedented highly compressed development cycle for exposure tools and photoresists is a significant challenge for the semiconductor industry.

From our perspective as designers of 193-nm photoresists, the ability to use existing 248-nm exposure tools with new 193-nm resist formulations has eased some of the burden of bringing up a new tool with a new and unproven resist process. In other words, a new 193-nm resist formulation can be evaluated on a high-NA 248-nm DUV tool, and compared with or benchmarked against a state-of-the-art DUV resist. This dual-wavelength nature of 193-nm resist technology brings about a further opportunity: the possibility of a single resist/process for multiple exposure tools. Additionally, the commercial availability of 193-nm tools, in the form of small field “mini-steppers” (e.g., the ISI ArF Microstep, 0.6 NA) has accelerated progress in the evaluation of new 193nm resist materials over the last year.

Nevertheless, the design of positive 193-nm single layer resists is a significant challenge. The field has advanced remarkably in the past few years due to the efforts of resist designers around the world(1-2). In a few brief years, the status of these chemically amplified positive resists has progressed from the discovery of etch-resistant, non-aromatic (alicyclic) materials with very poor imaging performance, reported in 1992 by workers at Fujitsu (3), to the development of a 'non-etch resistant' acrylic terpolymer resist in 1993 by the IBM/MIT team (4), with imaging performance commensurate with requirements for prototype tool-testing purposes. These early advances were followed by perhaps the first demonstration of combined etch resistance and imaging performance in 1995 by workers from IBM, who focused on the control of the thermal properties of the alicyclic-modified (acrylic) resist as a key in obtaining a high performance resist. This work introduced the use of steroidal (cholanoate) additives for thermomechanical and dissolution modification, and improved etch resistance (5). In very short order following these three critical findings, several acrylic (or acrylic-cyclic olefin hybrid) resist approaches have been published which combine imaging performance with plasma etch resistance, most notably from Bell Laboratories, Lucent (6), NEC (7) and Fujitsu (8).

The majority of recently published work on 193-nm SLRs involves the design of new etch resistant polymers, focusing on new backbone polymer chemistry (6,9), alicyclic pendant groups (7), and acid-labile protecting groups (8,9). For example, workers at Lucent Technologies have prepared new resist materials from the free radical alternating copolymerization of norbornene and maleic anhydride (6). This work on resist design using alternating copolymerization has been substantially broadened by workers from Samsung (10) and Hyundai (11). A joint effort from IBM, University of Texas and BF Goodrich has produced new resists with the potential for significant improvement in plasma etch resistance from the addition polymerization of norbornene derivatives (9, 12). Recently, investigations of protecting group chemistry at both IBM (9) and Fujitsu (7) have emphasized the importance of alicyclic protecting groups to provide *dual-function* (etch resistance and dissolution switching) in 193nm resists. Workers at NEC have disclosed a variation of this theme, with alicyclic esters bearing deprotectable side groups (7).

The impact of photoacid generator (PAG) structure has been largely ignored for 193-nm single layer resists due to the concentrated polymer-oriented work discussed above. Most published work to date has involved the use of triflic or metallic (antimonate or arsenate) photoacids. Many PAGs used in DUV (248 nm) resists are inefficient when formulated with (non-phenolic) polymers used in 193-nm resists, presumably due to the lack of the electron transfer sensitization pathway thought to be in use in phenolic systems (i.e., 248nm resists).

Our goal in this investigation was to develop an understanding of the impact of photoacid generator structure on properties important to the development of high performance 193-nm positive photoresists: the efficiency of deprotecting the high activation energy side groups employed in our resists (e.g. t-butyl and isobornyl esters) as a function of PAG structure, the consequence of volatility of the photoacid on the dissolution properties of the resists, and the impact of photoacid size on the fine feature lithography of some model resists.

Designing a 193nm Single Layer Resist

In the design of a standard DUV resist formulation, a polymer (PHS or Poly(hydroxystyrene)) is "protected" from unexposed dissolution through judicious choice of an acid-reactive protecting group. A suitable photoacid generator (PAG) is chosen, based on consideration of ease of preparation, dissolution inhibition, photoacid size, etc. The formulation is dissolved in a good casting solvent, quite possibly with leveling additives for good coating uniformity. The DUV resist formulation "menu" is shown schematically in Figure 2.

The design of analogous 193-nm SLR's is made much more complex because of the current lack of a "PHS-equivalent" for the 193-nm era. Multiple choices are available for a polymer backbone to "build" the resist around. None as yet have emerged as the natural drop-in replacement for PHS. Figure 3 illustrates the choices involved in the design of 193-nm SLR's. Many polymers and polymerization methods have been revealed very recently. Each polymer route has advantages and disadvantages. Additionally, the question of protecting group is more complex than in

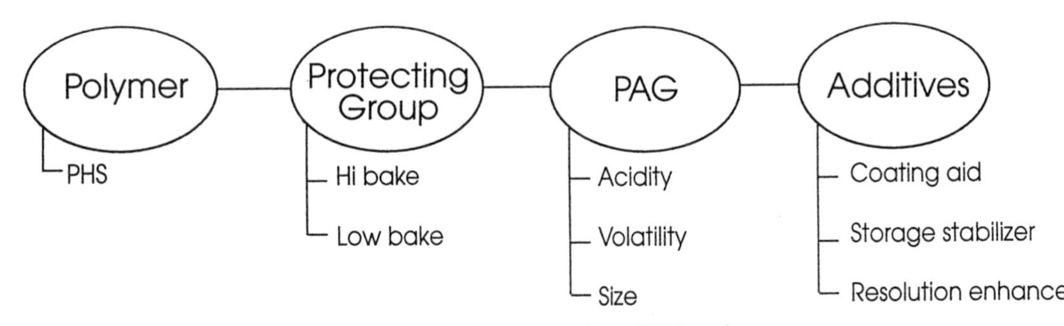

2. The design menu for DUV resists.

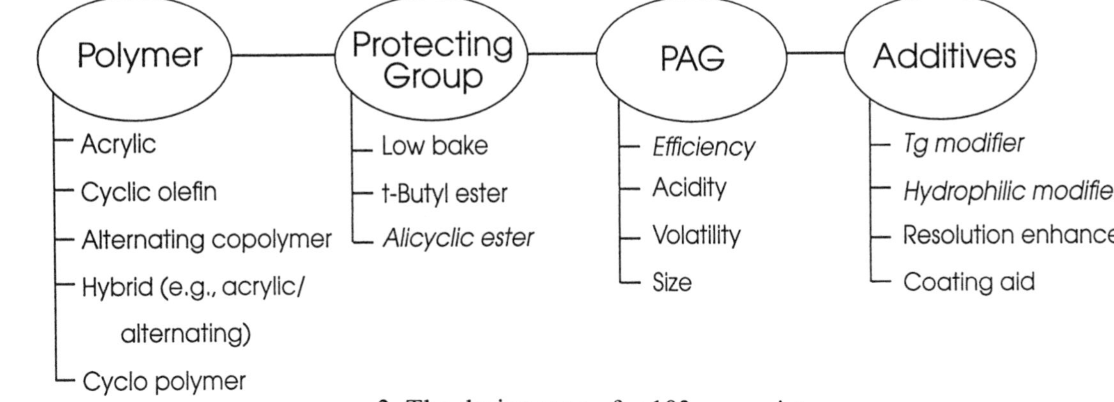

3. The design menu for 193-nm resists.

the case of DUV resists because the use of etch resistant but acid-reactive "protecting groups" is very popular for many 193-nm resist approaches. The selection of a proper PAG is difficult for 193-nm resists, because many PAGs used in 248-nm resist products are ineffective when formulated with 193-nm resist polymers. This is thought to be due to the lack of electron transfer sensitization in 193-nm polymers, presumably due to the absence of phenolic hydroxyl groups (attached to PHS) which fill this role for DUV resists. We have recently developed a class of high performance PAGs which provide a significant boost to the performance of 193-nm resists. These compounds (iodonium sulfonate derivatives) combine relatively high efficiency for "fast" photospeeds in non-phenolic polymers with the required transparency for 193-nm resists (e.g., acrylics), excellent thermal stability and strong dissolution inhibition, and tailorable photoacid size. Control of the photoacid size, polymer structure and resist process are important parameters in gaining high performance imaging in chemically amplified resists.

For example, the first "high resolution" imaging single layer resist for 193nm lithography was developed several years ago by IBM and the 193nm lithography group at MIT Lincoln Laboratory (4). This formulation, known as IBM Version 1 resist, was used as a "Tool Evaluation Resist" to evaluate the imaging performance of new 193nm exposure tools. Version 1A (initial formulation) resist was used extensively on the SVGL Micrascan 193 prototype at Lincoln Laboratory. In fact, the resist was integral to the optical characterization of the tool and accelerated optimization of the prototypes optical characteristics. The highest resolution obtained at 193nm (NA = 0.5) for Version 1A was 0.22 microns. A substantial improvement in performance will be demonstrated below, when the initial iodonium salt PAG used in V1A resist (TBIT, see Figure 1) was replaced with the newer PAG structures shown in the same Figure.

In the following section, we describe the tremendous impact of the structure of the photoacid generator on the behavior of 193-nm resists. Structural impact on three aspects of resist performance were investigated: efficiency of deprotection, acid volatility and diffusion, and finally, imaging resolution.

The PAG/Protecting Group Question

In our investigation of the nature of the acid-labile protecting group on 193-nm resist behavior (9,12) we uncovered *two separate regimes* of reactivity to photogenerated acid:

Low Bake Protecting Groups efficiently deprotect with post expose bake temperatures of less than 100°C (e.g., THP ester) and in some cases deprotect before the post-exposure bake (PEB) takes place (e.g., THF ester). These protecting groups have attractive reactivities and polarities, but do not tolerate carboxylic acid well, and preclude high temperature (above ca. 120°C) processing. The potential for contamination of exposure tool optical elements from volatilization of these reactive groups during the exposure may be substantially greater than with the less reactive (high-bake) protecting groups discussed below.

High Bake Protecting Groups demand post expose bake temperatures in excess of 100°C for sufficient chemistry to take place. This regime is typified by t-butyl esters and acid-cleavable alicyclics (e.g., isobornyl ester). Our 193-nm resists have thus far operated with these protecting groups of lower reactivity, thus allowing for high temperature bake processes, stability towards carboxylic acids, and straightforward preparation and storage. The reactivity difference of a "low bake" system (THF ester) vs a "high bake" (t-butyl ester) protecting group was illustrated in the following experiment, using a metallic (sulfonium antimonate) PAG. The extent of relative deprotection vs. dose (we refer to as chemical contrast curves (9, 14) was obtained as a function of post exposure bake temperature. The data clearly indicates that a PEB temperature of 100°C is insufficient in the case of the t-butyl ester protecting group, but is an appropriate temperature for the more reactive THF ester to efficiently deprotect. The t-butyl ester deprotects in an efficient manner at 150°C, well above the thermal deprotection limit of the much more reactive THF ester.

A consequence of the preference for the "high bake" protecting groups for the design of 193-nm resists is that no separation of the temperature regime of deprotection chemistry and acid diffusion is possible, placing a tremendous burden on the photoacid: efficient deprotection must occur, but the acid should not diffuse out of the film or into the "unexposed" regions at the temperatures required to cause this efficient deprotection chemistry to take place. Thus, a continual struggle in the "high bake" resist is balancing the need for efficient deprotection with the requirements of low acid diffusion. *The struggle to control deprotection chemistry and acid diffusion/evaporation (simultaneously) in a positive resist is the central problem in appropriate design of high bake photoresists.* Greater reflection problems that accompany shorter wavelength lithography bring about thin film interference effects that further compound this difficult situation. We anticipate that the use of reflection-suppression technologies should help ameliorate some of these problems (13). Key considerations include reactivity of the protecting group, photoacid size, strength and volatility, bake temperature and the resist glass transition temperature (14-15).

The Triflic Acid Problem

A "high-bake" copolymer of t-butyl methacrylate and methyl methacrylate containing no carboxylic acid (TBM-MMA) was formulated with two acid generators, an iodonium triflate (TBIT) and triphenylsulfonium hexafluoroantimonate. Films were prepared by spin-coating both formulations, and then exposing at a high dose (ca. 30 mJ/cm2) (250-nm filtered light) to achieve a high level of deprotection of the acid-cleavable group. In this experiment, the t-butyl ester copolymer exhibits *photoacid-dependent dissolution*. When the antimonate PAG is used, the exposed film remains insoluble in developer when PEB temperature is too low (120 °C), but becomes highly soluble in developer at high (150 °C) PEB temperatures. While the triflate PAG (TBIT) also requires a high bake temperature to drive the deprotection to high conversions, the triflate formulation fails to develop. More reactive esters (e.g., THP and THF), when incorporated into similar methacrylate copolymers, develop cleanly regardless of photoacid used (12). These effects are presumably due to photoacid volatility. Less reactive esters (e.g., t-butyl, isobornyl) require high temperature PEB to realize high deprotection efficiency, likely exacerbating volatility. We used a TBOC indicator film (14) to judge the relative volatility of several photoacids, including metallic, triflic, and higher alkyl and aryl sulfonic.

A variety of photoacid generators (PAGs) were formulated with a t-butyl ester-containing methacrylate terpolymer which has been previously described (4), and shown in Figure 4. The PAGs chosen included triflic acid generators: TBIT (bis-t-butylphenyliodonium triflate, and TPS (triphenylsulfonium) triflate. Also included was a traditional metallic PAG, triphenylsulfonium hexafluoroantimonate (TPS SbF6), and finally an iodonium salt containing a large, relatively high molecular weight perfluorohexylsulfonate anion (n=4; Figure 1).

Visual inspection of the TBOC indicator film suggested that of the photoacids studied, the resists containing the triflic acid-releasing PAGs showed extensive evaporation, as the line of demarcation between the exposed and unexposed area in the photoresist film was clearly transferred to the TBOC film only in the case of triflic acid (from both TBIT and triphenylsulfonium triflate). Different photoacids produced very different acid volatility. Triflic acid is indeed volatile (from photoresist films) in the conditions of this experiment, producing a substantial (6%) shrinkage in the TBOC film when either of the triflate salts were used. The sulfonium antimonate indeed formed a metallic-acid with far less volatility, producing no measurable shrinkage in the TBOC film. The film formulated with the iodonium perfluorohexane sulfonate also produced no signs of volatility in this experiment (14) demonstrating that it is possible to prepare PAGs with organic acids that exhibit very low volatility, but that also function efficiently to cause radiation-induced deprotection of 193-nm resists (read below).

A variety of iodonium salt derivatives of our "standard PAG", TBIT, were prepared and evaluated (see Figure 1). These iodonium salts were prepared using minor modifications to the synthesis as described by F. M. Beringer, where the counterions were obtained by metathesis with the appropriate sulfonic acid salt (17). Several derivatives exhibited very low activity when formulated into an acrylic terpolymer (Figure 4) and evaluated as prototype resists. The "active" PAGs, i.e., compounds that produce efficient deprotection of t-butyl ester containing acrylic polymers, were based on sulfonic acids of two types: "active" aromatics containing electron withdrawing substituents, and haloalkyl sulfonic acids. Figure 5 shows chemical contrast (deprotection vs. dose) curves for four iodonium sulfonic acids, including the triflic acid compound (TBIT) that has been used in our resists (IBM Version 1 and 2). Also included is a significantly larger haloalkylsulfonic acid (high activity), an "activated" arylsulfonic acid (high activity) and a non-activated aryl sulfonic acid (very low activity). Exposed films of these were baked (PEB) at 140 °C. The larger haloalkyl and activated aromatic sulfonic acids shown in Figure 5 exhibit near-zero volatility when subjected to the TBOC sandwich experiment discussed above.

A test to compare non-volatile iodonium haloalkylsulfonate with triflate was conducted by formulating two resists based on a methacrylate terpolymer (Figure 4) of a composition tailored to produce zero-thinning (no unexposed development) through reduction in the methacrylic acid concentration (to ca. 15 mole%). In this way, no "information" (the impact of acid diffusion/volatility) is lost during development. Two model resists were formulated from this hydrophobic polymer (14,15). Resist "A" was formulated with TBIT, the triflic acid generator which was shown above to produce significant photoacid volatility. Resist "B" was formulated with a perfluoroalkylsulfonate compound where n=4 (perfluorohexanesulfonate, see Figure 1).

$$\left(CH_2-\underset{\underset{\underset{O-C(CH_3)_3}{|}}{C=O}}{\overset{CH_3}{C}}\right)_x\left(CH_2-\underset{\underset{\underset{O-CH_3}{|}}{C=O}}{\overset{CH_3}{C}}\right)_y\left(CH_2-\underset{\underset{\underset{OH}{|}}{C=O}}{\overset{CH_3}{C}}\right)_z$$

4. Structure of the methacrylate terpolymer used in Version 1 photoresist.

Chemical Contrast of Selected PAGs
193nm Exposure IBM Version 1 Resist

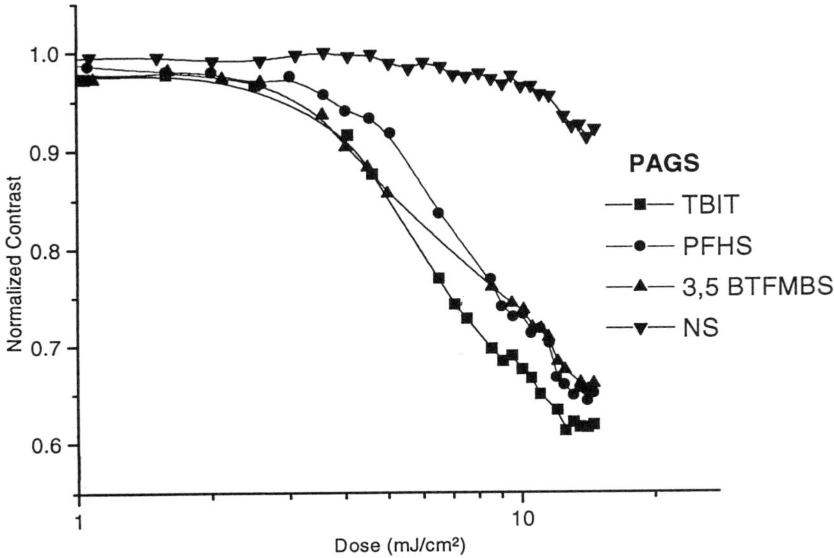

5. Chemical contrast of PAGs in V1 resist. PFHS refers to the perfluorohexanesulfonate salt (Figure 1, n=4), and NS refers to naphthylsulfonate.

Clean development was observed using the non-volatile photoacid (resist "B"), while the more volatile triflate resist "A" showed severe webbing at the top of the pattern, presumably due to triflic acid volatilization which causes a reduced photoacid concentration at the resist/air interface *(less photoacid = less deprotection = webbing)*. This experiment identifies one possible aspect of the difficulty in formulating an acrylic resist hydrophobic enough to withstand strong developers used in semiconductor processing using a triflate PAG.

Recently, the Version 1 resist, based on the methacrylate terpolymer structure shown in Figure 4, with a relatively high carboxylic acid concentration (ca. 25 mole%) was significantly reformulated (IBM Version 1C) with a larger, less mobile photoacid, using the perfluorobutanesulfonate (n=2; Figure 1) anion. The improved PAG leads to a tremendous boost in imaging performance, and a substantial change in the optimum processing temperatures. High quality images down to 0.175 microns have been printed in the SVG Micrascan 193 prototype at MIT Lincoln Laboratory. Additionally, substantially smaller features have been printed in Version 1C using the ISI Microstep (0.6 NA) 193nm tool at the IBM Almaden Research Center. We have been able to print an array of high quality 0.13 micron features through a substantial dose/focus range using Version 1C resist (through 0.7 microns of focus at 32 mJ/cm^2) see Figure 6.

Features as small as 0.12 microns have been successfully printed at 193nm using this Version 1C resist process (Figure 7). It is important to note that this represents standard quartz-on-glass mask technology, with no optical enhancements employed. Significant process latitude using a variable pitch mask (lines and larger spaces) has been observed at 0.12 microns using the Version 1C resist.

While the optimized process for the triflate-based Version 1 resist (using TBIT as PAG) employs a high post-apply bake PAB (for annealing (16)), best results are found using a low PEB, presumably to limit diffusion and evaporation of the triflic photoacid. Using this process in combination with the larger PAGs discussed above produces severe standing waves. The "best process" when using the less volatile PAGs found in Version 1C resist uses a moderate PAB and a higher PEB to simultaneously drive diffusion and deprotection chemistry. Experiments to determine the linewidth change as a function of bake temperature are in progress. We anticipate that the tolerance to temperature variation will be substantially improved using the larger PAG molecules in combination with higher PEB temperatures.

Literature Cited

1. Allen, R.D.; "Progress in 193 nm Photoresists", *Semiconductor International*, September **1997**, 72-80.

2. Allen, R. D.; Conley; W. E., Kunz, R. R.; *"Deep UV Resist Technology"*, Chapter 4 in **Handbook of Microlithography,** Ed. P. Rai-Choudhury, SPIE Optical Engineering Press, Bellingham, WA, 1997.

3. Kaimoto, Y.; Nozaki, K.; Takechi, S.; Abe, N; *Proc. SPIE* **1992**, *1672*, 66.

4. Allen, R. D.; Wallraff, G. M.; Hinsberg, W. D.; and Kunz, R. R.; *"Methacrylate Terpolymer Approach in the Design of a Family of Chemically Amplified Positive*

IBM V1C Photoresist Performance

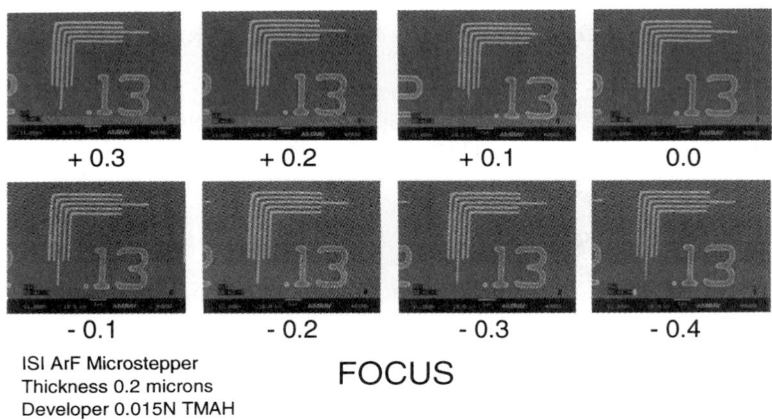

ISI ArF Microstepper
Thickness 0.2 microns
Developer 0.015N TMAH
Dose 32 mJ/cm²

6. High resolution 0.13 micron patterns from the processing of V1C Resist exposed at 193nm on the ISI ArF Microstep.

IBM Version 1C 193nm Photoresist
ISI ArF Microstepper 0.6 NA

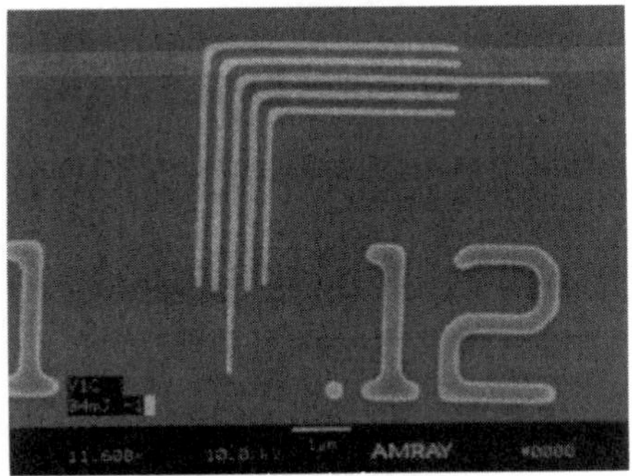

Thickness: 0.2 microns
Developer: 0.015N TMAH
Dose: 34 mJ/cm²

7. High resolution 0.12 um features printed in V1C resist exposed at 193nm on the ISI ArF Microstep using no optical enhancements and a standard chrome-on-quartz mask.

Resists", Chapter 11 in ACS Symposium Series, No. 537, **Polymers for Microelectronics,** L. Thompson, C. G. Willson, and S. Tagawa, ACS, Washington, DC, 1994; Kunz, R. R.; Allen, R.D.; Hinsberg, W. D.; Wallraff, G. M.; *Proc. SPIE,* **1993**, *1925*, 167.

5. Allen,R. D.; Wallraff, G. M.; DiPietro, R. A.; Hofer, D. C.; Kunz, R. R.; *Proc. SPIE* **1995**, *2438*, 474.

6. Wallow, T. I.; Houlihan, F. M.; Nalamasu, O.; Chandross, E.; Neenan, T. X.; Reichmanis, E.; *Proc. SPIE,* **1996** *2724*, 355; Houlihan, F.; Wallow, T.; Nalamasu, O.; Reichmanis, E.; *Macromolecules,* **1997**, *30*, 6517.

7. Iwasa, S.; Maeda, K.; Nakano, K.; Ohfuji, T.; Hasagawa, E.; *J. Photopolym. Sci. Technol.*, **1996**, *9(3)*, 447.

8. Nozaki,K.. Watanabe,K. Yano,E. Kotachi,A. . Takechi,S and Hanyu,I. *J. Photopolym. Sci. Technol.*, **1996**, *9(3)*, 509.

9. Allen, R. D.; Sooriyakumaran, R.; Opitz, J.; Wallraff, G.; DiPietro, R.; Breyta, G.; Hofer, D.; Kunz, R.; Jayaraman, S.; Schick, R.; Goodall, B.; Okoroanyanwu, U.; Willson, C. G.; *Proc. SPIE,* **1996**, *724*, 334.

10. Wallraff, G. M.; Opitz, J.; Breyta, G.; Ito, H.; Fuller, B.; *Proc. SPIE* **1996**, *2724*, 149.

11. Kihara, N.; Saito, S.; Ushirogouchi, T.; Nakase, M.; *J. Photopolym. Sci. Technol.* **1995**, *8(4)*, 561.

12. Allen, R.; Sooriyakumaran, R.; Opitz, J.; Wallraff, G.; DiPietro, R.; Breyta, G.; Hofer, D.; Kunz, R.; Okoroanyanwu, U.; Willson, C. G.; *J. Photopolym. Sci. Technol.*, **1996**, *9(3),* 465.

13. Kunz, R.; and Allen, R.; *Proc. SPIE,* **1994**, *2195*, 447.

14. Allen, R.; Opitz, J.; Larson, C.; Breyta, G., DiPietro, R.; Hofer, D.; *Proc. SPIE,* **1997**, *3049*, 44.

15. Allen, R.; Opitz, J.; Larson, C.; Wallow, T.; Breyta, G., DiPietro, R.; Sooriyakumaran, R.; Hofer, D.; *J. Photopolym. Sci. Technol.*, **1997**, *10(3),* 503.

16. Ito, H.; Breyta, G.; Hofer, D.; Sooriyakumaran, R.; Petrillo, K.; Seeger, D.; *J. Photopolym. Sci. Technol.*, **1994**, *7(3),* 433.

17. Beringer, F. M.; *Jo. Amer. Chem. Soc.*, **1959,** ***81,*** 352.

Chapter 18

Calixarene and Dendrimer as Novel Photoresist Materials

Osamu Haba, Daisuke Takahashi, Kohji Haga, Yoshimasa Sakai, Tomonari Nakayama, and Mitsuru Ueda

Department of Human Sensing and Functional Sensor Engineering, Graduate School of Engineering, Yamagata University, Yonezawa, Yamagata 992, Japan

Negative-working alkaline developable photoresists based on calix[4]-resorcinarene (**1**) or calixarene dendrimer (**2**), a cross linker, and a photo acid generator have been developed. Compound **2** was prepared by the condensation of compound **1** with 3,5-diallyloxybenzylbromide, followed by the removal of allyl groups. The resist consisting of **1** (70wt%), a photo acid generator, diphenyliodonium 9,10-dimethoxyanthracene-2-sulfonate (DIAS) (10wt%), and 4,4-methylenebis[2,6-bis(hydroxymethyl)-phenol] (MBHP) (20wt%) as a cross-linker showed a sensitivity of 2.2 mJ·cm^{-2} and a contrast of 3.1 when it was exposed to 365 nm light and postbaked at 130 °C for 3 min, followed by developing with a 0.1% aqueous tetramethylammonium hydroxide (TMAH) solution. On the other hand, the resist formulated by mixing **2** (70wt%), DIAS (10wt%), and the cross-linker, 2,6-bis(hydroxymethyl)phenol (BHP) produced a clear negative patternes by the exposure of 365 nm (10 mJ·cm^{-2}) UV light, postbaked at 110 °C for 3 min, and developed with a 0.3% TMAH aqueous solution.

The microelectronics technology has shown an astounding progress for the past decade, where radiation-sensitive polymeric materials called "resist" play an important role to produce circuit patterns in substrates (*1*). The classical diazonaphtoquinone /novolac resist is still the "workhorse" of the microelectronics industry. The lithographic performance of the resist is profoundly affected by the molecular weight distribution of novolac resin (*2*). The optimum resist performance was obtained with a complex mixture of different molecular weights novolac fractions. It is thus interesting to investigate the photoresist performance of monodisperse analogues of novolac resins. In this chapter, we will deal development of two novolac alternatives for photoresist materials.

One of the analogues is calix arenes which are oligomeric cyclic analogues of novolac resin derived through the condensation of *p*-alkyl phenols with formaldehyde. Quite recently, *p*-methylcalix[6]arene hexaacetate was successfully used as a high-resolution negative resist in electron beam lithography (*3*). Calixarenes are generally high-melting compounds, insoluble in water and only sparingly soluble in organic solvents. However, a variety of calix[4]resorcinarenes, derived from resorcinol and aldehydes, have been reported to have a good solubility in both organic solvents and aqueous alkaline solution (*4*). Thus we selected calix[4]resorcinarene (**1**) as a matrix.

Dendrimers are polymers with a new molecular architecture, which are characterized by possessing central polyfunctional core, from which arise successive layers of monomer units with a branch occurring at each monomer unit. They are monodisperse materials like the calixarenes, and their molecular weight reaches some thousands like the novolac resin. Thus the dendrimers possessing phenolic shell and calixarene core, is another analogues of novolac resin. We designed a new dendrimer (**2**) which contains phenol groups in the exterior for solubility in aqueous alkaline solution and calix[4]resorcinarene in the interior to increase the number of the phenol group even in the lower generation.

In this chapter, we report new negative working alkaline developable resists consisting of **1** and **2**, a photoacid generator and a crosslinker.

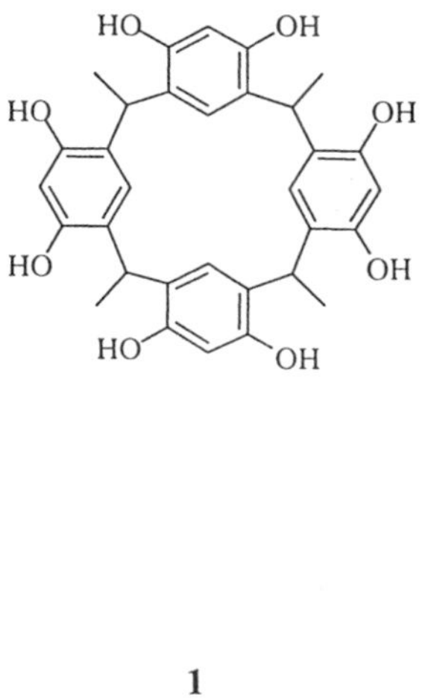

1 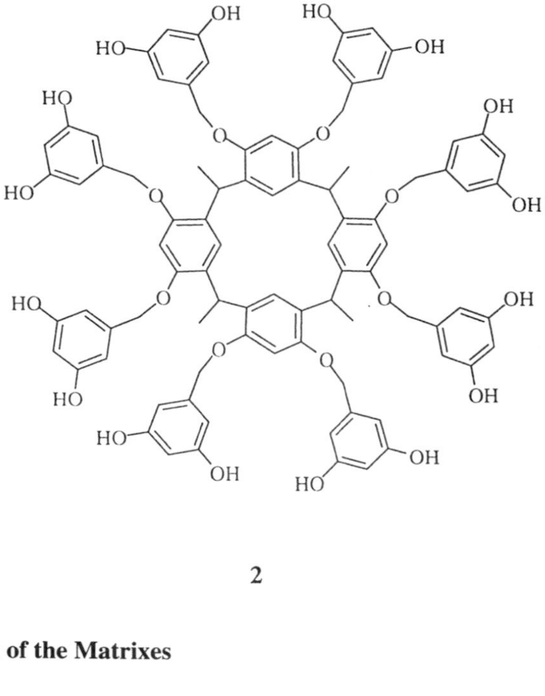 **2**

Syntheses and Characterization of the Matrixes

Syntheses. Calixarene **1** was prepared by condensation of resorcinol and acetaldehyde in water according to the reported procedure (*5*). The resulting material showed good solubility in dipolar aprotic solvents, acetone and especially in alcohols at room temperature. Synthesis of **2** was carried out starting from methyl 3,5-dihydroxybenzoate (**3**) according to Scheme 1 modifying the "convergent-growth" method (*6*). The hydroxyl

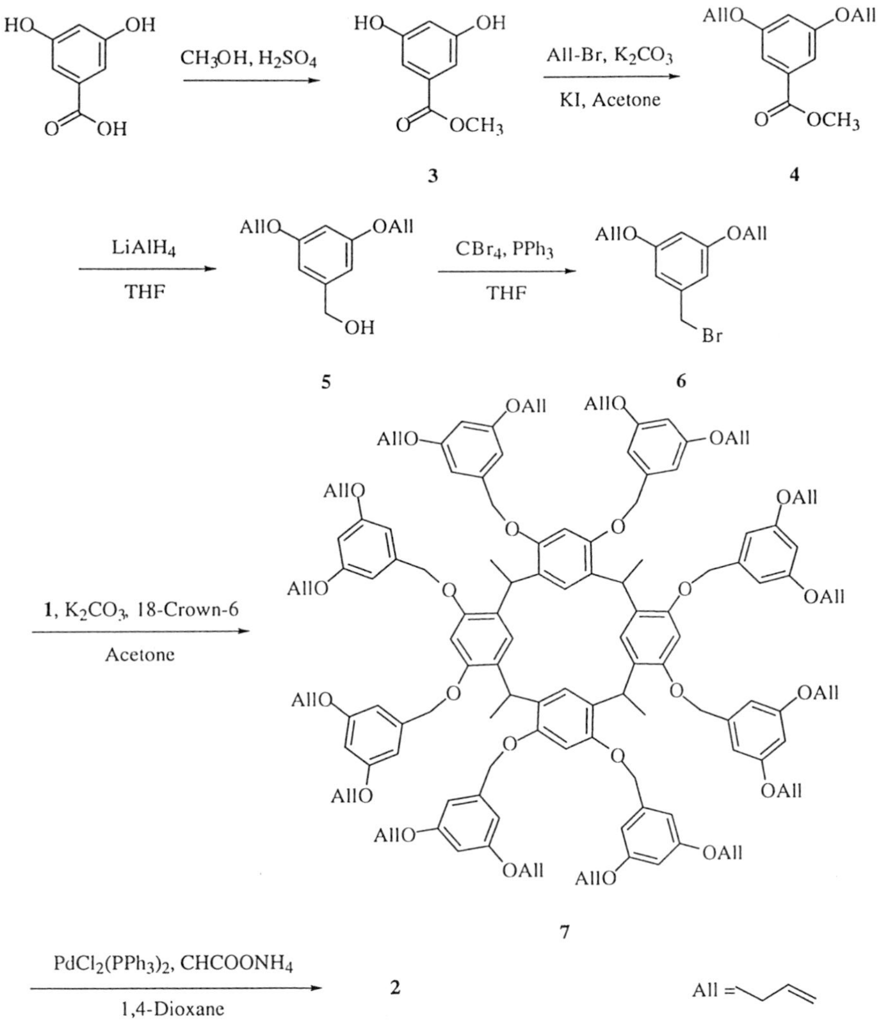

HO OH

O OH

CH_3OH, H_2SO_4

HO OH

O OCH_3

3

All-Br, K_2CO_3

KI, Acetone

AllO OAll

O OCH_3

4

$LiAlH_4$

THF

AllO OAll

OH

5

CBr_4, PPh_3

THF

AllO OAll

Br

6

1, K_2CO_3, 18-Crown-6

Acetone

7

$PdCl_2(PPh_3)_2$, $CHCOONH_4$

1,4-Dioxane

2

All =

groups were protected with allyl ether, then reduction of the ester group gave 3,5-diallyloxybenzyl alcohol (**5**). The alcohol **5** was reacted with CBr_4/PPh_3 to give bromide (**6**) in good yield. The reaction of **6** and **1** was carried out with potassium carbonate and 18-crown-6 in acetone to give allyl protected first-generation dendrimer (**7**). The cleavage of the allyl ether of **7** was done by bis(triphenylphosphine)palladium(II) dichloride and ammonium formate in 1,4-dioxane to give the dendrimer with exterior hydroxyl groups (**2**). Dendrimer **2** showed good solubility in many organic solvents, such as acetone, alcohol and aprotic polar solvents as well as aqueous alkaline solution.

Characterizations. Figure 1 shows ^1H NMR spectra of **1**, **7** and **2** measured in DMSO-d_6. The supectrum of **1** gave well resolved doublet peaks at 1.3 and 4.5 ppm and singlet peaks at 2.5, 6.2, 6.8 and 8.6 ppm, which are attributable to methyl, benzyl, aromatic and phenolic protons, respectively. Peaks at 5.2-5.4 ppm due to allyl groups seen in the spectrum of **7** completely disappeared in the spectrum of **2**. Thus the deprotection could be achieved completely. In the spectrum of **2**, four regions of 1.2-1.6, 4.0-5.2, 5.8-7.2 and 8.9-9.4 ppm should be assigned to methyl, benzyl methylene aromatic and phenolic hydroxy groups, respectively. Because each peak observed is broad or complex, the dendrimer **2** may consist of some conformers arising from the conformation of the calix[4]-resorcinarene core unit.

The formation and purity of dendrimers **7** and **2** were also confirmed by ultraviolet matrix-assisted laser disorption/ionization time-of-flight (MALDI-TOF) mass spectra (Fig-

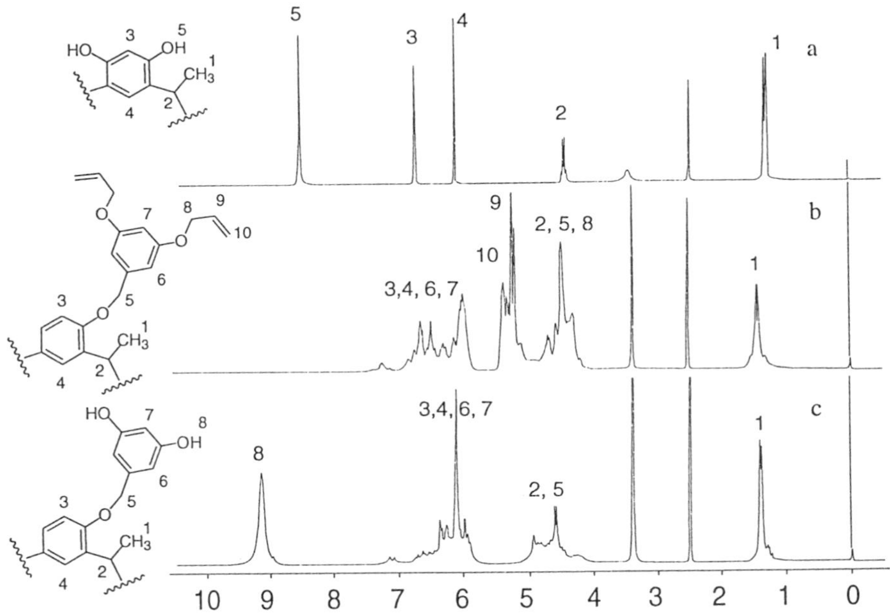

Figure 1. ^1H NMR spectra (270 MHz) of **1** (a), **7** (b) and **2** (c), measured in DMSO-d_6 at room temperature.

ure 2). In the spectrum of **7**, only one peak was found at 2180.8, which is very comparable to the theoretical molecular weight of **7** ([M+Na+]=2185.57). While the spectrum of **2** shows some noises, the main peak at 1545.2 well agree with the calculated molecular weight of **2** ([M+Na+]=1544.55).

Thermal Behavior. Figure 3 shows TG and DSC curves of **2**. The 10% weight loss temperature was 332 °C under nitrogen atmosphere. Thus the dendrimer **2** has high thermal stability. The DSC curve shows glass transition temperature (Tg) at 133 °C. In addition, no melting point was observed up to 300 °C in the DSC curve. This indicates **2** is amorphous which is preferable for film formation.

Film Formation and Dissolution Behavior. In order to study the formation of film and the dissolution behavior of films of **1** and **2** toward an alkaline solution, **1** was dissolved at 25 wt% in 2-methoxyethanol at room temperature, then spin-coated on Si wafer. The wafer was prebaked (100 °C for 5 min) to remove the residual solvent. The resulting clear and tough transparent film was obtained. Such films were also obtained from **2** by dissolving it in diglyme (30wt %) at room temperature, followed by spin-coating on Si wafer, and then prebaking at 80 °C for 10 min. The films of **1** and **2** dissolved completely in 0.1 % tetramethylammonium hydroxide (TMAH) aqueous solution at room temperature within 5 sec.

Design of Photoresist Material

Photo Acid Generator. Many onium salts, such as diaryliodonium and triarylsulfonium salts, are well known as photoacid generators (*7*). However, there are few onium salts that have an absorption band at wavelength longer than 300 nm (*8*). Recently, diphenyl iodonium 9,10-dimethoxyanthracene-2-sulfonate (DIAS) having an absorption at wavelengths longer than 300 nm was developed and shown to produce 9,10-dimethoxyanthracene-2-sulfonic acid upon irradiation with 365 nm light (*9*). Furthermore, DIAS as a photoacid generator was successfully applied to formulation of a positive photoresist composed of bisphenol A protected with a tert-butoxycarbonyl group and a novolac resist matrix (*10*).

UV spectra of **1** and **2** in acetonitrile (Figure 4a and b, respectively) indicate that **1** and **2** have two strong absorptions at 216 nm and 285 nm due to π-π^* transitions and were almost transparent around 250 nm and above 300 nm. Thus DIAS , which has a strong absorption in the range of 300-450 nm, can be used as the photo acid generator.

Cross-Linker. It is well known that polyfunctional benzylic alcohols act as good cross-linkers for poly(4-hydroxystyrene) (*11*). This acid-catalyzed cross-linking reaction was studied in detail, and the reaction was proposed to proceed via a direct C-alkylation as well as an initial O-alkylation, followed by a subsequent acid-catalyzed rearrangement to the final alkylated product. Furthermore, both a thermal cross-linking and an acid-catalyzed cross-linking process were proposed for this alkylation (*12*). Thus we decided to use 4,4'-methylenebis[2,6-bis(hydroxymethyl)phenol] (MBHP) and 2,6-bis(hydroxymethyl)phenol (BHP) in conjunction with **1** and **2**, respectively, on the basis of its avail-

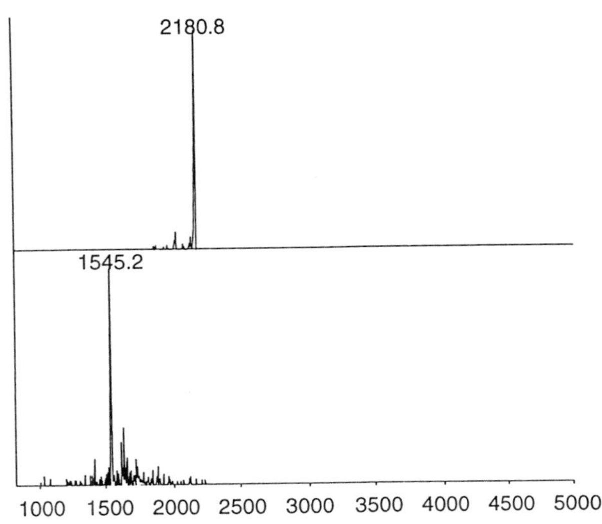

Figure 2. MALDI-TOF mass spectra of **7** (upper) and **2** (lower). 3,5-Dihydroxybenzoic acid (DHBA) was used as a matrix.

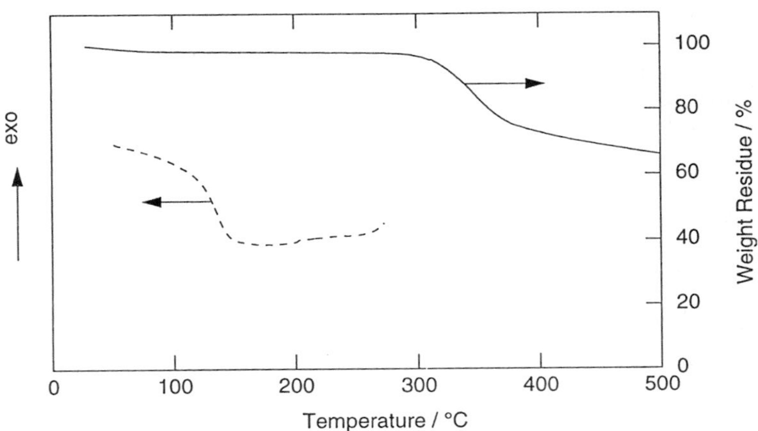

Figure 3. TG (solid line) and DSC (broken line) curves of dendrimer **2**.

ability and sensitivity. The cross-linking will occur as shown in Scheme 2 when BHP is used for **1** or **2**.

Sensitivity. We formulated a resist consisting of **1** (70 wt%), MBHP (20 wt%), and DIAS (10 wt%) in diglyme. The sensitivity curve for a 1-µm-thick film, shown in Figure 5a, indicated that the sensitivity ($D^{0.5}$) and the contrast ($\gamma^{0.5}$) were 2.2 mJ·cm^{-2} and 3.1, respectively, with 365 nm light, after post exposure bake (PEB) treatment at 130 °C for 3 min, followed by development with a 0.1% TMAH aqueous solution at room temperature.

The resist consisting of **2** (70wt%), MBH(20wt%), and DIAS (10wt%) in cyclohexanone was also forumulated. The sensitivity curve for a 1-µm-thick film (Figure 5b) indicates the sensitivity and the contrast were 2.3 mJ·cm^{-2} and 17, respectively, with 365 nm UV light when PEB was carried out at 110 °C for 3 min and a 0.3wt% aqueous TMAH solution was used as a developer.

SEM Images of the Resulting Pattern. In Figure 6a is presented a scanning electron micrograph (SEM) of a contact-printed negative image that was obtained using the resist consisting of **1**, prebaked at 80 °C for 10 min (thickness of 1 µm), exposed to 25 mJ·cm^{-2} of 365 nm UV radiation, post-baked at 95 °C for 3 min, and developed in a negative mode with a 0.1% TMAH aqueous solution at room temperature. The resist is capable of resolving a 1 µm feature when a 1-µm-thick film is used.

Figure 6b shows a SEM photograph of the contact-printed negative image that was obtained using the resist consisting of **2**, prebaked at 80 °C for 10 min (thickness of 1 µm), exposed to 10 mJ·cm^{-2} of 365 nm UV radiation, post baked at 110 °C for 3 min, and developed in a negative mode with a 0.3% TMAH aqueous solution at room temperature. A 3 µm line can be cleanly delineated in a 5 µm thick film.

Conclusions

Calix[4]resorcinare (**1**) and calix[4]resorcinarene dendrimer (**2**) were prepared as new photoresist materials. These compounds acted as excellent photoresist matrix with the photo acid generator, DIAS, and the cross-linker, MBP or BMHP. Both resists showed the high sensitivity and contrast such as 2.2 mJ·cm^{-2} and 3.1 for **1**, and 2.3 mJ·cm^{-2} and 17 for **2**.

Experimental Section

Reagents. Calix[4]resorcinarene (**1**) was prepared by a reported procedure (*5*). Closs-linkers, BHP and MBHP (*8*) were prepared by the reaction of phenol with formaldehyde in water. Photo acid generator, DIAS was prepared by the reaction of diphenyliodonium chloride with sodium 9,10-dimethoxyanthracene-2-sulfonate, which was obtained by the reduction of sodium anthraquinone-2-sulfonate with zinc and aqueous sodium hydroxide solution, followed by the methylation with dimethyl sulfate (*9*). 3,5-Dihydroxybenzoic acid (Merck) and 18-crown-6 (Tokyo Kasei Co.) were commercially available and used

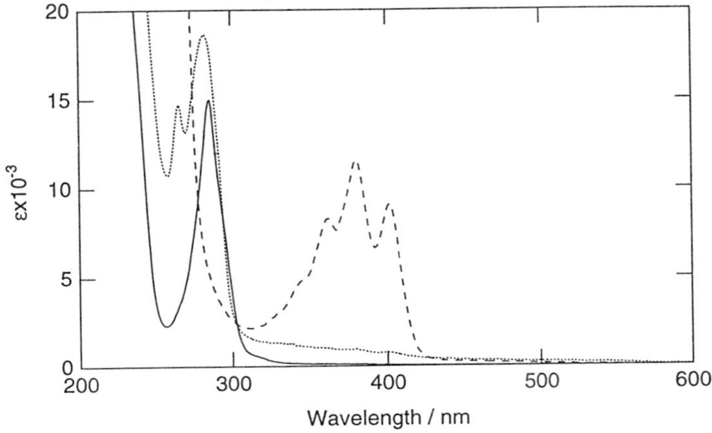

Figure 4. UV spectra of **1** (solid line), **2** (dotted line) and DIAS (broken line) in acetonitrile solution. Sample concentration was 1.0×10^{-5} mol·l^{-1} for **2** and DIAS, 6.5×10^{-5} mol·l^{-1} for **1**. Path length was 1.0 cm.

DIAS

$h\nu$

BHP

$-H_2O$

Δ

Cross-linking

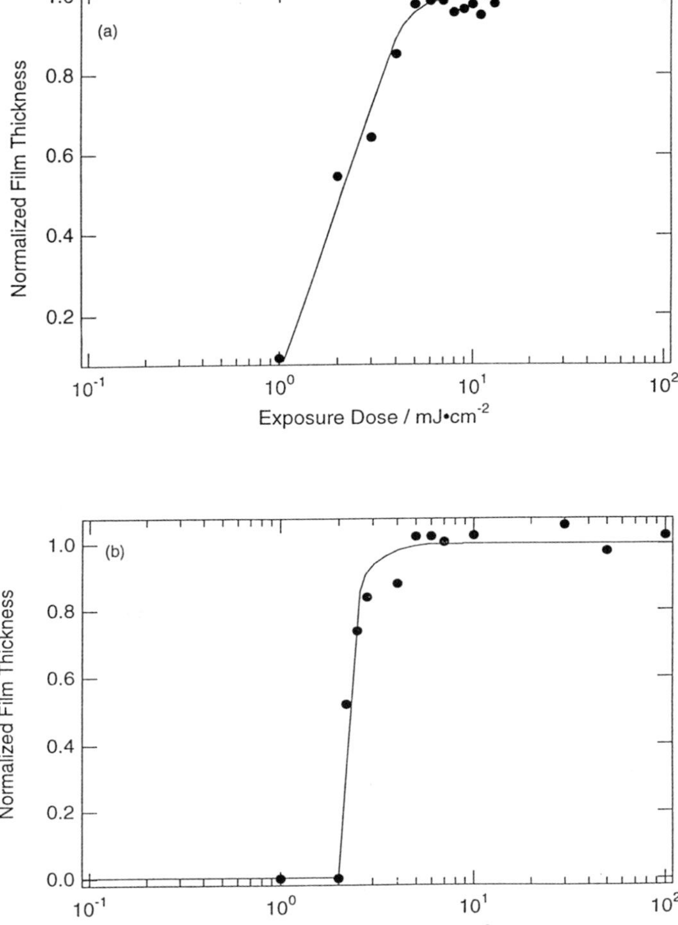

Figure 5. Exposure characteristic curves for the systems of **1**, MBHP, DIAS (a) and of **2**, BHP, DIAS (b).

(a)

(b)

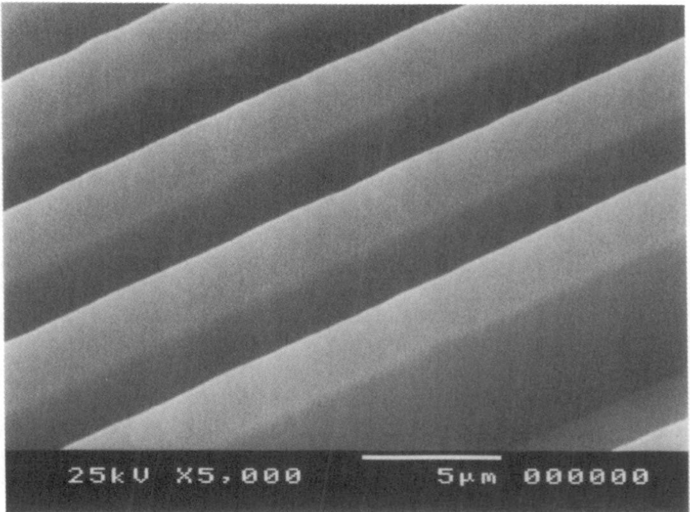

Figure 6. Scanning electron micrographs of the negative image printed in the systems consisting of **1** (a) and **2** (b).

as received. THF and 1,4-dioxane were refluxed over sodium-benzophenone ketyl and distilled just before use.

Methyl 3,5-Diallyloxybenzoate (4). A mixture of methyl 3,5-dihydroxybenzoate (**3**) (7.0 g, 41.7 mmol), 3-bromopropene (10.8 ml, 125 mmol), K_2CO_3 (17.3 g, 125 mmol) and KI (0.21 g, 1.25 mmol) in 150 ml of acetone was stirred at 60 °C for 24 hours under nitrogen atmosphere. The reaction mixture was cooled to room temperature and filtered. The filtrate was evaporated under reduced pressure. The residue was dissolved in diethyl ether, and washed with saturated aqueous Na_2CO_3 solution. The organic layer was dried over $MgSO_4$, filtered and evaporated. Yellow oil was obtained. Yield 9.31 g (37.5 mmol, 90%). IR (KBr): ν (cm^{-1}) =1720 (C=O). ^{1}H-NMR (270 MHz, DMSO-d_6): δ (ppm) 7.07-7.08 (d, 2H, Ar<u>H</u>), 6.82-6.83 (t, 1H, Ar<u>H</u>), 5.97-6.11 (m, 2H, -C<u>H</u>=C), 5.23-5.44 (dd, 4H, -CH=C$\underline{H}_2$), 4.61-4.63 (d, 4H, -OC$\underline{H}_2$-), 3.84 (s, 3H, -CH_3).

3,5-Diallyloxybenzyl Alchohol (5). A mixture of **4** (5.0 g, 20 mmol) and THF was added dropwise to a stirred suspension of lithium aluminium hydride (1.5 g, 40 mmol) in 100ml of THF. The mixture was refluxed for 5 hours under nitrogen atmosphere and then left overnight at room temperature. Water (9 ml) was added dropwise. The mixture was filtered and evaporated under reduced pressure. The residue was dissolved in diethyl ether and washed with saturated aqueous NaCl solution. The organic layer was dried over $MgSO_4$ and evaporated. Colorless oil was obtained. Yield 3.25g (14.77 mmol, 73%). ^{1}H-NMR (270 MHz, DMSO-d_6): δ (ppm) 6.50-6.51 (d, 2H, Ar<u>H</u>), 6.38-6.40 (t, 1H, Ar<u>H</u>), 5.96-6.10 (m, 2H, -C<u>H</u>=C), 5.22-5.42 (dd, 4H, -C=C$\underline{H}_2$), 5.16-5.20 (t, 1H, -CH_2O<u>H</u>), 4.51-4.53 (d, 4H, -OC$\underline{H}_2$-C).

3,5-Diallyloxybenzyl Bromide (6). To a solution of **5** (2.8 g, 13 mmol) and carbon tetrabromide (5.9 g, 18 mmol) in THF was slowly added triphenylphosphine (4.7 g 18 mmol). The mixture was stirred at room temperature for 4 hours under nitrogen atmosphere. The reaction mixture was filtered, evaporated under reduced pressure and the product purified by column chromatography with n-hexane/ethyl acetate (12/1, v/v) as an eluent to give an yellow oil. Yield 2.2 g (7.6 mmol, 60%). ^{1}H-NMR (270 MHz, DMSO-d_6): δ (ppm) 6.63-6.64 (d, 2H, Ar<u>H</u>), d 6.49-6.50 (t, 1H, Ar<u>H</u>), d 5.96-6.08 (m, 2H, -C<u>H</u>=C), d 5.24-5.43 (dd, 4H, -C=C$\underline{H}_2$), d 4.60 (s, 2H, -C$\underline{H}_2$Br), d 4.53-4.55 (d, 4H, -OC$\underline{H}_2$-C), d 4.60 (s, 2H, -C$\underline{H}_2$Br).

Protected Dendrimer (7). A mixture of calix[4]resorcinarene (0.45 g, 0.83 mmol), **5** (2.1 g, 7.3 mmol), 18-crown-6 (0.19 g, 0.72 mmol) and K_2CO_3 (1.3 g, 9.1 mmol) in dried acetone was stirred at 60 °C for 48 hours under nitrogen atmosphere. The reaction mixture was cooled to room temperature and filtered. The filtrate was evaporated under reduced pressure. The residue was dissolved in diethyl ether, and then washed with saturated aqueous K_2CO_3 solution. The organic layer was dried over $MgSO_4$, filtered and evaporated. The product was purified by column chromatography with *n*-hexane/ethyl acetate (5/2, v/v) as eluent to give an yellow oil. Yield 1.3 g (0.61 mmol, 74%).^{1}H-NMR (270 MHz, DMSO-d_6): δ (ppm) 6.13-6.79 (m, 32H, Ar<u>H</u>), 5.98-6.13 (m, 2H, -C<u>H</u>=C), 5.19-5.35 (dd, 4H, -C=C$\underline{H}_2$), 4.66-4.68 (q, 4H, -C<u>H</u>-), 1.41-1.44 (d, 12H, -C$\underline{H}_3$).

Deprotected Dendrimer (2). To a mixture of bis(triphenylphosphine) palladium(II) dichloride (0.15 g, 0.21 mmol) and ammonium formate (2.6 g, 41 mmol) was added a solution of **7** (1.4 g, 0.65 mmol) in 1,4-dioxane (20 ml). The reaction mixture was stirred at 110 °C for 24 hours under argon atmosphere. After cooling, 10 wt% aqueous NaOH solution (80 ml) was added, and the resulting aqueous solution was washed with diethyl ether. The aqueous layer was acidified with concentrated hydrochloric acid to pH 3 and then extracted with ethyl acetate. The organic layer was dried over $MgSO_4$, filtered and evaporated under reduced pressure. The residue was dissolved in 5 ml of methanol and the solution was poured into cold water (100 ml). Precipitation was collected and dried *in vacuo*. Recrystallization from water/acetone gave a white powder. Yield 0.76 g (0.50 mmol, 77%)

Photosensitivity. **1** was dissolved in 20wt% diglyme at 40 °C, and to this solution were added MBHP and DIAS. Films spin-coated on silicon wafers were prebaked at 80 °C for 10 min and exposed to a filtered super-high-pressure mercury lamp. Imagewise exposure through a mask was carried out in a contact-printing mode.

Measurement. Infrared spectra were recorded on a HORIBA FT-210 spectrophotometer. ^{1}H-NMR spectra were obtained using a JEOL EX 270 spectrometer. UV-spectra were recorded on a Shimadzu UV-2200. Film thickness was measured with a Dectak 3030 system (Vecco Instruments Inc.). Ultraviolet matrix-assisted laser disorption/ionization time-of-flight (MALDI-TOF) mass spectra were obtained by means of the KRATOS KOMPACT MALDI III mass spectrometer (SHIMADZU CO.) operated in the linear-positive mode. The samples were dissolved in tetrahydrofuran and mixed with 2,5-Dihydroxybenzoic acid as the matrix.

Literature Cited

1. Willson, C. G., In *Introduction to Microlithography 2nd ed.*; Thompson, L. F.; Willson, C. G.; Bowden, M. J. Eds.; ACS: Washington DC, 1994, pp. 139.
2. Allen, R. D.; Rex Chen, K. J.; Gallagher-Wetmore, P. M. *Proc. SPIE* **1995**, *2438*, 250.
3. Fujita, J.; Ohnishi, Y.; Ochiai, Y.; Matsui, S. *Appl. Phys. Lett.* **1995**, *68*, 2438.
4. Gutsche, C. D. *Aldrichimica Acta* **1992**, *28*, 3.
5. Högberg, A. G. S. *J. Org. Chem.* **1980**, *45*, 4498.
6. Hawker, C. J.; Fréchet, J. M. J. *J. Am. Chem. Soc.* **1990**, *112*, 7638.
7. Crivello, J. V. In *UV Curing: Science and Technology*; Pappas, S. P. Ed.; Technology Marketing Corp.: Stanford, 1978; Chapter 2.
8. Wallraf, G. M.; Allen, R. D.; Hinsberg, W. D.; Willson, C. G.; Simpson, L. L.; Webber, S. E.; Sturterant, J. L. *J. Imaging Sci. Technol.* **1992**, *36*, 468.
9. Nitoh, K.; Yamaoka, T.; Umehara, A. Chem. Lett, **1991**, 1869.
10. Nitoh, K.; Yamaoka, T.; Umehara, A. *Polym. Adv. Technol.* **1992**, *3*, 117.
11. Lee, S. M.; Fréchet, J. M. J.; Willson, C. G. *Macromolecules* **1994**, *27*, 5154.
12. Lee, S. M.; Fréchet, J. M. J. *Macromolecules* **1994**, *27*, 5160.

Chapter 19

Calixarene Resists for Nanolithography

Yoshitake Ohnishi[1], Naoko Wamme[2], and Jun-ichi Fujita

Fundamental Research Laboratories, NEC Corporation, 34 Miyukiga-oka, Tsukuba 305, Japan

Calixarenes have been developed as negative electron resists for nanolithography. These cluster-like, or roughly ball-shaped molecules form very flat and hard films by spin-coating. The high resolution of these resists down to several nm is due to the smallness of the molecules and having no entanglements between molecules as is in ordinary chain polymers. As etching resistance of calixarenes is sufficient in plasma-etch processes, nanofabrication of metal or semiconductors is easily carried out by conventional resist processes.

Calixarenes, coined by C. D. Gutsche, are the cyclic oligomers produced by condensation of phenols and formaldehyde (*1*). In other words, they are cyclic phenol resins, but their physical and chemical properties are much different from those of linear phenol resins. Many of the calixarenes are crystalline and generally have poor solubilities in either water or in organic solvents. Their melting points are generally high, while ordinary phenol resins, novolaks, soften below 150°C.

Calixarene chemistry is a comparatively new field. It became popular after the "one-step synthesis" method was established in the late 1970s. Researchers interests are mainly concentrated on the characteristics of these molecules as organic hosts to guest molecules or atoms. However, calixarenes can be regarded as new

[1]Current address: 2-1-14, Daita, Setagaya, Tokyo 155, Japan.
[2]Current address: 4-22-2, Higashiterao, Tsurumi, Yokohama 230, Japan, nee Mita.

prospective candidates for electronic materials because of their excellent physical and chemical properties.

We previously reported on 5,11,17,23,29,35-hexamethyl-37,38,39,40,41,42-hexaacetoxycalix[6]arene (hereafter abbreviated as MC6AOAc) (*2,3*). This compound is soluble in organic solvents such as toluene, xylene and chlorobenzene, but poorly soluble in alcohols and ketones. In solid state, it is stable either chemically or mechanically up to 320°C. This molecule is, roughly speaking, a round molecule having a diameter of about 1 nm. Although there is no entanglement between molecules as is in linear polymers, we verified making calixarene films from solution by spin-coating. This film works as a good insulating film in semiconductor devices (*3,4*). As an electron resist, MC6AOAc showed ultra high resolution and enabled us to fabricate nano-structures with conventional resist processes (*5*).

We have also reported several examples of fabricating nano-scale features. A fine Ge line of 7 nm width was made from 10 nm width MC6AOAc resist pattern (*5*). The smoothness of the resist side wall enables the line width to be narrowed by over-etching. Making of Al lines was also demonstrated (*6*).

Some fabrications of "mesoscopic" devices using calixarene resists were also carried out. We have reported a nano-dot array fabrication on Si substrate having dots 15 nm in diameter and 35 nm pitch, and discussed its resolution limit (*7*). An SEM photo micrograph of a nano-dot array is shown in Figure 1.

Recently, Spector et al. showed a fine results for production of sub-20 nm soft X-ray zone plates (*8*). Sakamoto et al. reported a fabrication of a 30-nm gate-length electrically variable shallow junction MOSFET (*9*). They also reported a fabrication of single electron transistors, very recently (*10*). A part of Si SET (poly-Si: 20 nm thick, electron box: 15 nm width, 20 nm height, junction: 7 nm) is shown in Figure 2.

Synthesis, Identification and Characterization

MC6AOAc was synthesized according to the route shown in Scheme 1 (*2,3*). *p*-Methylcalix[6]arene (MC6A) was synthesized by a one-step method. A slurry of 5.2g (0.05 mol) of *p*-cresol, 2.5g (0.11 mol) of paraformaldehyde and 3mL of 5N KOH (0.015 mol) in 50mL xylene was refluxed for 4 hrs in an inert atmosphere. The precipitate was collected and washed, in succession, with water and ethanol. The yield was 74.3%. It is noteworthy that only hexamer was produced and neither tetramer nor octamer was found. Acetylation was carried out quantitatively using excess acetic anhydride and *p*-methylcalix[6]arene in pyridine at r.t. for 2 hrs. The product was poured into a large excess of water. The precipitated solid was filtered, washed and dried. The dried product (**1a**) was identified as hexaacetate of

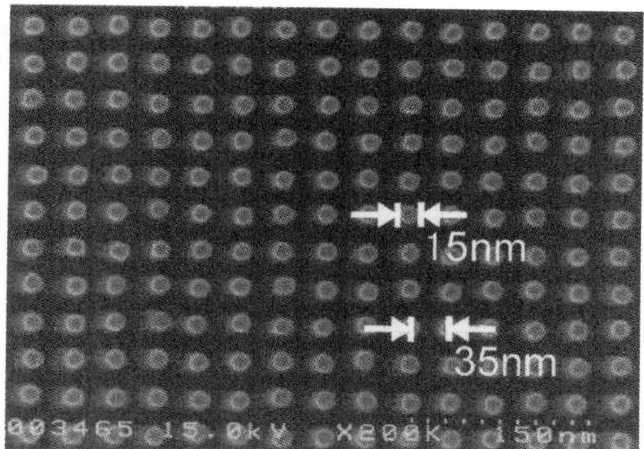

Figure 1. An SEM photo of a dot array pattern made with MC6AOAc. (Reproduced with permission from *Kobunshi*, **1996**, *45*, 666. Copyright 1996 Soc. Polymer Sci. Japan)

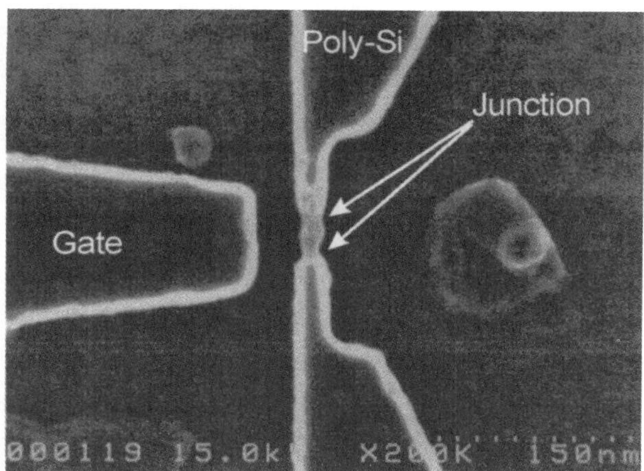

Figure 2. An SEM photo of a part of Si SET.

p-methylcalix[6]arene by elemental analysis, IR, NMR and mass spectra (MS) (m/e = 972).
An acetone solution of **1a** was vigorously stirred for several minutes to result in the appearance of a white precipitate. The precipitate was recovered and dried. This precipitate (**1b**) is a conformational isomer of the initially dissolved MC6AOAc(**1a**), also identified by the measurements described above.

5,11,17,23,29,35-Hexachloromethyl-37,38,39,40,41,42-hexamethoxycalix[6]-arene (hereafter CMC6AOMe) was synthesized also as an electron resist with a higher sensitivity for electron beam. This compound was first reported by Ungaro's group at Parma University (*11*). Synthesis was carried out according to their article as shown in Scheme 2, and the result was exactly the same as they reported (*12*). This compound has a melting point of 273°C, is soluble in organic solvents and also forms good homogeneous films by spin-coating from solution.

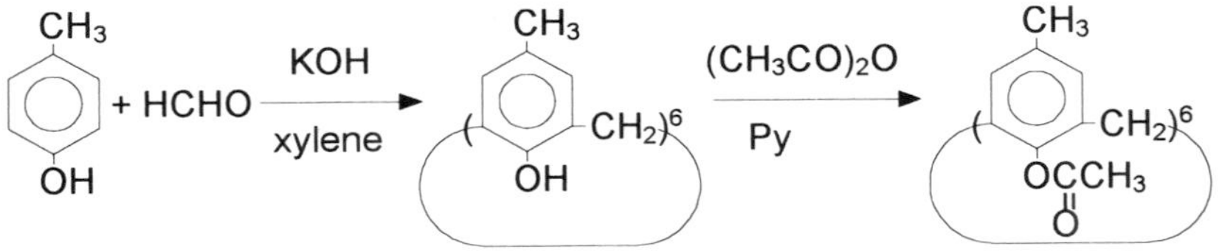

CH_3 + HCHO, KOH / xylene → $(CH_2)_6$ OH; $(CH_3CO)_2O$ / Py → $(CH_2)_6$ $OCCH_3$ (C=O)

Scheme 1. Synthesis of MC6AOAc

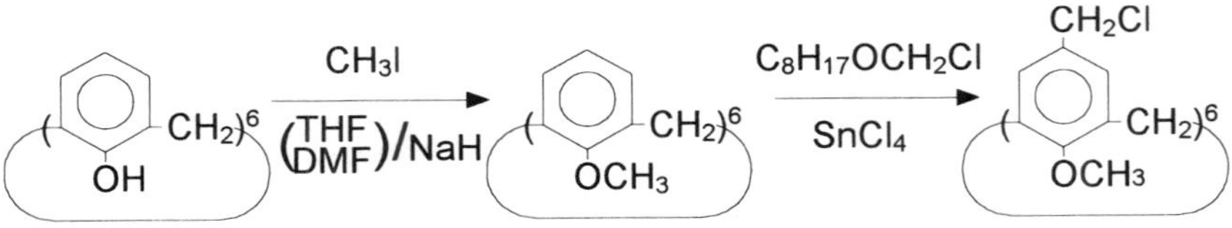

$(CH_2)_6$ OH; CH_3I / (THF/DMF)/NaH → $(CH_2)_6$ OCH_3; $C_8H_{17}OCH_2Cl$ / $SnCl_4$ → CH_2Cl $(CH_2)_6$ OCH_3

Scheme 2. Synthesis of CMC6AOMe

Conformational Problems

We found three conformational isomers of MC6AOAc (**1**) so far. **1a** has a melting point of around 254°C. The DSC measurement (Mettler TA 3000) of **1b** is shown in Figure 3. It is also confirmed by melting point observations under a microscope that **1b** is stable and shows no softening or flow up to 320°C. At around this point the acetoxy group dissociates and decomposition of this compound occurs at about

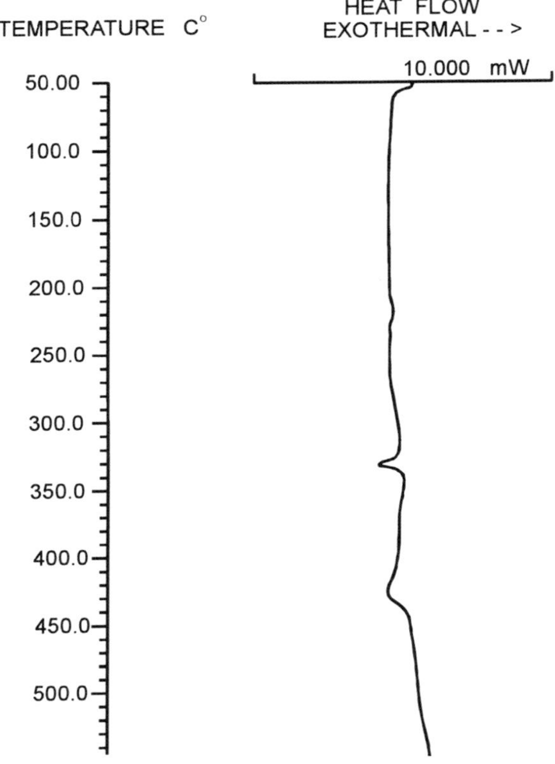

Figure 3. DSC chart of MC6AOAc(**1b**). Heating rate: 10°C/min.

420°C. TG-DSC measurements (Shimazdu TG-40M and Perkin-Elmer DSC-7) proved that Tg occurs at 127.3°C. As there is no softening or thermal instability at the point, it seems to be a slight conformational change in the molecule, and it affects nothing to processing films of **1b**. Isomer **1c** is also found from the first precipitate from an acetone solution of **1a** and has a high melting point of 363°C and a tendency to form microcrystals.

From the consideration of thermal stability and amorphousness, we selected **1b** as the nano-resist. Obtaining a material which is amorphous is one of the essential requirements, because crystalline materials forms microcrystals in films and must be avoided when coating uniform resist films. The wide angle X-ray diffraction measurements of **1a**, **1b** and **1c** were carried out by a Rigaku RU-200 (Cu Kα). This diffractmetry apparently indicates that **1b** is an amorphous substance as shown in Figure 4, while the spectra of **1a** and **1c** showed many sharp peaks indicating they are existing as microcrystalline materials.

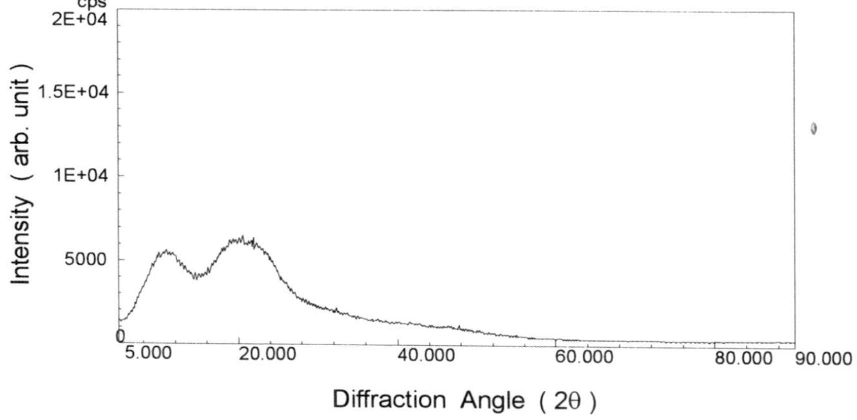

Figure 4. Wide angle X-ray diffraction spectrum of MC6AOAc(**1b**).

NMR Spectra. ^{1}H-NMR, ^{13}C-NMR, COSY and NOESY measurements of MC6AOAc (**1b**) were carried out using a Bruker AMX-400. Figure 5 shows ^{1}H-NMR spectra (400 MHz) in $CDCl_3$ and in toluene-d_8 solutions. The solvent effect of the ^{1}H and ^{13}C chemical shifts, which is caused by the interaction between the carboxyl group of the solute and the benzene ring of the solvent, shows that each oxygen atom in the acetoxy group projects outward from the molecule cylinder. In NOESY spectra, the cross-peak between H in benzene ring and H in acetyl group was observed. We conclude that the "up-down" relation between methyl groups, which project outward from the average plane of the molecule, is alternate in the neighboring unit (*13*). This conformation is expected to enhance the solubility of MC6AOAc in organic solvents because the acetoxy groups projected to the outward from its calix.

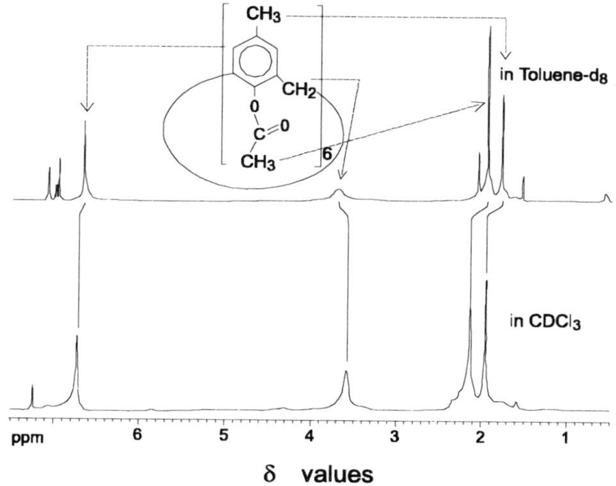

Figure 5. ^{1}H-NMR spectra (400 MHz) of MC6AOAc (**1b**) in $CDCl_3$ and in toluene-d_8.

Mass Spectra. The conformation presented on the basis of the NMR data is also corroborated from MS data of MC6A and MC6AOAc (Figure 6). Spectra were measured using a Shimadzu 9020DF GC/MS, with an ionization voltage of 15 and 70 eV, acceleration 2.0 kV. The mass spectra of MC6A show peaks corresponding not only hexamer but also of the pentamer and tetramer. The ratio among these peaks varied by ionization voltage. On the other hand, the spectra of MC6AOAc show a peak of hexamer and few peaks of which the acetyl group or other small fragments are eliminated from the hexamer, and show no peaks for the pentamer and tetramer. For all mass spectra taken, we found no peaks higher than for the hexamer.

The reason why the mass spectra of MC6A show peaks corresponding pentamer and tetramer is given below. It is known that the most susceptible part under electron beam irradiation in phenol resins is the methylene part (*14*). MC6A has a calix shape, then methylene units reside at the surface of the molecule and are attacked by electron beam. Monomer units in MC6A are eliminated one by one giving peaks corresponding pentamer and tetramer. On the other hand, MC6AOAc has no more calix shape and the acetoxy units at the surface are attacked and fragmented by electron beam.

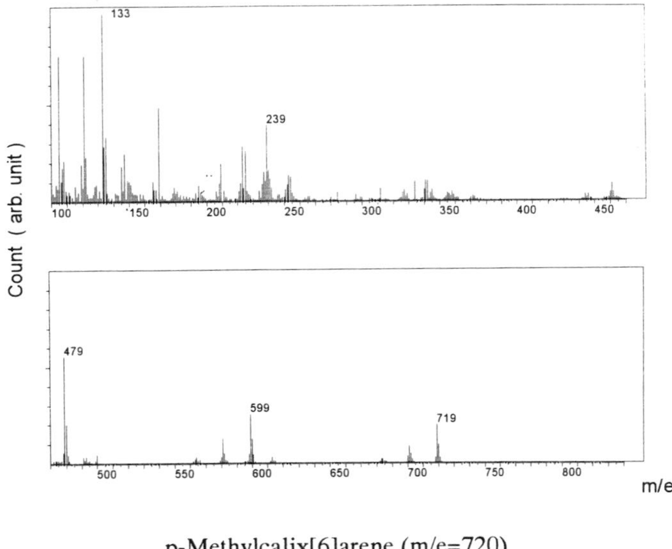

p-Methylcalix[6]arene (m/e=720)

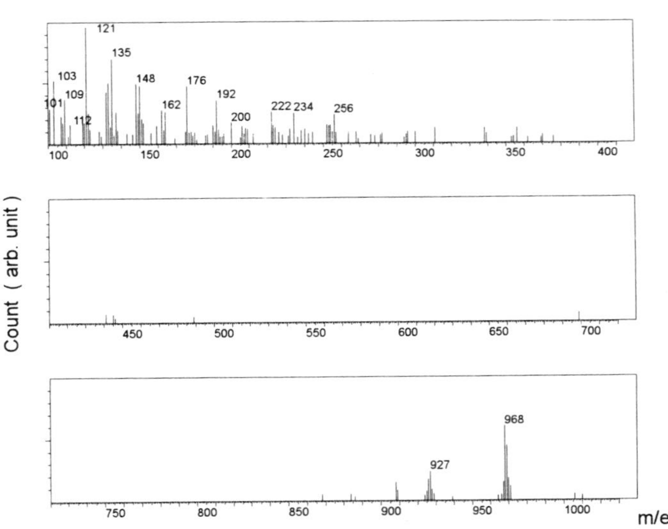

Hexaacetarte of p-methylcalix[6]arene (m/e=972)

Figure 6. Mass spectra of MC6A and MC6AOAc(**1b**). Ionization voltage : 70 eV

Film Properties.

Although there have been several reports on calixarene–polymer composite membrane, no other study on making and characterizing calixarene films has been reported except one which describes monolayers from mercurated calixarenes (*15*).

We found that calixarene films are easily made on substrate from solution by either spin or spray coating. For example, a 10 nm thick uniform film is made from a solution of 0.5 wt.% MC6AOAc in toluene by spin coating at 3000 rpm. An 800 nm thick film is made from 14 wt.% MC6AOAc in a 3:1 mixture of chlorobenzene and chloroform. The spin speed used in this case was 1000 rpm. In every case, after the calixarene film was spun onto a substrate, a post-bake (100°C, 30 min in N_2) was carried out to remove the residue of solvents. Adhesion to Si, Ge, GaAs and Al substrates was good.

Unfortunately, MC6AOAc films thicker than 800 nm tend to crack. This might be caused by high hardness of the film and/or a lack of entanglement between molecules as is in ordinary polymer films. Microhardness measurements showed that the hardness of MC6AOAc films are 5 to 10 GPa (i.e., 150 to 300 Vickers hardness). This value is more than ten times higher than that of ordinary polymers, even harder than a Cr film.

Roughness of the calixarene film surface is about 2 nm (R_{max}, peak to peak value, measured by AFM).

Calixarene films work also as heat stable insulating layers. MC6AOAc film was successfully applied to the local insulation for laser-direct-writing circuit restructuring in large-scale integrated circuits (*4*).

Resist Properties.

MC6AOAc is essentially insensitive to light, but exposure to high energy beams result in cross-linking the resist producing negative tone patterns. Using a JBIL-150 (JEOL) focused ion beam direct writing system (Be^{++} 260 KeV), a practical dose to crosslink the resist is 4 x 10^{13} ions / cm^2, and patterns are developed with ethanol (*3*). Using a JBX-5FE (JEOL) EB writing system (50 KeV, spot size ca. 5 nm), the gel dose is 0.8 mC/cm^2 and a practical dose is 7 mC/cm^2 (*5*). Fine lines down to 10 nm were also made with a line dose of 20 nC/cm, delineated by an SEM S-5000 (Hitachi) with a beam current of 100 pA at 30 kV acceleration voltage (*5*). Also nano-dots were successfully made using the JBX-5FE with a spot dose of $1x10^5$ electrons/dot (*7*). It must be noted that these nano-patterns have very smooth side wall. The side wall roughness is as small as the surface roughness, i.e., ca. 2nm. The etching rate under CF_4 plasma is nearly the same as that of Si, i.e., one fourth to that of poly(methyl methacrylate) (PMMA) (*5*). This durability allows us to transfer the pattern to a semiconductor or metal substrates.

Sensitivity of CMC6AOMe was increased ten times that of MC6AOAc, as shown in Figure 7. The practical dose is about 0.7 mC/cm^2. This figure corresponds to 40 $electrons/nm^2$, which implies that the sensitivity is close to the limit for nano-fabrications considering statistical fluctuation of electron beam. CMC6AOMe also showed high resolution capability and etching durability. This material is recommendable for practical use.

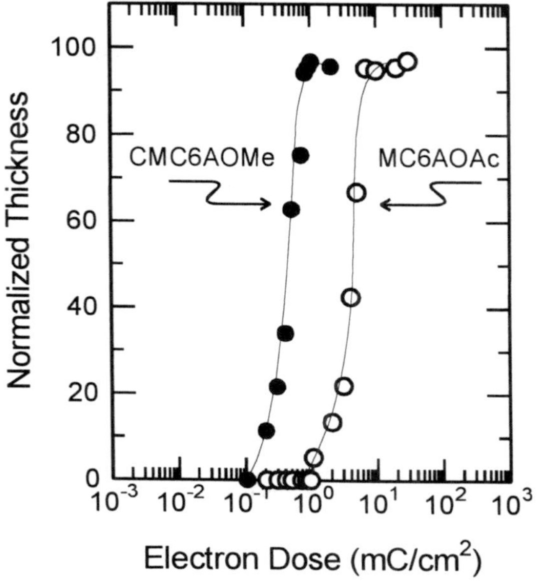

Figure 7. Sensitivity characteristics for MC6AOAc and CMC6AOMe. (Reproduced with permission from ref. 7. Copyright 1996 Am. Vacuum Soc.)

Discussion.

Although PMMA has been known as a positive resist with a high resolution capability, negative resists are more convenient for making nano-dots or fine lines.. As a negative resist based on a cross-linking reaction, the ultimate resolution is limited by the size of the resist molecule itself. It is easily predictable that a small, cluster-like molecule may give a high resolution which ordinary chain polymers cannot do. Some small molecules, for example, some derivatives of C60 can be candidates for nano-resists. Merits of calixarenes are: 1) many derivatives can be easily synthesized by substitution on phenol rings; 2) as the chemical structures are cyclic phenol resins, these materials are expected to be safe in toxicity or pollution; and 3) starting monomers are very inexpensive and high yields are obtained with simple synthetic processes.

The smallness of the resist molecules also provide the flatness of the film surface and smoothness of the side wall, which is crucial in nano-fabrication (*16*).

Sensitivity for calixarenes are comparatively low compared to today's standard electron resists. This is natural because small molecules require many cross-links per unit volume to make a gel state (*17*). To improve the sensitivity, a chloromethyl group was selected because its reaction is a step-wise reaction. Introducing a reactive group which undergoes a chain reaction, such as an epoxy group, or employing a diffusion mechanism in resist reaction was deliberately avoided to maintain the high resolution capability of calixarenes (*18,19*). Additionally, it must be remembered that "10 nm width" is only ten times of the calixarene molecular size, therefore mixed systems of calixarenes and some other reactive compounds were also avoided.

To realize finer fabrications having the dimensions down to few nm, calixarenes are no more appropriate resists as they have around 1nm diameter. It will be required to employ smaller size particles having diameter around few tenth nm, i.e., atoms.

Some Requirements and Consideration on Calixarenes.

Here we consider some requirements on calixarenes as a nano-resist. The first to be considered is solubility, of course, to make films from solutions. Although Shinkai and his coworkers developed many water-soluble calixarenes, most of the calixarenes have poor solubilities either in water or in organic solvents and many of them are highly crystalline (*1*). An attempt to enhance solubility to organic solvents by substituting the upper rim with higher alkyl groups was in vain.

Some calixarenes are, though, highly soluble. For example, 5,11,17,23-tetra-*t*-butyl-25,27-dimethoxy-26,28-dihydroxycalix[4]arene (*20*) is highly soluble in organic solvents, but films thereof show microcrystals in the films, as the calixarene is a highly crystalline substance. To avoid such inhomogeniety, having a material which is amorphous must be considered. We found that among calix[n]arenes, n = 6 (hexamer) are comparatively less crystalline. Tetramers (n = 4) and octamers (n = 8) are generally highly symmetrical and easy to crystallize. For example, in 5,11,17,23,29,35-hexa-*t*-butyl-37,38,39,40,41,42-hexaacetoxycalix[n]arene, n = 4 and 8 are highly crystalline but n = 6 is rather difficult to crystallize even to make a small crystal for X ray crystallographic analysis. Then we have synthesized several derivatives in which the *t*-butyl group was replaced with hydrogen, methyl,

ethyl etc.. The *t*-butyl and methyl derivatives showed high solubilities in organic solvents, while other derivatives were poorly soluble. Although the *t*-butyl derivative has a high melting point of 360°C (*21*), it shows softening and micro-crystallization at about 200°C. The methyl derivative is thermally stable and amorphous, probably because of its conformational structure. As described in the previous part, in this molecule the acetoxy groups are pushed out from its calix. Therefore, we selected the methyl derivative as a prospective candidate for a nano-resist.

The size of a calixarene molecule must be also considered. A calixarene molecule having a bulky substituent is not an appropriate one, as its molecular size becomes much bigger than 1 nm.

Conclusion.

Calixarenes are prospective candidates of materials for microelectronics. As electron resists, their small and round shape gives ultra-high resolution. Synthesis is comparatively easy starting from very inexpensive substances. Films of calixarenes were made with excellent quality, high heat resistivities and flat surfaces. Pattern fabrications are carried out easily with conventional processes. Thus, calixarene resists provide a practical means to fabricate nanos-tructures down to around 10 nm. It would be a great help to resarch and development of "mesoscopic devices".

Acknowledgments.

The authors wish to thank Drs. Y. Ochiai and S. Matsui for discussions, Drs. Y. Nakahara and K. Tateishi for NMR, MS measurements, and Dr. M. Todoki of Toray Research Center for stimulating discussion on thermal properties.

Literature Cited.

1. Gutsche, C. D. *Calixarenes*, Royal Soc. Chem., Cambridge, 1989.
2. Mita, N. *Soluble Calixarene Derivatives and Films Thereof,* U.S. Patent No. 5,143,784, **1992.**
3. Wamme N.; Ohnishi, Y. *Proc. Am. Chem. Soc. PMSE* **1992**, *67*, 451.
4. Seki, Y.; Morishige, Y.; Wamme, N.; Ohnishi, Y.; Kishida, S. *Appl. Phys. Lett.* **1993**, *62*, 3375.
5. Fujita, J.; Ohnishi, Y.; Ochiai, Y,; Matsui, S. *Appl. Phys. Lett.* **1996**, *68*, 1297.
6. Ohnishi, Y.; Fujita, J.; Ochiai, Y.; Matsui, S. *Microelec. Eng.* **1997**, *35*, 117.

7. Fujita, J.; Ohnishi, Y.; Ochiai, Y.; Nomura, E.; Matsui, S. *J.Vac.Sci.Technol.*, **1996**, *B 14*, 4272.
8. Spector, S.J.; Jacobsen, C. J.; Tennant, D. M. *J. Vac. Sci. Technol. B,* to be published.
9. Sakamoto, T.; Kawaura, H.; Baba, T.; Fujita, J.; Ochiai, Y. *J. Vac. Sci. Technol. B* to be published.
10. Sakamoto, T.; Kawaura, H.; Baba, T. *Abst. Silicon Nanoelectronics Workshop* **1997**, 66 Kyoto Japan.; *Appl. Phys. Lett.* to be published.
11. Arduini, A.; Casnati, A.; Pochini, A.; Ungaro, R. *5th Int'l Symposium on Inclusion Phenomena and Molecular Recognition*, **1988**, H 13, Orange Beach, Ala..
12. Almi, M.; Arduini, A.; Casnati, A.; Pochini. A.; Ungaro, R. *Tetrahedron* **1989,** *45*, 2177.
13. Nakahara, Y.; Mita, N.; Ohnishi, Y.; Tateishi, K. *Abst. 203rd ACS Nat'l Mtg.* **1992,** *Anal. Chem. No. 62*, San Francisco, CA.
14. Tanigaki, K.; Iida, Y. *Macromol. Chem., Rapid Commun.* **1986**, *7*, 485.
15. Markowitz, M. A.; Janout, V.; Castner, D. G.; Regen, S. L. *J. Am. Chem. Soc.* **1989**, *111*, 8192.
16. Yoshimura, T.; Shiraishi, H.; Yamamoto, J.; Okazaki, S. *Appl. Phys. Lett.* **1993**, *63*, 764.
17. Charlesby, A. *Atomic Radiation and Polymers*, Pergamon, London, 1960.
18. Ohnishi, Y.; Itoh, M.; Mizuno, K.; Gokan, H.; Fujiwara, S. *J. Vac. Sci. Technol.* **1981**, *19*, 1141.
19. Ohnishi, Y.; Tanigaki, K.; Furuta, A. In *Polymer in Electronics;* Davidson, T., Ed.; ACS Symposium Series No. 242, Am. Chem. Soc., Washington D. C., **1984**, 191.
20. Gutsche, C. D.; Dhawan, B.; Levine, J. A. *Tetrahedron* **1983**, *39*, 409.
21. Gutche, C. D.; Dhawan, B.; No, K. H.; Muthukrishnan, R. *J. Am. Chem. Soc.* **1981**, *103*, 3782.

Chapter 20

Design and Preliminary Studies of Environmentally Enhanced Water-Castable, Water-Developable Positive Tone Resists: Model and Feasibility Studies

Jennifer M. Havard[1], Dario Pasini[1], Jean M. J. Fréchet[1,3], David Medeiros[2], Shintaro Yamada[2], and C. Grant Willson[2]

[1]Department of Chemistry, University of California, Berkeley, CA 94720–1460
[2]Departments of Chemistry and Chemical Engineering, University of Texas, Austin, TX 78712–1962

The design of water soluble positive tone resists has been explored using water-soluble poly(2-isopropenyl-2-oxazoline) as the substrate. The overall chemically amplified design incorporates two successive solubility changes to achieve the desired image tone. The initial change in solubility affecting the entire resist film is achieved during the pre-exposure thermal "bake" step, by addition of an appropriately designed carboxylic acid modifier to the matrix. If a diacid is used, crosslinking occurs leading to insolubilization. Alternatively, a monocarboxylic acid may be used to insolubilize the poly(oxazoline) film through a simple polarity switch. The second change in solubility affecting only those areas exposed to radiation is achieved by the photogeneration of acid within the polymer film. Upon post-exposure baking, the photogenerated acid cleaves the carboxylic acid modifier in a process that restores solubility to the polymer matrix. The preparation of a variety of carboxylic acid modifiers and the demonstration of the individual steps of the overall process has been accomplished confirming the validity of this general approach to fully water-soluble positive-tone resists.

A great deal of interest has recently evolved in the environmental enhancement of lithographic processes. Resist materials have traditionally been cast from organic solvents and developed in aqueous solutions of organic bases, such as tetramethylammonium hydroxide. The replacement of these formulations with new materials cast from and developed in pure water would in many cases have a variety of benefits. Using water to replace traditional casting solvents would help to lower volatile organic compound (VOC) emissions from fabrication centers allowing increased production, while reducing the potential for exposure to these chemicals. The replacement of the currently favored aqueous base developer solutions by pure

[3]Corresponding author.

water could also lead to significant cost savings and facilitate the treatment of waste streams from fabrication facilities. Obviously, the overall performance of the water soluble resists would ultimately determine their suitability for various applications.

This study explores the feasibility of developing positive tone resists that can be cast and developed from water alone. Its purpose is to determine which approaches might be suitable for the design of a positive tone water-soluble resist. Each of the components of an eventual resist are explored separately with the help of model reactions to develop guidelines for the design of an eventual positive-tone water-soluble resist system. Since neither the matrix resin (Novolac) nor the photoactive compound (diazonaphthoquinone) components of classical i-line resists (*1*) are water soluble, resists incorporating chemical amplification (*2,3*) were targeted.

A variety of chemically amplified completely water soluble negative tone resists have recently been described (*4,5*). These materials can be cast from water and are designed to undergo a crosslinking reaction upon irradiation. This crosslinking insolubilizes the film in the irradiated regions, leading to the formation of negative tone images upon postbaking and development. However, it was found that due to their crosslinked design, these materials suffered from swelling problems during development. Although the crosslinked areas are physically prevented from dissolving, they still retain a high affinity for water and thus are easily swelled in aqueous solutions.

The design of a positive tone fully water soluble resist is much more problematic since the irradiated regions must remain soluble in water, and the non-irradiated regions must somehow become insoluble in water. In order to achieve this goal, we have explored the novel concept of a dual solubility switch in which two distinct chemical transformation, one thermal, and one photoinduced, take place at different stages of the imaging process. We describe below experiments designed to validate this concept through the preparation of various components and the use of model reactions.

Results and Discussion

Figure 1 outlines our dual solubility switch approach in which the original water-soluble polymer film undergoes a solubility change during the prebake step which is also used to remove as much of the solvent (water) as possible. This solubility change can be obtained through two different but related routes.

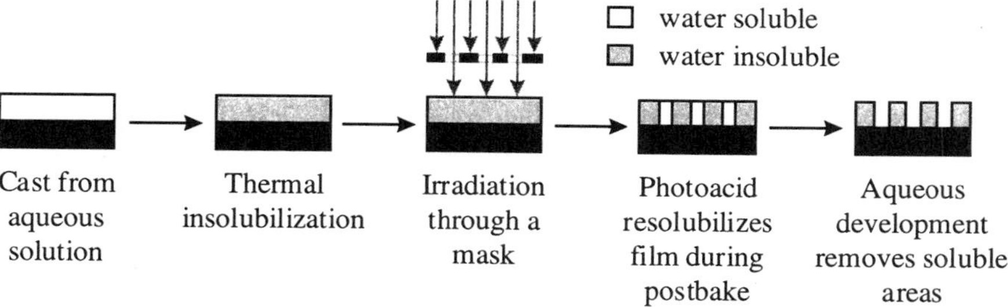

Figure 1: Dual solubility switch design for positive tone water soluble imaging

The first route relies upon crosslinking of the film, making it completely insoluble. The second method makes use of a solubility modifier that reacts with functionalities present in the original film to render it insoluble in water while preserving its solubility in other solvents. In both cases, the film is then irradiated through a mask to produce a latent image. In keeping with the well established processes developed for chemically amplified resists, this latent image consists of photogenerated acid dispersed in the exposed areas of the polymer film. The second solubility switch results from thermal activation of the latent image during the post-exposure bake. The reaction that takes place during this second stage of the imaging process is simply selected for its ability to reverse the insolubilization of the film, returning the irradiated regions to their initial water solubility. Development in pure water washes away these areas to give the desired positive tone images.

The Choice of a Matrix Resin. As was mentioned earlier, the common matrix resins for today's lithography are phenolic resins such as Novolac and poly(4-hydroxystyrene). Though some of our early work had involved simple water soluble alcohols such as poly(vinyl alcohol), schemes for their reversible *in situ* insolubilization were sometimes complicated by irreversible processes or side-reactions. As a result we chose to test the water-soluble linear polymer that is obtained by free-radical polymerization of 2-isopropenyl-2-oxazoline, 1. Monomer 1 can be polymerized through a variety of techniques, as shown in Scheme 1 (*6,7*). Both radical or anionic polymerization conditions lead to a polymer containing pendant oxazoline rings, while a more complex structure is obtained under cationic conditions as both the vinyl and the oxazoline moieties are reactive.

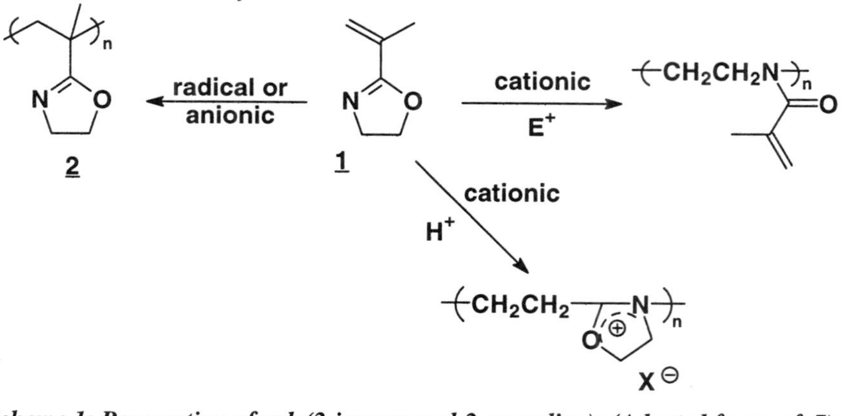

Scheme 1: Preparation of poly(2-isopropenyl-2-oxazoline)s (Adapted from ref. 7)

In practice, the intrinsic reactivity of the oxazoline ring with cationic species may render the poly(2-isopropenyl-2-oxazoline) resin 2 too reactive for use in a fully water-soluble resist. However the ability of oxazoline ring to react with a variety of simple molecules makes polymer 2 well-suited for this study focused on the exploration of new designs and the demonstration of key concepts. We are currently evaluating several other types of reactive matrix resins that may ultimately be better suited than 2 for the design of fully functional positive tone water-soluble resists.

The Thermal Insolubilization Step. Nishikubo *et al* have used the ring-opening modification of oxazolines and other similar cyclic iminoethers to perform crosslinking reactions in solution. In particular, they observed that crosslinking occurred when polymers containing cyclic iminoethers were heated in solution with bifunctional nucleophilic compounds such as diacids or bisthiols (*8*). Subsequent work has also shown that monofunctional nucleophiles may be used to modify oxazoline rings in solution (*9*).

Therefore, it is expected that films containing polymer **2** and a dicarboxylic acid should crosslink upon heating, as shown in Scheme 2. Alternatively, the thermal reaction of **2** with a suitably designed monofunctional carboxylic acid acting as a solubility modifier, should also lead to a product insoluble in water as a result of the modification of both the oxazoline ring and the carboxylic acid function. This type of thermal reaction, which may be done in the pre-bake step, is particularly attractive as it meets our first objective of insolubilization of the entire film prior to irradiation.

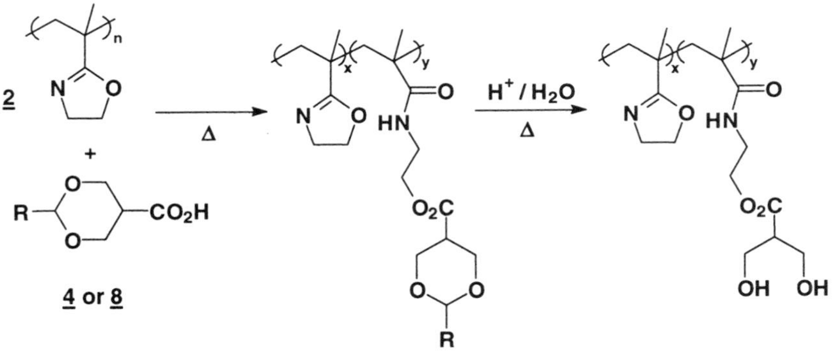

2 + 4 or 8 —Δ→ … —H^+/H_2O, Δ→ …

Scheme 2: The chemistry of the dual solubility switch design

The second part of the dual switch design requires the selective, radiation-induced modification of this film back to a water soluble state. This goal may be achieved through incorporation of an acid cleavable site within the carboxylic acid crosslinker or solubility modifier. Radiation-induced generation of a catalytic amount of acid in selected areas of the resist film would then allow the selective destruction of the insolubilizing linkages.

Acetal units, already used extensively in acid catalyzed chemically amplified resist chemistry (*10,11*), are an attractive choice for these cleavable units. Upon hydrolysis, acetals generate an aldehyde and two alcohol moieties, that may be designed to contribute to an increase in the water solubility of the irradiated fragments.

Design and Preparation of Radiation Cleavable Crosslinkers and Solubility Modifiers. In order to achieve the thermal insolubilization of our water soluble films, as outlined in Figure 1, a series of crosslinkers, compounds **4a-c** and **6**, were initially prepared. Compounds **4a-c** were obtained through the saponification and decarboxylation of intermediate bis(diethylmalonate)acetal products, **3a-3c**, as shown in Scheme 3. Compounds **3a** and **3b** were prepared from the bisaldehyde and diethyl bis(hydroxymethyl) malonate in refluxing toluene. Compound **3c** was formed by the transacetalization of 1,3-bis(ethoxymethyl)benzene (*15*) with diethyl bis(hydroxymethyl)malonate to give **3c**. Saponification with KOH and subsequent decarboxylation in refluxing pyridine yielded products **4a-c**.

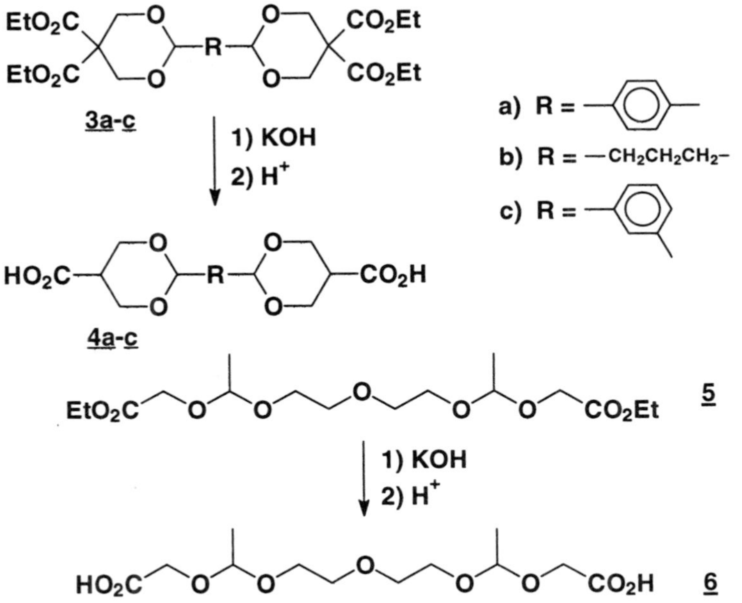

Scheme 3: Preparation of crosslinkers 4a-c and 6

These materials (**4a-c**) were found to be soluble in lower alcohols, but had little aqueous solubility. A structurally related compound, **6**, designed to have improved water solubility, was then prepared as shown in Scheme 3. Ethylene glycol divinyl ether was condensed with ethyl glycolate using pyridinium *p*-toluenesulfonate to give **5**. This intermediate was then saponified with KOH to give the final water-soluble product, **6**. Unfortunately the thermal stability of compound **6** was not sufficient to allow its use in a resist formulation.

A similar design incorporating a cleavable acetal group as well as a carboxylic acid moiety was used in the preparation of water soluble polarity modifiers. Compounds **8a-c** prepared as outlined in Scheme 4, are completely soluble in alcohols, but only **8a** has sufficient water solubility to allow its use in our original resist design.

Small Molecule Model Compounds. Model reactions were performed to test the effectiveness of these solubility modifiers. The reaction of compounds **8a-c** with 2-ethyl-2-oxazoline in refluxing toluene led to **9a-c**. After isolation and characterization, both **9a** and **9b** were found to be water soluble while **9c** was insoluble in water. This suggested that the insolubilization step could be applied if compound **8c** or a derivative thereof were to be used as the solubility modifier.

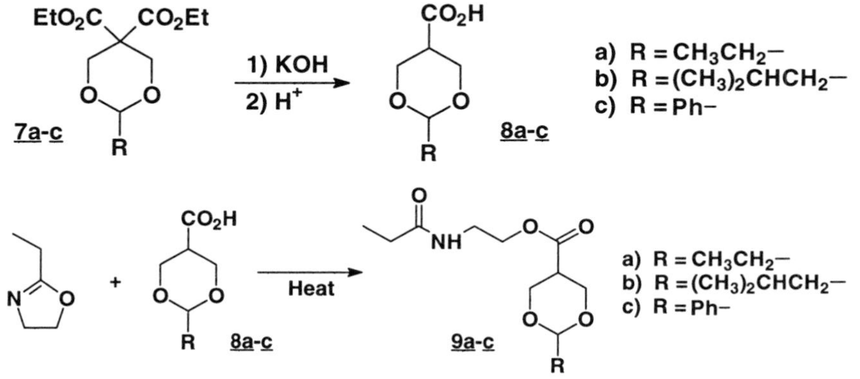

1) KOH
2) H^+

a) $R = CH_3CH_2-$
b) $R = (CH_3)_2CHCH_2-$
c) $R = Ph-$

Heat

a) $R = CH_3CH_2-$
b) $R = (CH_3)_2CHCH_2-$
c) $R = Ph-$

Scheme 4: Preparation of of solubility modifiers 8a-c and model compounds 9a-c

Initial Film Studies. Resist films containing **8c** with **2** in *n*-butanol were spin-coated onto silicon wafers and baked at 100-120°C for varying amounts of time. By varying the amount of **8c** relative to the number of pendant heterocyclic rings we were able to determine the correct amount of reagent needed to bring about the desired solubility change upon baking. Therefore, it is necessary to incorporate 30 mole % of **8c** (relative to oxazoline rings) with a prebake of 3 minutes at 120°C to render the resist film insoluble in water.

The chemical reaction taking place in the film during the prebake step (the first step of Scheme 2, where R=Ph), was monitored using infrared (IR) spectroscopy. While a film containing only the poly(2-isopropenyl-2-oxazoline) matrix resin and no carboxylic acid shows a band at 1653 cm^{-1}, indicative of C=N stretching in the oxazoline ring, a film containing 30 mole % **8c** baked at 120°C for 5 minutes shows clearly that the expected ring-opening reaction has taken place. A new band at 1753 cm^{-1} characteristic of ester C=O stretching is observed, as is a new N-H stretch at 3418 cm^{-1}. Additional confirmation that the reaction has taken place is obtained through a comparison with the IR spectrum of model compound **9c**, which correlates well with the spectra of the baked resist films.

Wholly Water Soluble Formulations. Having shown the validity of the approach with **8a** and **2**, wholly water soluble resist solutions containing varying amounts of **8a** in a matrix of **2** were tested. These attempts were unsuccessful, and the films did not become insoluble in water upon heating. Therefore a method was developed to render compound **8c** "temporarily" water soluble to allow its use in the spin-coating of a resist film from water solution. This is easily done by transforming compound **8c** into its water-soluble ammonium carboxylate derivative **10** by reaction with ammonia. Thermogravimetric analysis of **10** shows that it reverts easily to **8c** at temperatures above 120°C with concomitant release of gaseous ammonia, as illustrated in Scheme 5. IR analysis of the product formed after this heating confirms that this thermally triggered reversion to free carboxylic acid indeed took place. This finding enables the formulation of a completely water soluble resist solution that will undergo the necessary polarity switch during the prebake.

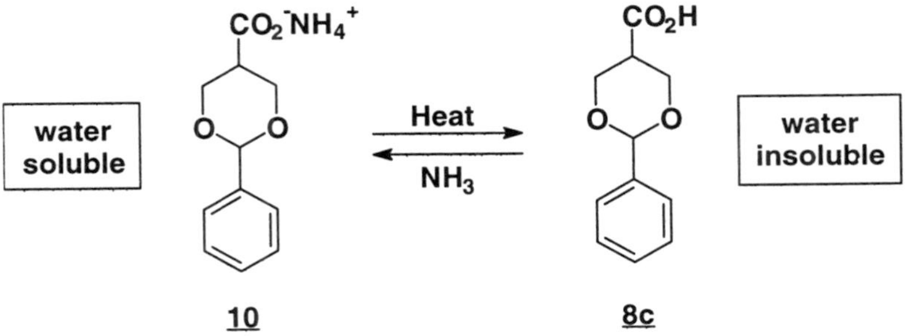

water soluble | $CO_2^-NH_4^+$ | Heat → / ← NH_3 | CO_2H | water insoluble

10   8c

Scheme 5: Aqueous solubilization of compound 8c via ammonium carboxylate 10

It is also possible to solubilize the crosslinkers **4a-c** by forming their ammonium carboxylates as shown for the solubility modifiers in Scheme 5. These water-soluble ionic compounds are also thermally unstable at the temperatures used for postbake and therefore they could also be used in the preparation of completely water soluble resists.

Imaging Studies. Aqueous solutions of poly(2-isopropenyl-2-oxazoline) with 30 mole % of **10** and 5 wt% of the water soluble photoacid generator (PAG) dimethyl 2,4-dihydroxyphenylsulfonium triflate **11** were spin-coated onto silicon wafers. After prebaking at 120°C for 3 minutes, the films were found to have undergone irreversible crosslinking rather than the expected reversible solubility switch. This problem was attributed to a crosslinking side-reaction involving the pendant oxazoline rings of the matrix resin, probably through an acid catalyzed ring-opening process. Since no exposure of the PAG had taken place, several model an control experiments were carried out. These pointed to the PAG itself since its structure includes two activated phenolic moieties that may be acidic enough in aqueous medium to cause crosslinking of the oxazoline rings at the elevated pre-bake temperature. Indeed, a 0.1M solution of the PAG in deionized water was found to have a pH of 3.4.

Therefore, a new water soluble photoacid generator had to be developed. Reaction of 1-methoxy-4-(methylthio)benzene with iodomethane and silver triflate

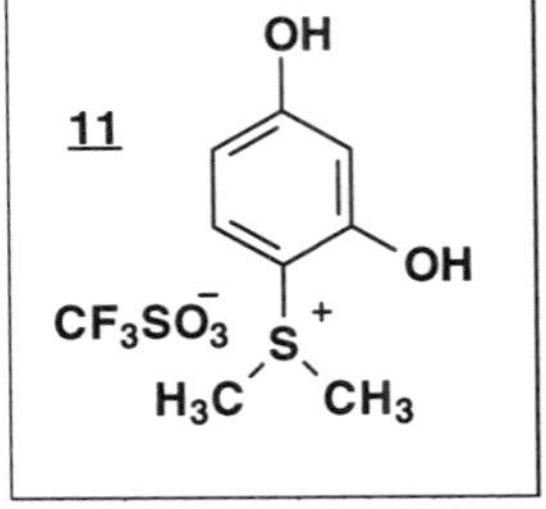

11: OH, OH, $CF_3SO_3^-$, S^+, H_3C, CH_3

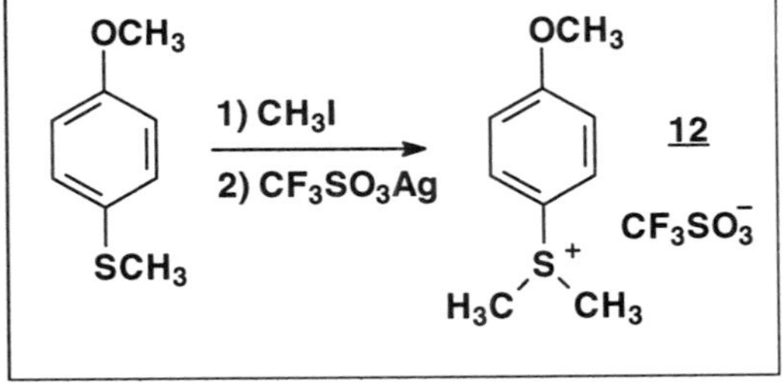

OCH_3, SCH_3 — 1) CH_3I; 2) CF_3SO_3Ag → 12: OCH_3, $CF_3SO_3^-$, S^+, H_3C, CH_3

yielded the new water soluble PAG dimethyl-(4-methoxyphenyl)-sulfonium triflate 12. Imaging of resists incorporating 12 confirmed that this photoacid generator was more suitable than 11 as no crosslinking took place in the pre-exposure bake. However, the desired positive tone images were not obtained after exposure and postbake because the oxazoline rings appeared to be more prone to acid catalyzed crosslinking than had been expected. Indeed our preliminary work with negative tone imaging of 2 by a crosslinking process had suggested that a reasonable window existed between the dose required for *in situ* acid generation and the dose required for crosslinking to afford the negative tone image. This finding suggested that an alternate to matrix resin 2 would be more suitable for the desired positive tone imaging process.

Polymer Model Studies. In order to complete our design studies of the solubility switch approach to positive tone imaging of water soluble resists, a model homopolymer 14 was designed to help ascertain the validity of our acetal cleavage concept, the last step in our overall imaging concept (Scheme 6).

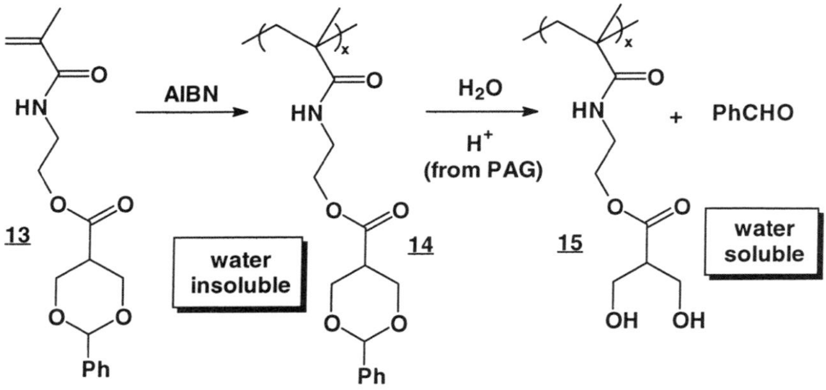

Scheme 6: Synthesis and imaging of homopolymer 14

The homopolymer was obtained by free-radical polymerization of the corresponding monomer 13, itself obtained by reaction of 8c with 2-isopropenyl-2-oxazoline 1. Films of a test resist consisting of a PGMEA solution of polymer 14 and triphenylsulfonium hexafluoroantimonate used as the PAG were spin-coated onto silicon wafers and tested in positive tone imaging experiments. Following a prebake step of 2 minutes at 100°C, the films were exposed to 254nm radiation through a mask and then postbaked at 90°C for 2 minutes. Though no obvious changes occurred in the films after this postbake step, benzaldehyde release was immediately observed if the film was sprayed with water during post exposure bake. This release of benzaldehyde which corresponds to the cleavage of the acetal groups of polymer 14, was confirmed by spectroscopic studies that showed the appearance of the hydroxyl groups of diol polymer 15. In addition, image development using water could be carried out as the exposed regions of the film dissolved away. Thus, this model study demonstrates that the second step of the dual solubility/polarity switch design is indeed feasible reversing the thermal insolubilization of the film and affording a fully water soluble material.

The Origin of Crosslinking in Resists Based on Matrix Resin 2. In an attempt to ascertain the exact cause of the failure of polymer 2 to image satisfactorily in positive-tone mode, despite the demonstration of all of the individual imaging steps, model copolymer 16 shown in Scheme 7 was prepared and used in imaging and control experiments.

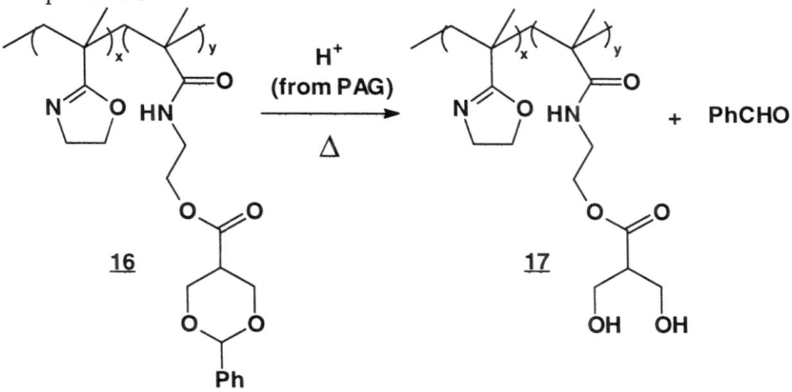

H^+ (from PAG) Δ

16 → 17 + PhCHO

Scheme 7: Attempted imaging/cleavage of copolymer 16 using photogenerated acid.

Once again a resist was prepared from a solution of polymer 16 and water-soluble onium salt 12 used as the PAG. Following irradiation at 254nm and post-exposure bake, no IR evidence for the formation of the desired diol-containing polymer 17 was obtained. Instead, crosslinking of the polymer film occurred suggesting once again the direct reaction of the photogenerated acid with the oxazoline pendant groups of 16. These findings suggest that the nitrogen atom of the oxazoline heterocycle is too basic for use in this positive-tone resist design. Photogenerated acid is trapped by the oxazoline moieties and is therefore unavailable for reaction with the pendant acetal functionalities that need to be cleaved for the solubility/polarity switch to occur. As demonstrated in our work with negative-tone water-soluble resists based on polymer 2 and analogs, the protonated oxazoline rings can undergo crosslinking during the post-exposure bake step, leading to insolubility.

Conclusion

This model and mechanistic study has clearly demonstrated that our novel approach to the design of fully water soluble positive tone imaging materials is valid. The oxazoline matrix resin we selected for our initial demonstration is not suitable for use with photogenerated acids due to its intrinsic basicity that exceeds that of the acetal imaging component of the system. Despite this drawback, the use of the oxazoline matrix resin has allowed us to gain in-depth understanding of the factors that determine the operation of a water soluble resist based on the dual solubility/polarity switch concept. Several other new materials incorporating this fundamental design are now under study in our laboratories and at least one has already demonstrated its ability to afford positive-tone images in a fully aqueous resist formulation. These results will be described in forthcoming publications.

Experimental

Instrumentation and Equipment. ^{1}H- and ^{13}C-NMR spectra were obtained using an IBM-Bruker AF-300 spectrometer. Shifts for ^{1}H-NMR and ^{13}C-NMR spectra were measured relative to the appropriate solvent signal. All shifts are reported in ppm from TMS. Infrared spectra were recorded on a Nicolet IR-44 spectrometer (DTGS detector) using KBr pellets or films cast on NaCl discs. Differential scanning calorimetry (DSC) was performed on a Seiko EXSTAR6000. Thermogravimetric analysis was performed on a Seiko SSC/5200. Molecular weight determinations and molecular weight distributions were measured by gel permeation chromatography using viscometric detection with universal calibration by polystyrene standards. Deep-UV exposures were performed using an Optical Associates Inc. exposure system with a low pressure Hg lamp. The output of the Hg lamp was filtered through a 254 nm narrow bandwidth filter (Oriel).

Materials. Terephthaldicarboxaldehyde (Aldrich), diethyl bis(hydroxymethyl) malonate (Aldrich), *p*-toluenesulfonic acid (Aldrich), potassium hydroxide (85 wt%, Fisher), pyridine (Fisher), glutaric dialdehyde (50wt% aqueous solution, Aldrich), ethylene glycol divinyl ether (Aldrich), ethyl glycolate (Aldrich), isophthalic dicarboxaldehyde (Lancaster), propionaldehyde (Eastman Chemical), isovaleraldehyde (Aldrich), 2-ethyl-2-oxazoline (Acros), ammonia (7M in methanol, Acros), 1-methoxy-4-(methylthio)benzene (Aldrich), silver trifluoromethanesulfonate (Aldrich), and methyl iodide (Aldrich), were used as obtained. 2,2'-Azobisisobutyronitrile (AIBN) (Aldrich) was recrystallized from methanol prior to use.

2-Isopropenyl-2-oxazoline, **1**, and poly(2-isopropenyl-2-oxazoline), **2**, were prepared as described in the literature (*12*). 2-Phenyl-1,3-dioxane-5-yl-carbonic acid, **8c**, was prepared through the modification of a literature procedure (*13*). Dimethyl 2,4-dihydroxyphenylsulfonium triflate, **11**, was prepared as described in the literature (*14*). 1,3-bis(diethoxymethyl)benzene was prepared through a modification of a literature procedure (*15*).

Synthesis of 3a: Terephthaldicarboxaldehyde (6.10g, 45.5mmol), diethyl bis(hydroxymethyl) malonate (20.0g, 90.8mmol), and *p*-toluenesulfonic acid (0.87g, 4.6mmol) were refluxed in toluene with azeotropic removal of water. Upon completion, the mixture was cooled to room temperature and filtered though a plug of silica gel eluted with methylene chloride. After evaporation of the solvents under reduced pressure, a white solid was obtained. The crude material was purified by recrystallization from 1:1 hexane:toluene to give 18.8g of white crystals (77%). ^{1}H NMR ($CDCl_3$): δ 7.42, 5.48, 4.83, 4.32, 4.19, and 1.18 ppm. IR (KBr): 3100-2800, 1737, 1720, 1270, 1160, 1120, 1110, 1060, 940 cm^{-1}.

Synthesis of 3b: Procedure similar to **3a**. Glutaric dialdehyde used in place of terephthalaldehyde. Crude solids were recrystallized from 8:2 toluene:hexane to give white flakes (60%). ^{1}H NMR ($CDCl_3$): δ 4.70, 4.50, 4.30, 4.19, 3.87, 1.60, 1.55, and 1.15 ppm. IR (KBr): 2982, 2877, 1732, 1468, 1370, 1259, 1217, 1153, 1106, 1042, 938, 877, 858 cm^{-1}.

Synthesis of 3c: A mixture of 1,3-bis(diethoxymethyl)benzene (*15*) (3.18g, 11.3mmol), diethyl bis(hydroxymethyl) malonate (4.98g, 22.6mmol), and *p*-toluenesulfonic acid (44mg, 0.23mmol)was heated to 100°C with removal of ethanol via distillation. Upon completion, the mixture was cooled to ambient temperature and filtered through a plug of silica gel eluted with ethyl acetate. After evaporation of the solvents under reduced pressure, a brown syrup was obtained, which crystallized upon sitting to give a clear waxy crystalline solid (4.40g, 88%). ^1H NMR ($CDCl_3$): δ 7.51, 7.42, 7.36, 5.48, 4.84, 4.32, 4.20, and 1.30 ppm. IR (KBr): 3060-2800, 1738, 1460, 1385, 1256, 1109, 711, 634 cm^{-1}.

Synthesis of 4a. The compound **3a** (5.00g, 9.28mmol) was refluxed in ethanol (50mL) with crushed KOH (2.10g, 37.4mmol). After 17 hours, the ethanol was distilled off, and the mixture was then refluxed in pyridine (50mL) for 15 hours. The pyridine was distilled off, and the solids dissolved in water (50mL). The aqueous solution was washed with ethyl acetate and then acidified to pH 1 with dilute HCl. **4a** precipitated out and was collected via filtration and rinsed with water to give 2.07g of a powdery white solid (66%). ^1H NMR (DMSO-d_6): δ 12.3-12.2, 7.35, 5.50, 4.30, 3.90, and 2.95 ppm. IR (KBr): 2998, 2878, 2568, 1697, 1383, 1153, 1091, 985, 796 cm^{-1}.

Synthesis of 4b. Procedure similar to **4a**. **3b** used instead of **3a**. A white solid was obtained (36 %). ^1H NMR (DMSO-d_6): δ 12.65, 4.41, 4.15, 3.62, 2.80, and 1.5-1.3 ppm. IR (KBr): 3000, 2868, 2800-2600, 1695, 1406, 1297, 1141, 1095, 1067, 1032, 936, 883, 748 cm^{-1}.

Synthesis of 4c. Procedure similar to **4a**. **3c** used instead of **3a**. A white solid was obtained (47%). ^1H NMR ($CDCl_3$): δ 12.79, 7.49, 7.40, 5.50, 4.31, 3.92, and 3.00 ppm. IR (KBr): 3100-2800, 1693, 1383, 1169, 1150, 1093, 987, 794, 746, 710 cm^{-1}.

Synthesis of 5. Ethylene glycol divinyl ether (0.50g, 3.16mmol), ethyl glycolate (0.66g, 6.34mmol) and pyridinium *p*-toluene sulfonate (0.043g) were dissolved in dry THF (50mL) and the solution was stirred at room temperature overnight. The solvent was removed under reduced pressure and the residue purified by silica gel flash chromatography (eluent: hexanes/EtOAc 5/5) to afford **5** as a colorless oil (0.88g, 76%). ^1H NMR ($CDCl_3$) δ 4.85, 4.20-4.05, 3.74-3.52, 1.32, and 1.22 ppm. IR(KBr): 3000-2800, 1757, 1391, 1205, 1141, 1096, 935, 858, 723, 588 cm^{-1}.

Synthesis of 6. Compound **5** (850mg, 2.32mmol) and KOH (336mg, 6.00mmol) were suspended in 1:1 ethanol:water. The solution was heated at reflux overnight, after which time a greenish precipitate was formed. The solution was filtered to remove solid impurities, the solvent was removed under reduced pressure and the solid residue was then dissolved in H_2O. After careful acidification to pH 5 with dilute HCl, the solvent was removed under reduced pressure and the residue was triturated in ethanol. After stirring overnight, the suspension was filtered and the solid dried under *vacuo* for 18 hours. ^1H NMR (D_2O): δ 4.85, 3.93, 3.74-3.59, and 1.32 ppm. IR (KBr): 3600-3200, 2950-2850, 1598, 1425, 1316, 1138, 957, 715, 610 cm^{-1}.

Synthesis of 7a. Propionaldehyde (4.81g, 0.083mol), diethyl bis(hydroxymethyl)malonate (18.22g, 0.083mol) and *p*-toluenesulfonic acid (0.81g, 0.004mol) were refluxed in toluene (100mL) with azeotropic removal of water. After cooling to room temperature, the solvent was removed under reduced pressure, and the oily residue was filtered through a short plug of silica gel eluted with ethyl acetate. Removal of the solvent under reduced pressure afforded **7a** as a colorless liquid (20.7g, 96%). An analytical sample was further purified by silica gel flash chromatography (eluent: hexanes/EtOAc: 8/2). ^{1}H NMR ($CDCl_3$): δ 4.68, 4.43, 4.30, 4.18, 3.92, 1.60, 1.26, and 0.91 ppm. IR(KBr): 3000-2850, 1731, 1468, 1365, 1256, 1224, 1154, 1051, 930, 865, 698, 647, 550 cm^{-1}.

Synthesis of 7b. Procedure similar to **7a**. Isovaleraldehyde used in place of propionaldehyde. **7a** was afforded as a colorless liquid (4.31g, 64%). ^{1}H NMR ($CDCl_3$): δ 4.80, 4.55, 4.35-4.12, 3.93, 1.78, 1.48, 1.28, and 0.89 ppm. Material used in subsequent reaction (synthesis of **8b**) without further purification.

Synthesis of 8a. A solution of compound **7a** (19.55g, 0.075mol) in ethanol (70mL) was heated under reflux with crushed KOH (9.26g, 0.165mol) for 16 hours. The solvent was then removed under reduced pressure to give a crude solid. Pyridine (40mL) was added to the solid, and this suspension was heated under reflux for another 16 hours, after which time the pyridine was removed by distillation. The remaining solid was dissolved in H_2O and acidified to pH 1 with dilute HCl. The aqueous layer was extracted with ethyl acetate and the combined extracts were dried over $MgSO_4$. After removal of the solvent under reduced pressure, the resulting oily solid was purified by silica gel flash chromatography (eluent: hexanes/EtOAc=8/2) to afford a white solid that was recrystallized from acetone/hexanes to give 2.30g of crystals (23%). ^{1}H NMR (acetone-d_6): δ 4.38, 4.21, 3.72, 2.88, 1.53, and 0.84 ppm. IR (KBr): 3400-2900, 2800-2600, 1688, 1477, 1423, 1406, 1286, 1224, 1154, 1090, 940, 746, 681 cm^{-1}.

Synthesis of 8b. Procedure similar to **8a**. **8b** used instead of **8a**. The crude product was recrystallized from toluene to give white crystals (43%). ^{1}H NMR (DMSO-d_6): δ 10.96, 4.48, 4.10, 3.66, 2.78, 1.65, 1.35, and 0.83 ppm. IR(KBr): 3300-2400, 1684, 1410, 1290, 1230, 1160, 974, 834, 749, 524 cm^{-1}.

Synthesis of 9a. Compound **8a** (79mg, 0.49mmol) and 2-ethyl-2-oxazoline (70mg, 0.71mmol) were dissolved in acetonitrile (20mL) and the solution was heated at reflux for 48 hours. The solvent from the reaction mixture was then removed under reduced pressure, and the solid residue was purified by silica gel flash chromatography (eluent: CH_2Cl_2/MeOH: 95/5) to yield **9a** as a white solid (40mg, 31%). ^{1}H NMR ($CDCl_3$): δ 5.73, 4.38, 4.27, 4.15, 3.75, 3.52, 2.99, 2.20, 1.63, 1.14, and 0.92 ppm. IR (KBr): 3300, 3000-2900, 2875, 1724, 1647, 1553, 1149, 1089 cm^{-1}.

Synthesis of 9b. Procedure similar to **9a**, except **8a** was replaced with **8b** and toluene was used as the solvent. **9b** was obtained as a white solid (37%). ^{1}H NMR ($CDCl_3$): δ 5.97, 4.28, 4.12, 3.72, 3.48, 2.93, 2.17, 1.75, 1.13, and 0.89 ppm. IR(KBr): 3310, 3086, 2957, 1731, 1641, 1552, 1147, 1064, 968, 910, 692, 582 cm^{-1}.

Synthesis of 9c. Procedure similar to **9a**, except **8a** was replaced with **8c** and toluene was used as the solvent. Recrystallization (acetone/hexanes) yielded **9c** as a white solid (76%). ^{1}H NMR ($CDCl_3$): δ 7.45, 7.37, 5.71, 5.44, 4.47, 4.20, 4.02, 3.72, 3.17, 2.23 and 1.17 ppm. IR (KBr): 3320, 3000, 1731, 1647, 1554, 1387, 1189, 1093, 748, 694 cm^{-1}.

Synthesis of 10. Ammonia (1mL) was added to a solution of compound **8c** (0.50g, 2.4mmol) in isopropanol (10mL). The solution was placed in the freezer, allowing **10** to crystallize. The crystals were collected by filtration, rinsed with acetone and dried under vacuo to give 0.43g of white flakes (78%). ^{1}H NMR (DMSO-d_6): δ 7.35, 7.2-6.6, 5.35, 4.19, 3.77, and 2.60 ppm. IR (KBr): 3400-2800, 1550, 1379, 1158, 1085, 986, 786, 740, 697, 612 cm^{-1}.

Synthesis of 12. 1-Methoxy-4-(methylthio)benzene (1.00g, 6.50mmol), and silver trifluoromethanesulfonate (1.67g, 6.50mmol) were dissolved in dry THF (10mL). The mixture was cooled with an ice bath, and methyl iodide (0.937g, 6.60mmol) was added dropwise. After stirring for 30 minutes, the reaction mixture was filtered to remove the yellow silver iodide precipitate. The solvent was removed under reduced pressure and 20 mL of ether was added to the flask. After stirring for 1 hour, the product was filtered off, giving 1.24 g of a white powder after drying under *vacuo* (65 %). ^{1}H NMR (acetone-d_6): δ 8.05, 7.22, 3.91, and 3.41 ppm. IR (KBr): 3024, 1914, 1598, 1505, 1250, 1157, 1032, 837, 643, 573, 517 cm^{-1}. λ_{max} (H_2O): 244nm.

Synthesis of 13. Compound **1** (1.33g, 0.012mol), compound **8c** (2.50g, 0.012mol) and a small amount of a polymerization inhibitor (BHT) were refluxed in toluene (60mL) for 48 hours. After removal of the solvent under reduced pressure, the product was isolated via silica gel flash chromatography (eluent: hexanes/EtOAc 1/1) as a white solid (1.50g, 40%). ^{1}H NMR ($CDCl_3$): δ 7.47, 7.37, 6.14, 5.70, 5.44, 5.37, 4.45, 4.25, 4.00, 3.59, 3.18 and 1.97 ppm. IR (KBr): 3292, 3080, 2982, 2858, 1728, 1654, 1618, 1551, 1383, 1094, 750, 698 cm^{-1}.

Synthesis of 14. A solution of **13** (0.784g, 2.45mmol) and AIBN (8.0mg, 0.049mmol) in 3mL of dry toluene (distilled from CaH_2) was heated at 70°C for 23 hours. The polymer was twice precipitated into ether, collected by filtration and dried under vacuo to yield a white powder (0.647g, 83%). ^{1}H NMR ($CDCl_3$): 7.5-7.2, 6.5-6.1, 5.5-5.4, 4.5-4.3, 4.2-3.8, 3.5-3.0, 2.0-1.6, and 1.2-0.8 ppm. IR (KBr): 3423, 2980, 2870, 1732, 1662, 1523, 1389, 1292, 1193, 1092, 1029, 750, 700 cm^{-1}.

Synthesis of 15. Compound **1** (0.25g, 4.5mmol), the compound **13** (0.48g, 3.0mmol) and AIBN (7.0mg, 0.042mmol) were dissolved in dry THF (1.1g). The solution was heated at 65°C for 48 hours. After removal of the solvent under reduced pressure, the crude solids were dissolved in a mixture of methanol and acetone and twice precipitated into ether. After filtration and drying under *vacuo*, a white solid was obtained (300mg, 41%). ^{1}H NMR (acetone-d_6): δ 7.3, 5.5, 4.5-3.0, and 2.0-0.7 ppm. IR (KBr): 3600-3200, 3000, 1735, 1666, 1384, 1090, 1020, 990, 756, 698 cm^{-1}.

Acknowledgments

Support of this research carried out in part at Cornell University, the University of California, and the University of Texas was provided by the Semiconductor Research Corporation, Sematech, and, in the initial stages, the Office of Naval Research. This support is gratefully acknowledged. In addition JMH thanks AMD for Fellowship support.

Literature Cited

1. MacDonald, S. A.; Willson, C. G.; Fréchet, J. M. J. *Acc. Chem. Res.* **1994**, *27*, pp. 151-158.
2. Fréchet, J. M. J.; Ito, H.; Willson, C. G. *Proc. Microcircuit Eng., Grenoble* **1982**, *82*, p. 260.
3. Lin, Q.; Simpson, L.; Steinhausler, T.; Wilder, M.; Willson, C. G.; Havard, J. M.; Fréchet, J. M. J. *Proc. SPIE, Metrology, Inspection and Process Control for Microlithography X* **1996**, *2725*, pp. 308-318.
4. Vekselman, A. M.; Darling, G. D. In *Advances in Resist Technology and Processing XIII*; Kunz, R. R., Ed.; Proceedings of the SPIE; The International Society for Optical Engineering: Bellingham, WA, 1996, Vol. 2724; pp. 296-307.
5. Havard, J. M.; Fréchet, J. M. J.; Pasini, D.; Mar, B.; Yamada, S.; Medeiros, D.; Willson, C. G. In *Advances in Resist Technology and Processing XIV*; Tarascon-Auriol, R. G., Ed.; Proceedings of the SPIE; The International Society for Optical Engineering: Bellingham, WA, 1997, Vol. 3049; pp. 437-447.
6. Tomalia, D. A.; Thill, B. P.; Fazio, M. J. *Polymer J.* **1980**, *12*, pp. 661-675.
7. Miyamoto, M.; Sano, Y.; Kimura, Y.; Saegusa, T. *Makromol. Chem.* **1986**, *187*, pp. 1807-1817.
8. Nishikubo, T.; Tokairin, A.; Takahashi, M.; Nosaka, W.; Iizawa, T. *J. Polym. Sci. Polym. Chem.* **1985**, *23*, pp. 1805-1817.
9. Nishikubo, T.; Kameyama, A.; Tokai, H. *Polymer J.* **1996**, *28*, pp. 134-138.
10. Hiro, M.; Fréchet, J. M. J. In *"Irradiation of polymers, Fundamentals and Technological Applications"* (R. Clough and S. Shalaby, Editors) ACS Symp. Ser. 620; American Chemical Society: Washington, DC, **1996**; Vol. 620, pp. 381-6. Ibid., *Polym. Prep.,* **1994**, 35, pp. 948-9.
11. Jiang, Y.; Bassett, D. R. In *Polymers for Microelectronics,* ACS Symp. Ser. 537; American Chemical Society: Washington, DC, 1992; Vol. 537, pp. 40-52.
12. Lalk, J. W. *et al*, US Patent 4376861, **1983**.
13. Scriba, G. K. E. *Arch. Pharm. (Weinheim)* **1993**, *326*, pp. 477-481.
14. Fréchet, J.M.J., and Shim, S.Y., US Patent 5,648,186, **1997.**
15. Ehrlichmann, W.; Friedrich, K., *Chem. Ber.* **1961**, *94*, pp. 2217-2220.

Chapter 21

Molecular Design for New Positive Electron-Beam Resists

Yukio Nagasaki

Department of Materials Science and Technology, Science University of Tokyo, Noda 278, Japan

Two types of new polymers were prepared for a positive resist toward an electron-beam (EB) exposure. Poly(silamine) telechelics, possessing alternating organosilyl and amino units in the main chain and sec-amino groups at both ends, were mixed with triisocyanate compound followed by coating on silicon wafer using spinner, submicron thickness of the poly(silamine) crosslinked film was obtained. On electron-beam irradiation of the poly(silamine) films, degradation of the film took place. However, the developed depth was only 30% even at a 1,000 μC cm^{-2} EB-dose and no thickness change occurred at higher doses, indicating that the charged polymer due to the EB-dose may prevent further cleavage of the polymer film. Actually, when $LiClO_4$ was doped into poly(silamine) gel, the film was developed almost completely at 2,000 μC cm^{-2}, supporting the charge-up mechanism. As another positive EB resist, 2-phenylallyl-ended poly(α-methylstyrene) (PAMS) was prepared via an anionic living polymerization technique. The obtained polymer was susceptible against an EB to form active species at the 2-phenylallyl chain end. The formation of the active species initiated a depolymerization of poly(α-methylstyrene) chain. Therefore, PAMS could be one of the candidates for the positive EB resist. The lithographic performance of the PAMS film was estimated by sensitivity characteristic curve. The Dp and the γ of the PAMS film were ca. 420 μC cm^{-2} and 4.2, respectively.

The miniaturization of microelectronic devices has been progressing rapidly for the last twenty years. For example, the accumulation of integrated circuits in an ultra-large-scale-integration (ULSI) progresses four times per every 3 - 4 years. (*1*)

For the fabrication of a further fine pattern, a high performance resist polymer must be developed. The primary demand for the high performance resists is, of course, to improve resolution. One of the important factors to improve resolution is the type of irradiation sources. In other words, a resist having a high sensitivity toward a shorter wavelength radiation (UV → deep UV → X-ray, electron beam(EB)) is required for fabrication of future integrated circuit devices.

One of the other important points to improve the resist performance is to investigate a novel polymeric system which is susceptible to the irradiation sources as stated above. We have recently been studying synthesis of organosilicon-containing polymers for negative EB-resist having high RIE resistivity. (*2*) One of the organosilicon-containing polymers prepared so far, poly[4-bis(trimethylsilyl)methylstyrene] showed high resolution parameter ($\gamma = 4 - 8$) and moderate sensitivity. (*3*) We have started very recently a new molecular design for development of high performance positive EB-resists.(*4*)

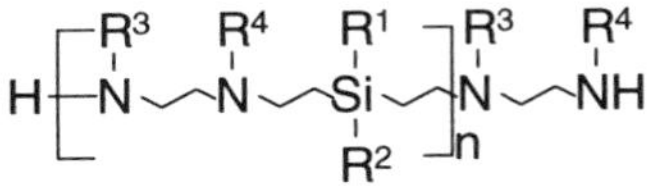

For development of a new high performance resist, new polymers possessing a new structure must be examined. Poly(silamine), (*5*) which was synthesized through anionic polyaddition reactions between divinylsilanes and diamines in the presence of a lithium catalyst, has unique repeating units, *viz.,* alternating organosilicon and amine units in the main chain. Recently, we found that poly(silamine) degrades when exposed to EB. This is attributable to susceptibility of lone pairs at amino groups toward electron-beams. Another important point of poly(silamine) is its high silicon content. Organosilicon-containing polymers(*6,7*) are being considered as one of the candidates for materials with high resistance to O_2 reactive ion etching (O_2 RIE), the mechanism of which is as follows: (*8,9*) When the etching starts, the surface of the Si-containing polymer reacts immediately with oxygen to form a thin SiO_2 layer on the surface. The thin SiO_2 layer thus formed prevents further etching by O_2 RIE. Such organosilicon-containing polymer resists, however, have several problems. For example, the organosilicon group itself does not show a high sensitivity to EB exposure. To improve this sensitivity, the introduction of EB-sensitive groups into the polymer by certain techniques such as copolymerization must be done. Such a design often decreases other characteristics such as the resolution and etching resistance. Thus, we started to study poly(silamine) as a positive EB resist, which wasexpected to provide high sensitivity and high etching resistance.

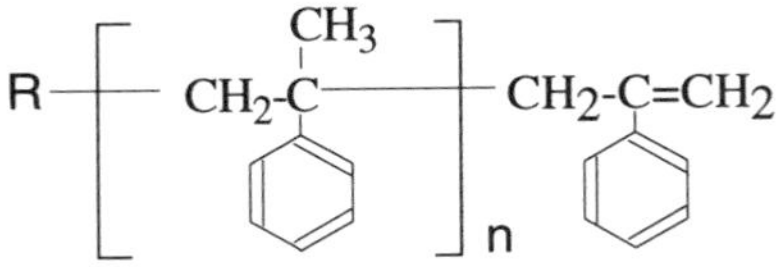

H-[N(R³)–N(R⁴)–Si(R¹)(R²)]ₙ–N(R³)–NH(R⁴)

Poly(silamine)

One of the other approaches to prepare positive resists is to utilize depolymerization tendency. Polymers having a low ceiling temperature (T_c) have been investigated for the positive type resist utilizing their

R–[CH₂–C(CH₃)(Ph)]ₙ–CH₂–C(Ph)=CH₂

2-Phenylallyl-ended poly(α-methylstyrene)

depolymerization characteristics.(*10*) For example, poly(phthalaldehyde) (PPHA), which shows the T_c of ca. -40 °C, can be depolymerized by acid treatment due to the high sensitivity of acetal repeating units toward acid hydrolysis. Thus, the PPHA film can be positive photo-resist when it is coupled with a photo-acid generator. (*11*) Poly(α-methylstyrene) (PMS) is one of the other well-known polymers which undergoes depolymerization at ambient temperature due to the low T_c (ca. 0 °C). (*12*) PMS is suitable as a resist because of its low thermal flow due to its high glass transition temperature (Tg = 130 °C) However, its sensitivity against EB-exposure is not high enough. Ito and his coworkers pointed out easier depolymerization of a cationically synthesized PMS, when the polymer coupled with an acid generator was irradiated with UV. (*13*) They explained that such an easier depolymerization tendency of the cationically synthesized PMS was attributable to the ω-end group. It is known that the ω-end group of the cationically synthesized PMA became not only unsaturated group but also benzyl alcohol and ether by chain transfer reaction and/or termiation reaction. These end groups are susceptible to acid attack to form cationic species at the end of polymer chain. These reactions made the polymer depolymerize much easier that that prepared by the anionic polymerization.

Thus, a structure of polymer end group often drastically change the characteristics of the polymer, e.g. depolymerization tendency in the case of PMS. Our idea was to synthesize uniform size PMS possessing a EB-susceptible double bond at the end of the polymer chain through the anionic polymerization technique. According to this idea, we prepared several types of uniform size PMS with end reactive group such as 2-phenylallyl, 2-tolylallyl and 2-phenyl-3,3-difluoroallyl groups.(*14*) The end-reactive PMS derivatives thus prepared were susceptible against certain stimuli such as heating, acid and base treatments. For example, the onset temperature of the thermal degradation of PMS derivatives with a 2-phenylallyl end group was more than 50 °C lower than that of hydrogen-terminated polymer.(*15*) In this review, two approaches to positive EB resists are described.

Experimental

Synthesis of Poly(silamine). Synthesis of poly(silamine) was described in our previous papers.(*4*). Briefly, in a 100-mL round-bottomed flask equipped with a three-way stopcock, tetrahydrofuran (THF; 12.23 mL), 3,6-diethyl-3,6-diazaoctane (1.35 g; 15 mmol), butyllithium (0.48 mL; 0.75 mmol) and dimethyldivinylsilane (3.7 g; 15 mmol) were placed via syringe. After the mixture was allowed to react at 50 °C for 1 d, 5 mmol of DAO was added to the solution and reacted for one more day. After a small amount of methanol was added to quench the reaction, the low boiling materials were removed by evaporation, then the remaining viscous liquid was washed three times with 100 mL of dimethylformamide (DMF) each to remove the low molecular weight (MW) compounds completely. Yield was ca. 90% The polymer thus obtained was analyzed by gel permeation chromatography (GPC), ^{1}H NMR, differential scanning calorimetry (DSC) and termogravimetric analysis (TGA).

Synthesis of 2-Phenylallyl-ended Poly(α-methylstyrene). Synthesis of poly(α-methylstyrene) having an end double bond was described in our previous papers(15,16). Briefly, to the reactor (100 mL flask equipped with three-way stopcock), THF (6.6 mL), sec-butyllithium (0.34 mL; 0.26 mmol) and α-methylstyrene (1.18 g; 10 mmol) were added in this order via a syringe at -78 °C. The color of the mixture turned brownish red immediately. The reaction was allowed to continue for 20 min. For an end-modification reaction, a 2 - 5 times excess molar amount of the end coupling agent, 2-phenylallyl bromide, was added to the living polymer solution and allowed to react for 20 min. The resulting mixture was analyzed by GPC and then poured into a large excess of methanol. The precipitate was purified by three successive reprecipitations from THF solution into methanol. The obtained polymer was freeze-dried with benzene to remove the solvents employed. Yield of the polymer was ca. 95%.

Resist Processing. Resist processing was described in detail in our previous papers. *(4,16,17)* Briefly, the 10 wt.% polymer solution was spin-coated at 1,500 r.p.m. for 30 sec on a silicon wafer surface. EB was exposed to the obtained polymer film using a JEOL JSM-5200 scanning electron microscope (SEM) equipped with a Tokyo Technology L & S Pattern Generator LSPG1-1S at a probe current of 10 pA and accelerating voltage of 20 kV. A dose that ranged from 1 - 2,000 μC cm^{-2} was employed. The exposed film was developed by soaking in a suitable solvent, then the remaining film thickness was measured with a Tencor ALPHA STEP 300.

The sensitivity and resolution parameter were calculated from the sensitivity characteristic curve as Dp and γ values. The Dp value denotes the dose at which the exposed area is fully developed to the substrate with minimal in the unexposed area.*(18)* The γ value was calculated from the following equation.

$$\gamma = [\log(Dp/Dp^0)]^{-1}$$

where Dp^0 is the dose at which the developer first begins to attack the irradiated film and, as before, is determined by extrapolating the linear portion of the film thickness remaining versus dose curve.

Measurements. GPC measurements were performed on a Shimadzu LC 6A Liquid Chromatograph with an RID-6A IR detector and TSK-Gel G4000HXL8 + G3000HXL8 + G2500HXL8 columns. THF containing 2wt% of triethylemine was used as eluent (1.0 mL min^{-1}) ^{1}H and ^{13}C NMR spectra (^{1}H: 399.65 MHz; ^{13}C: 100.53 MHz) were measured on a JEOL EX400 spectrometer using $CDCl_3$ as a solvent at room temperature. Chemical shifts relative to $CHCl_3$ (^{1}H: δ = 7.26) and $CDCl_3$ (^{13}C: δ = 77.0) were employed. The glass transition temperature of the polymer samples was determined using a differential scanning calorimeter (DSC) (Mettler TA4000 system) at a heating rate of 20 °C/min from -170 to +200 °C. The thermal decomposition of the polymer samples was evaluated using thermal gravimetric analysis (TGA) (Mettler M3) at a heating rate of 10 °C/min from +35 to +500 °C in an argon atmosphere.

Results and Discussion

Poly(silamine) as Positive EB-Resist

Synthesis of Poly(silamine) Derivatives. As we reported previously, (*5*) the reaction between dimethyldivinylsilane (DVS) and 3,6-diazaoctane (DAO) in the presence of a lithium catalyst gives the polyadducts [poly(silamine)], the MW of which is controllable from a few hundred to ten thousands (Scheme 1). When piperazine (PIP) was used as a diamine source instead of DAO, the anionic polyaddition reaction proceeds also without any side reaction. (*6*)

Scheme 1

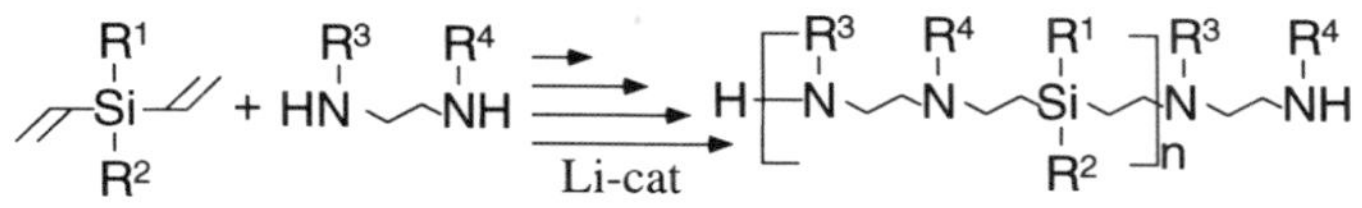

R¹ R³ R⁴ R³ R⁴ R¹ R³ R⁴
Si + HN NH —Li-cat→ H—[N N Si N]$_n$ NH
R² R²

To obtain poly(silamine) having a phenylene linkage in each repeating unit, the anionic polyaddition reactions between 1,4-bis(vinyldimethylsilyl)benzene (VSB) and DAO and/or PIP were carried out. Because the nucleophilic reactivity of VSB was almost the same as that of DVS, (*4*) the polyaddition proceeded smoothly in a homogeneous solution. From the GPC analysis of the polymer obtained by the reaction between VSB and DAO, the Mn of the obtained polymer was 1,600 with a distribution factor (Mw / Mn) of 2.33. The molecular weight distribution (MWD) was close to the value from Flory's theory of polyaddition reactions. (*19*) After the polymer was washed three times with DMF, the Mn of the polymer became 4,300 due to exclusion of the low MW fractions.

To clarify that the obtained polymer was a true polyadduct but not a vinyl polymer, ^{1}H NMR analysis of the polymer was carried out (Fig. 1). Using VSB, DAO and poly(silamine) **I** (see Fig. 2) as reference compounds, the assignments of the signals were carried out and described in the figure. Around 2.4 - 2.5 ppm, multiplet signals assignable to methylene protons adjacent to nitrogen appeared, while a singlet signal was shown at 7.5 ppm originating from the aromatic protons. The ratio of these signals was 12 / 4, indicating that the obtained polymer possesses a complete alternating 1,4-bis(1,1-dimethyl-1-silatrimethylene)phenylene and N,N'-diethylethylenediamine structure.

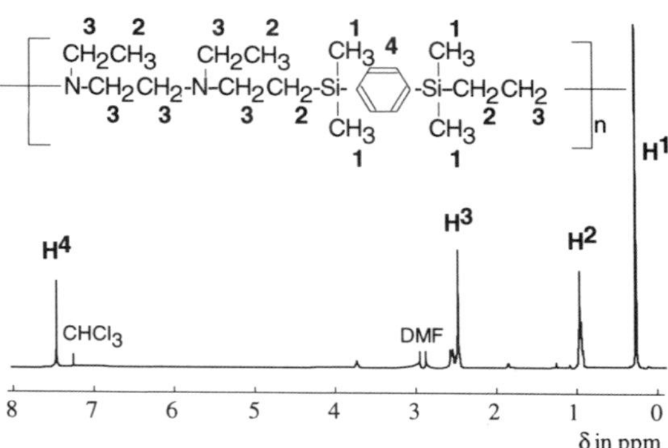

Fig. 1. ^{1}H NMR spectrum of poly(silamine) **II**.(Reproduced from ref. 4 by courtesy of publishers, Hütig and Wepf, Zug, Switzerland)

The reaction between VSB and PIP in THF gave

a polymer having a relatively higher MW. For example, when the reaction was carried out under the conditions of $[VSB]_0$ / $[PIP]_0$ / $[\text{lithium piperide}]_0$ = 1.0 / 1.0 / 0.05 mol L^{-1} in THF at 50 °C for 24h, the Mn of the obtained polymer was 5,200 with a MWD of 2.02. Though lithium piperide is insoluble in THF, the addition reactivity toward vinylsilanes is very fast. The higher reactivity of lithium piperide may be one of the reasons why the obtained poly(silamine) **IV** has a higher MW compared with those of the other poly(silamine)s. The obtained poly(silamine) **IV** was precipitated in acetone. The Mn of the polymer obtained after the precipitation in acetone was 7,200 with a MWD = 1.49, indicating that the low MW fractions in the crude polymer was excluded. The repeating units of the obtained polymer were identified in the same way as mentioned above using ^1H NMR spectroscopy as well as elementary analysis.

On the basis of these results, it is concluded that poly(silamine) derivatives which possess phenylene linkages in the main chain can be obtained by the reaction of VSB with diamine compounds such as DAO and PIP. Fig. 2 summarizes poly(silamine)s which were synthesized so far by the anionic polyaddition reactions between two kinds of divinylsilanes (DVS and VSB) and two kinds of diamines (DAO and PIP).

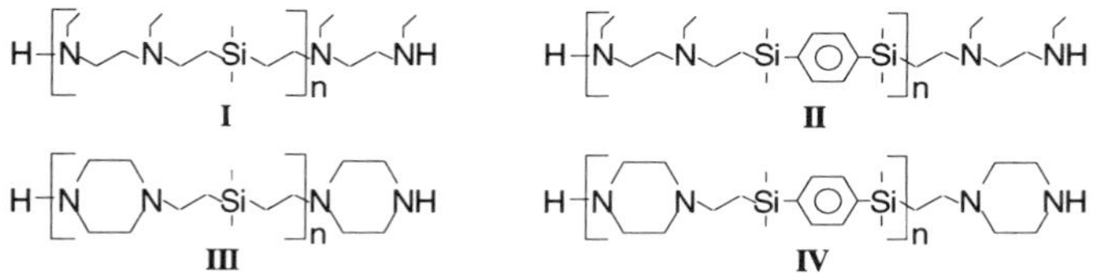

Fig. 2. Poly(silamine) derivatives synthesized by the anionic polyaddition reactions between divinylsilanes (DVS and VSB) and diamines (DAO and PIP)

Characteristics of poly(silamine)s. Thermal analyses of poly(silamine)s **I - IV** were carried out to investigate their features, the several results of which are listed in Table 1. As we already reported, (*20*), poly(silamine) **I** has an extremely flexible chain due to the organosilicon linkage. For example, Tg of **I** was - 88.1 °C, which is one of the most flexible polymers known and is similar to silicone. Poly(silamine) **II**, which possesses a phenylene group in each organosilicon unit instead of DVS, the Tg was ca. -30 °C higher than that of poly(silamine) **I**, which was similar to a polyamine possessing a phenylene linkage in each repeating unit. (*21*). (cf. Tg of poly[(1,4-diethylenephenylene)-*alt*-(N,N'-diethylethylenediamine)] was - 58 °C.) When a DAO unit was replaced by PIP unit in each repeating unit, poly(silamine)s **III** and **IV**, the Tg increased up to -20 and +15 °C, respectively. These two samples were obtained as a white powder after purification by precipitation in acetone. From TGA of the poly(silamine) samples, it was found that the poly(silamine) samples were fairly stable at high temperatures. Actually, the temperatures at which 5 % weight loss occurs for these poly(silamine)s were more than 380 °C.

Table 1. Characteristics of poly(silamine) derivatives

Sample	10^{-3} • Mn[b]	Mw / Mn[b]	Tg / _C[c]	$d_{5\%}$ / °C[d]
I[a]	2.6	1.67	-88.1	380
II[a]	3.0	2.91	-55.3	380
III	5.1	1.80	-27.4	400
IV	7.2	1.49	15.0	400

a) Poly(silamine) samples **I** and **II** were prepared according to our previous papers[6].
b) Mn and Mw/Mn determined from GPC.
c) determined by DSC.
d) Temperature at which 5 % weight loss occurs was determined by TGA.

As shown in Fig. 3, exothermic and endothermic peaks appeared in the DSC profile in the case of poly(silamine) **III**. These exothermic and endothermic signals were attributable to crystallization and melting, respectively. The heat of fusion of poly(silamine) **III** was determined to be 10.0 kJ mol^{-1} using the DSC melting peak. It should be noted that most of polymers show their heat of fusion in the range between 2 - 10 kJ mol^{-1} *(22)*, indicating that poly(silamine) **III** has fairly high crystallinity due to the rigid piperazine units. It was rather surprising for us that poly(silamine) **IV** did not show any crystalline structure under the same experimental conditions. The alternating rigid phenylene and piperazine units may prevent facile crystallization.

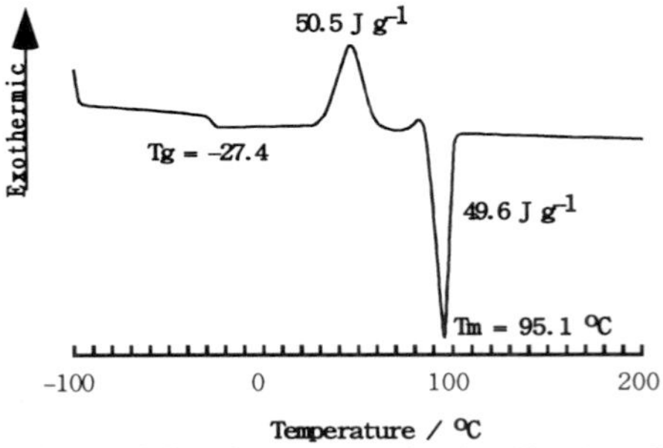

Fig. 3 DSC heating curve of poly(silamine) **III** in the range from -100 °C to 200 °C (5 °C min^{-1}) (The sample was annealed at 200 °C and then cooled at 120 °C min^{-1}) (Reproduced from ref. 4 by courtesy of publishers, Hütig and Wepf, Zug, Switzerland)

Evaluation of poly(silamine)s as EB-resist. Matrix assisted laser desorption ionization - time of flight (MALDI-TOF) mass is known as one of the most powerful analytical instruments for examining polymer characteristics such as MW, end-groups and repeating units. *(23)* One of the features of this technique is that no fragmentation takes place during the electron beam dose due to the minimum energy transfer from the matrix used. For example, the molecular weight of each polymer molecule of linear poly(ethylene oxide) having a primary amino group at one end and a hydroxyl group at the other end can be completely determined by this technique. *(24)* When the poly(silamine) samples were analyzed using the MALDI-TOF mass technique, however, the MW was significantly lower than that expected from the GPC and the ^{1}H NMR results, *(25)* indicating the fragmentations of the poly(silamine) molecules even though the minimum energy was obtained from the matrix. These fragmentation can be explained by Scheme 2. If this kind of cleavage reaction proceeds not only under MALDI-TOF mass but also under the normal EB-dose, poly(silamine) can be anticipated to be a highly sensitive positive-type resist with high O_2 RIE resistance.

Scheme 2 **EB**

$$XCH_2\text{-}CH_2\ddot{N} \longrightarrow XCH_2\text{-}CH_2\overset{+\bullet}{N} \longrightarrow X\dot{C}H_2 \quad CH_2{=}\overset{+}{N}$$

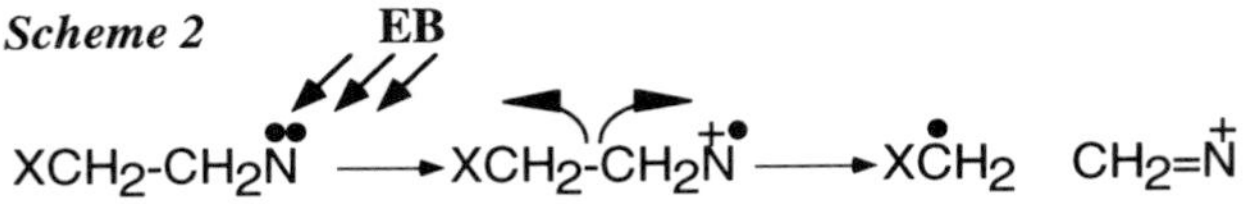

Because poly(silamine) **IV** exhibited the highest Tg among the four samples examined in this study, we investigated the sensitivity of **IV** toward EB-exposure. Poly(silamine) showed a fairly good film forming property by spin-coating from a THF solution. When a 10 wt.% THF solution of poly(silamine) **IV** was spin-coated at 7,000 rpm, a 5,000 Å thick film was obtained. Fig. 4 shows the sensitivity characteristic curve of poly(silamine) **IV**. With increasing EB-exposure, the film started to decrease in thickness 50 μC cm^{-2}, indicating that effective degradation took place due to an EB dose of more than 50 μC cm^{-2}. The remaining film thickness decreased to ca. 40 % with increasing EB dose. At more than 150 μC cm^{-2}, however, the film thickness did not change at all, indicating that some side reactions may take place along with the cleavage reaction.

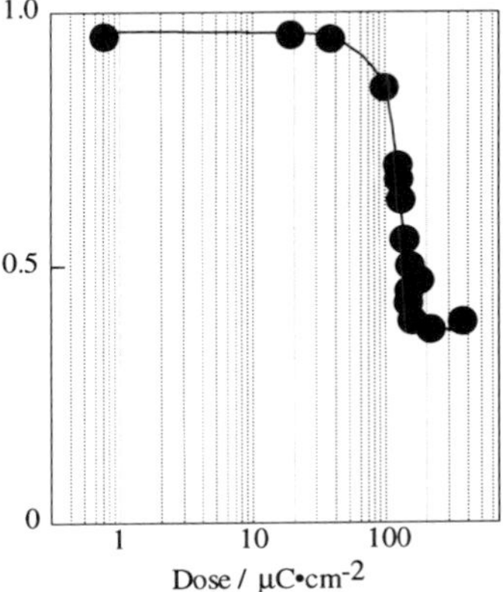

Fig. 4 Sensitivity characteristic curve of poly(silamine) **IV** developed by methanol. (Thickness: 5,000 Å) (Reproduced from ref. 4 by courtesy of publishers, Hütig and Wepf, Zug, Switzerland)

Because poly(silamine) **IV** has only one linkage which can be cleaved by the EB-dose in each repeating unit (Si-CH_2CH_2-N), the sensitivity was not very high. In addition, the film could not be completely developed by the EB exposure due to some side reactions. Therefore, the increase in processability, *viz.*, film forming property, caused a decrease in sensitivity. Poly(silamine) **I** can be expected to show a higher sensitivity to the EB-dose than that of polymer **IV** due to all the N-CH_2CH_2-X repeating units in the main chain. Since poly(silamine) **I**, however, shows a very low Tg, the obtained polymer was an oily material. Thus, it was impossible to investigate the resist characteristics.

Since poly(silamine) **I** was prepared through the polyaddition reaction between DVS and DAO, the end groups of the obtained polymer can be controlled by the initial monomer ratio, *viz*., the initial mole ratio of $[DAO]_0$ / $[DVS]_0$ > 1 gave the polymer having a sec-amino-group at both ends. The obtained telechelic poly(silamine) **I** can be crosslinked by tri-functional isocyanate. For example, when telechelic poly(silamine) **I** was mixed with Colonate HL® (Japan Polyurethane Co. Ltd.) in THF, the mixture gelled within 1 min. This reaction can be applied to the poly(silamine) **I** gel preparation on a Si-wafer. Within one minute after the polymer **I** was mixed with the tri-isocyanate, the mixture was spin-coated on the Si-wafer to form the poly(silamine) **I** thin film gel, the thickness of which was ca. 2,000 - 7,000 Å.

Using the obtained crosslinked poly(silamine) **I** thin film, the characteristics toward EB-exposure were investigated. After the EB-exposure followed by methanol development, the thickness of the film decreased, indicating positive characteristics. It is rather surprising to us that water was a much better developing solvent than methanol, indicating that the fragmented polymer became charged by the EB-dose making it a water soluble material. Figure 5 shows sensitivity characteristic curve of the poly(silamine) **I** gel film which was developed by methanol. As can be seen in the figure, the film thickness started to decrease with an EB-dose less than 10 μC cm^{-2}, indicating higher sensitivity of the poly(silamine) **I** film against EB than that of **IV**. However, the developed thickness attained was only 30% even at a 1,000 μC cm^{-2} EB-dose. In the dose region higher than 1,000 μC cm^{-2}, the remaining film thickness increased again, indicating that the charged polymer due to the EB-dose may prevent further cleavage of the polymer film. Actually, when $LiClO_4$ was doped into poly(silamine) **I** gel, the film was developed almost completely at 2,000 μC cm^{-2}, supporting the charge-up mechanism.

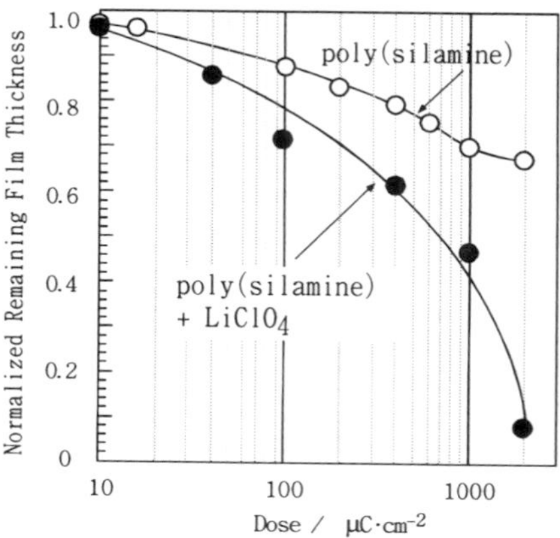

Fig. 5 Sensitivity characteristic curve of closslinked poly(silamine) **I** developed by water (Open Plot: without $LiClO_4$ dope; Close Plots: Doped with $LiClO_4$ ($[N]_0/[Li]_0$= 40); Film thickness = 4,000 Å)

Therefore, developed poly-(silamine) homologues can be one of the promising candidates for positive type lithographic material.

2-Phenylallyl-ended Poly(α-methylstyrene) as Positive EB-Resist.

Preparation of End-Functionalized Poly(α-methylstyrene). It is well known that MS can be polymerized by the anionic polymerization technique without any side reactions such as termination or chain transfer reactions.(*26*) In our previous paper, we revealed that organosilicon-containing α-methystyrene derivative, 4-bis(trimethylsilyl)methyl-α-methylstyrene (SMS), can be polymerized in the same manner as MS to form a living polymer.(*10*) The end-modification of the living polySMS by 2-phenylallyl bromide (PAB) proceeded almost quantitatively.

The anionic polymerization and the end modification of MS by PAB proceeded similarly to that of SMS. Figure 6 shows a GPC curve of the PMS obtained by the end-modification reaction with PAB. The obtained PMS exhibited a very low MWD (Mw/Mn = 1.06) even after the end modification reaction. It should be noted that the end modification agent, PAB possesses both carbon-halogen and carbon-carbon double bonds in the same molecule, which has several possibilities of side reactions such as dimerization of the prepolymers and oligomerizations of the coupling agent with the prepolymers when it treated with the prepolymer. The Mn of the obtained polymer was 19,000.

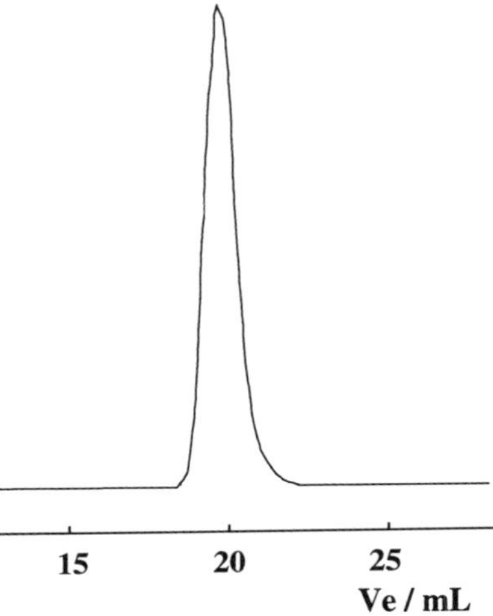

Fig. 6 GPC profile of 2-phenylallyl-ended poly(α-methylstyrene) (PAMS)

The extent of the end-modification was analyzed by ^{1}H NMR as shown in Fig. 7. As can be seen in the figure, the two singlet signals appearing at 5.0 and 5.3 ppm were assignable to two olefinic protons at the end of the polymer. By

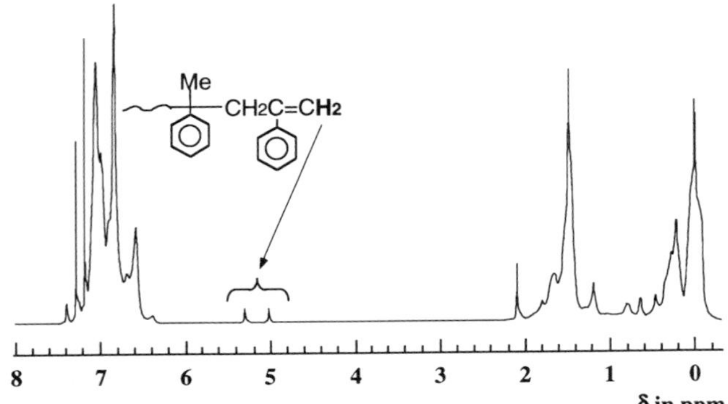

Fig. 7 ^{1}H NMR spectrum of 2-phenylallyl-ended poly(α-methylstyrene)

comparison of these signals with phenyl protons in each repeating unit, 2-phenylallyl group was confirmed to have been introduced almost completely to the PMS end.

Degradation Tests of PAMS. The degradation behavior of 2-phenylallyl-ended PMS, PAMS, was investigated by TGA. The hydrogen-terminated PMS started to degrade at 310 °C as shown in Fig. 8. When the 2-phenylallyl group was introduced into the end of the polymer chain, the onset temperature of the degradation shifted to the lower temperature side. This fact can be explained in the following way. The high temperature induced the cleavage reaction of the end-double bond to form a radical species, which initiated the depolymerization reaction of the poly(α-methylstyrene) derivatives as shown in Scheme 3. The whole TGA profile of PAMS shifted to lower temperature field, indicating quantitative end-modification of the polymer which agreed well with the NMR data.

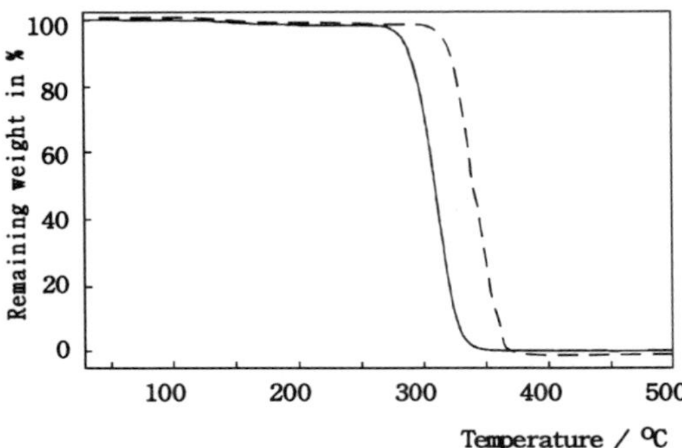

Fig. 8 Thermogravimetric analysis of 2-phenylallyl-ended poly(α-methylstyrene) (solid line) and hydrogen-ended poly(α -methylstyrene) (dotted line)

Scheme 3

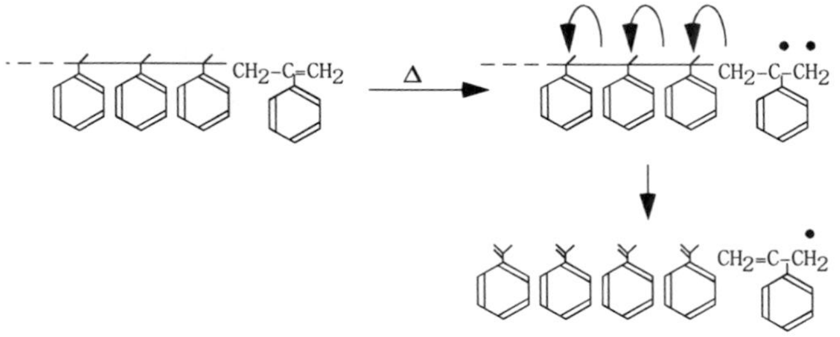

CH_2–C=CH_2 Δ → CH_2–C–CH_2 ↓ CH_2=C–CH_2

The alkaline treatment of PMS with 2-phenylallyl end group also induced a depolymerization in ogranic solvent. When PAMS was mixed with n-BuLi in THF at ambient temperature, almost all of the polymer disappeared as shown in Fig. 9. A low molecular weight oligomeric mixture was observed instead of the starting polymer. This is in sharp contrast to the hydrogen-terminated polymers, which remained almost completely after the reaction with n-BuLi. (*11*) This result indicates that butyllithium adds to the double bond at the end of the polymer chain to form the active carbanion, followed by the depolymerization reactions.

Evaluation as EB resist. As stated above, the 2-phenylallyl end group in PAMS worked effectively to generate active species at the end of polymer chain for both thermal and alkaline treatments to result in the initiation of the depolymerizaiton. If an EB exposure works also for generation of active species at the phenylallyl end group in PAMS, the sensitivity of PMS homologues must be increased by the EB exposure. Figure 10 shows sensitivity characteristic curve for PAMS (Mn = 19,000; Mw/Mn = 1.06) with EB-dose along with that for hydrogen terminated PMS (Mn = 21,000; Mw/Mn = 1.04). As can be seen in the figure, both polymers showed positive character against EB dose. As reported previously, H-terminated PMS showed very low sensitivity against EB exposure. Only 30% of the film decreased at 1,000 μC cm^{-2} dose under our conditions (developed by methanol/methyl isobutyl keton 2/3 vol./vol.). Contrary to H-terminated PMS, PAMS showed much higher sensitivity. Actually, the film was completely developed at 420 μC cm^{-2} under the same conditions. The higher sensitivity of the PAMS film than that of H-terminated PMS may be attributable to the end 2-phenylallyl group in PAMS, *viz.*, the end double bond scission took place much easier than that of the main chain scissions when EB was exposed to the PAMS film. The end double bond scission generated the active species at the end of PAMS chain which induced the depolymerization of PAMS.

A γ value for PAMS was 4.1. A fairly high contrast

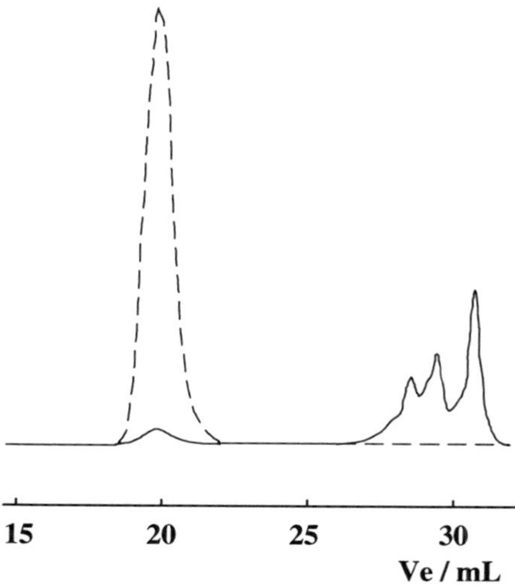

Fig. 9 GPC of PAMS before (dotted line) and after (solid line) the butyllithium treatment

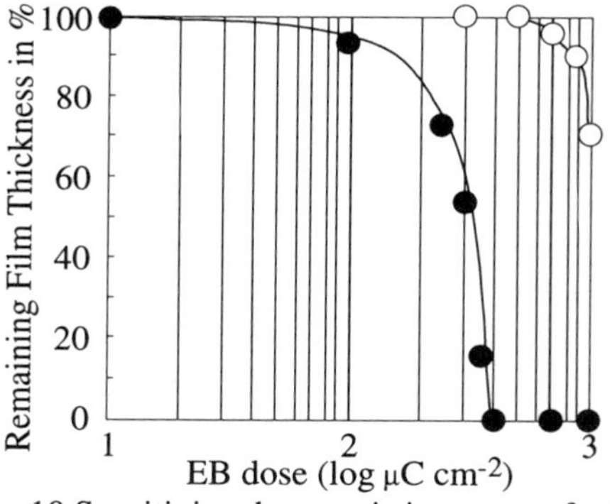

Fig. 10 Sensitivity characteristic curves of 2-phenylallyl-ended poly(α-methylstyrene) (closed circle) and hydrogen-ended poly(α-methylstyrene) (Thickness: 6,000 Å)

parameter (γ) may be attributable to both the low molecular motion in each segment and very narrow molecular weight distribution of the polymer.

Conclusion

Utilizing the anionic polymerization technique, two types of new polymer were prepared. Poly(silamine) was decomposed by the EB-exposure, which can be developed with water. Thus, poly(silamine) is anticipated as new types of positive water developing EB-resist having high etching resistance.

2-Phenylallyl-ended poly(α-methylstyrene) was susceptible not only to heating but also to alkaline treatment to induce degradation to monomer and/or oligomeric mixture. In the same manner, micropatterning can be obtained by the EB exposure. Since the glass transition temperature of PAMS is high, positive EB-resist having high resolution characteristic can be anticipated.

Acknowledgments

The authors would like to express sincere appreciation to *Prof. Masao Kato, Messrs. Hideo Kimura, Noriyuki Yamazaki, Ken-ichiro Deguchi and Naoki Yamazaki,* for carrying out a part of this study. A part of this study was supported financially by *Futaba Foundation.*

References & Comments

1 a) Ogawa, T. *J. Photopolym. Sci. Technol.* **1996**, 9, 379. b) Thompson, L. F. In *Introduction to Microlithography Second Edition*; Thompson, L. F.; Willson C. G.; Bowden, M. J.; Ed., ACS, Washington, DC, **1994**, Chap. 1, pp. 1-17.

2 a) Kato, N.; Nagasaki, Y.; Kato, M.; *Polym. Adv. Tech.* **1990**, 1, 341. b)Nagasaki, Y.; Kato, N.; Yamazaki, N.; Kato, M. In *Silicon-containing Polymers*, Jones, R. G., Ed, The Royal Society of Chemistry, Cambridge **1996**, pp. 44-50. (c) Aoki, H.; Tokuda, T.; Nagasaki, Y.; Kato, M. *J. Polym. Sci., Part A, Chem. Ed.* **1997**, 35, 2827.

3 Kato, N.; Takeda, K.; Nagasaki, Y.; Kato, M. *Ind. Eng. Chem. Res.* **1994**, 33, 417.

4 Nagasaki, Y.; Deguchi, K.; Yamazaki, N.; Kato, M.; *Macromol. Rapid. Commun.* **1996**, 17, 51

5 Nagasaki, Y.; Honzawa, E.; Ksato, M.; Kihara, Y.; Tsuruta, T. *J. Macromol. Sci. -Pure & Appl. Chem.* **1992**, A29, 457.

6 Jones, R. G. *Trends Polym. Sci,* **1992**, 1, 372.

7 Gozdz, A. S. *Polym. Adv. Tech.* **1994**, 5, 70.

8 Sugita, K.; Ueda, N. *Prog. Polym. Sci.* **1992,** 17, 319.

9 Saigo, K.; Watanabe, F. *J. Polym. Sci.*, **1989**, A27, 2611.

10 Wilson, C. G. In *Introduction to Microlithography Second Edition;* Thompson, L. F.; Willson C. G.; Bowden, M. J.; Ed., ACS, Washington, DC, **1994**, Chap. 3, pp. 139-268.

11 Ito, H.; Wilson, C. G. *Polym. Eng, Sci.* **1983,** 23, 1012.

12 Ivin, K. J.; Léonard, J. *Eur. Polym. J.* **1970**, 6, 331.

13 Ito, H.; England, W. P. *Polym. Prep.* **1990**, 31, 427. b) Ito, H.; England, W. P., Ueda, M. *J. Photopolym. Sci. Tech.* **1990**, 3, 219.

14 Nagasaki, Y.; Yamazaki, N.; Kato, M. *Macromol. Rapid Commun.* **1996**, 17, 123.

15 Nagasaki, Y.; Yamazaki, N.; Kato, M. *Polymer* **1996**, 37, 4321.

16 Nagasaki, Y.; Kimura, H.; Kato, M. *J. Photopolym. Sci. Tech.* **1997**, 10, 317

17 Nagasaki, Y.; Yamazaki, N.; Kato, M. *Angew. Makromol. Chem.,* **1997**, 247, 163

18 Thompson, L. F. In *Introduction to Microlithography Second Edition;* Thompson, L. F.; Willson C. G.; Bowden, M. J.; Ed., ACS, Washington, DC, **1994**, Chap. 4, pp. 269-376.

19 Flory, P. J., *Principles of Polymer Chemistry;* Cornell University Press, Ithaca, NY, 1953

20 Nagasaki, Y.; Kazama, K.; Kato, M.; Kataoka, K.; Tsuruta, T.; *Macromol. Symp.* **1996**, 109, 27.

21 Kataoka, K.; Koyo, H.; Tsuruta, T.; *Macromolecules*, **1995**, 28, 3336.

22 Brandrup, J., Immergut, E. H., *Polymer Handbook;* Third Edition, John Wiley & Sons, New York, 1989

23 Creel, H. S. *Trends Polym. Sci.* **1993**, 1, 336.

24 Kim, Y. J.; Nagasaki, Y.; Kataoka, K.; Kato, M.; Yokoyama, M.; Okano, T.; Sakurai, Y. *Polym. Bull.* **1994**, 33, 1.

25 The MW determined from GPC was calibrated by the polystyrene standards. To avid a possible adsorption of poly(silamine) on the surface of the gel in GPC column, 2 wt% of triethylamine was added to the eluent (THF). The MW determined from the GPC agreed well with that from 1H NMR, in which the MW was determined from the ratio of N-CH2 in the repeating units vs. that at the both ends.

26 Müller, A. H. E.; Fontanille, M., In *Comprehensive Polymer Science;* Vol. 3, Pergamon Press, Oxford, **1988**, pp. 387-433

LITHOGRAPHIC MATERIALS AND PROCESSES

Elsa Reichmanis

While VLSI design and fabrication continue to demand increasingly smaller and more precise device features, existing technologies continue to be extensively utilized for sub-critical dimension delineation. Within this scenario, full understanding of the materials issues is required in order to achieve maximum performance. This understanding, in turn, will allow the design and development of more effective materials technologies for all lithographic options. In addition, novel chemistries can often be used to enhance the performance of imbedded lithography toolsets, extending their lifetime and thus reducing the overall cost of operation.

For instance, understanding the complex interactions that lead to dissolution inhibition in conventional novolac based positive photoresists can be rationally expected to lead to the design of improved materials for this application. Molecular level understanding of the relevant principles underlying effective dissolution inhibitor design will also be applicable to the design of materials for advanced lithographic technologies such as 193 nm lithography and e-beam lithography. While most materials that are used in device fabrication today are positive resists, there remains a need to define effective negative acting materials alternatives in that such chemistries could prove advantageous for selected geometries. Research in these areas continues to be active.

Novel approaches to chemistry and processing can lead to materials technologies that may effect solvent free image development. An alternative to the traditional solution developed chemistries is the use of dry development techniques employing reactive ion etching pattern transfer. The driving force that is behind the development of these schemes is simply the demand for improved resolution which requires imaging features with increasingly higher aspect ratios and smaller linewidth variations over steep substrate topography. A number of schemes have been proposed to address this issue, namely; the use of polymer planarizing layers, anti-reflection coatings, and contrast enhancement materials. Notably, surface modification of resists can be used to selectively incorporate species into resist films to effect differential etching in an appropriate RF discharge. Such dry-process schemes could be envisioned to lead to "green" lithographic materials and processes.

Other areas of intense lithographic materials research include the design of improved processes to promote adhesion of polymeric resists to device substrates. As materials technologies evolve, it is not unexpected that the interactions of the relevant polymers to the various device substrates will change resulting in a critical need to fully understand the phenomena leading to effective adhesion.

Each of the many alternative resist materials and processes described above need to be further explored and evaluated for their applicability to VLSI device manufacturing. The future of microlithography is bright and contains many challenges in the areas of resist research and associated processing. There is no doubt that within the decade, many new materials will be commonplace within the manufacturing environment.

Chapter 22

The Influence of Structure on Dissolution Inhibition for Novolac-Based Photoresists: Adaption of the Probabilistic Approach

Christopher L. McAdams[1], Wang Yueh[1], Pavlos Tsiartas[2], Dale Hsieh[1], and C. Grant Willson[1-3]

[1]Department of Chemistry and Biochemistry and [2]Department of Chemical Engineering, The University of Texas at Austin, Austin, TX 78712

In this paper, several theories describing the molecular-level interactions between novolac and the dissolution inhibitors are discussed. Evidence is provided that small changes in the hydrogen bonding interactions between the dissolution inhibitor and the novolac resin can create large differences in resist dissolution rates. The strength of this interaction is shown to depend on the basicity of the hydrogen bond acceptor site in the inhibitor. This correlation is based on a linear free energy relationship in a series of benzophenones with different substituents. Also, the dissolution inhibition efficiencies of a series of sulfonyl esters are shown to be better than structurally similar carbonyl esters. These experiments reveal a set of criteria for the rational design of dissolution inhibitors.

The dissolution of novolac-based photoresists into aqueous base is a complex phenomenon dependent on many factors: inhibitor concentration (*1*), residual solvent concentration in the film (*2*), developer temperature (*3*), polymer structure (*4*), polymer molecular weight (*5*), developer pH (*6*), added salts (*7*) to the developer, and, the topic of this paper, inhibitor efficiency (*8*). Inhibitor efficiency describes how much influence an inhibitor has on the dissolution rate of novolac.

Many theories describing the dissolution of novolac and its inhibition by diazonaphthoquinone (DNQ) have been presented in the literature. These range from the azo-coupling reaction of the DNQ with the phenolic resin (*9*) to the molecular blocking of diffusion of developer into the resist. Recently, our group at the University of Texas (*10*) presented the Probabilistic Approach, which has provided insight on the topic. Using the Probabilistic Approach as a theoretical framework, we will present a hydrogen bonding molecular interaction model in which the dissolution inhibition is caused by a perturbation in the "effective pK_a" of the polymer.

[3]Corresponding author.

Stonewall Model. In the "Stonewall Model," Hanabata *et al.* (*9*) showed using gel permeation chromatography (GPC) that multifunctional DNQ can crosslink the novolac *via* a base catalyzed azo-coupling reaction (Figure 1). They explained that this reaction increases the polymer molecular weight, which results in a decrease in the dissolution rate in the unexposed areas of the resist film.

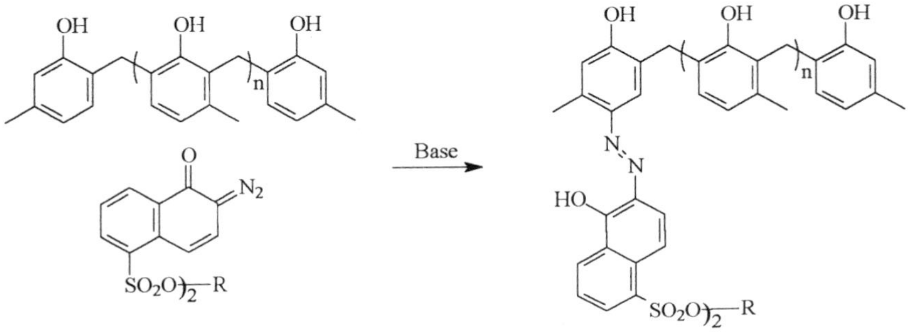

Figure 1. Crosslinking of novolac by a base-catalyzed azo-coupling reaction (Adapted from reference 9.)

Several limitations of this model show that it is not the main mechanism for dissolution inhibition in DNQ/novolac systems. First, the reaction does not occur with all inhibitors. Diazodiones such as Meldrum's Diazo do not undergo base-catalyzed azo-coupling, but they are effective inhibitors (*11*). Also, Murata *et al.* (*12*) showed that efficient dissolution inhibition can be observed with inhibitors that do not have diazo functionality. For example, phenyl naphthyl sulfonate is a better dissolution inhibitor than the corresponding phenyl DNQ sulfonate indicating that the dissolution inhibition ability can be attributed to the *sulfonate ester* of the DNQ, not the DNQ chromophore itself (Figure 2). Despite the shortcomings of the Stonewall Model, it was an important contribution to the understanding of dissolution inhibition, because it was the first microscopic, molecular-interaction model for inhibition.

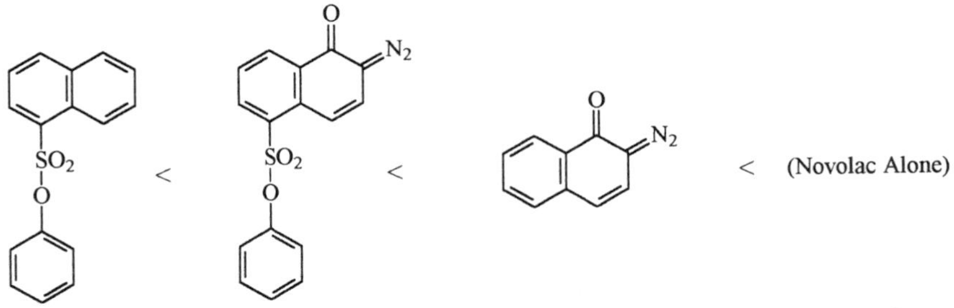

Figure 2. Relative dissolution rates of novolac and inhibited novolac samples (Adapted from reference 13.)

Diffusion Model. In 1986, Arcus presented the membrane model (*3*) for the dissolution of novolac resins. While this model does little to elucidate the molecular-level interaction of DNQ with novolac, it does provide a framework in which to discuss dissolution phenomenon. Arcus's theory is based on the premise that the rate of transport (diffusion) of the developer into the photoresist film controls the dissolution rate. Arcus postulated that the major mechanism for the developer discrimination between exposed and unexposed regions of a positive photoresist is the increased ability of the exposed regions to transport ions, base, and water through the photoresist film that is acting as a membrane.

Reiser expanded the diffusion model for dissolution of novolac (*13-24*) using percolation theory (*25*, *26*) as a theoretical framework. Percolation theory describes the macroscopic event, the dissolution of resist into the developer, without necessarily understanding the microscopic interactions that dictate the resist behavior. Reiser views the resist as an amphiphilic material: a hydrophobic solid in which is embedded a finite number of hydrophilic active sites (the phenolic hydrogens). When applied to a thin film of resist, developer diffuses into the film by moving from active site to active site. When the hydroxide ion approaches an active site, it deprotonates the phenol generating an ionic form of the polymer. In Reiser's model, the "rate of dissolution of the resin ... is predicated on the deprotonation process [and] is controlled by the diffusion of developer into the polymer matrix" (*27*).

The mathematical treatment of the percolation theory divides a percolation field into a grid comprised of cells. The hydroxide ion approaches the percolation grid from the top and must diffuse through to the bottom. Within each cell there is either an active site (a phenol) or no active site. When cells with active sites are adjacent to each other, they are considered to be in a "cluster" through which the developer may diffuse. The density of active sites, p, that is high enough to create a single percolation cluster is known as the percolation threshold, p_c. The dissolution rate, R, of the film is predicted by the scaling law:

$$R = constant\ (p - p_c)^n \qquad (1)$$

where the *constant* is a film dependent parameter and n is the number of degrees of freedom for a three dimensional matrix. When p is smaller than p_c the developer cannot diffuse through the film, and when p is greater than p_c, the percolation field may be penetrated resulting in the dissolution of the resist.

Percolation theory helps explain some dissolution phenomena. Unfortunately, it does not explain why the dissolution rate is molecular weight dependent nor why different monomeric species inhibit novolac with different efficiencies.

***Ortho-Ortho* Effect.** Furata *et al.* (*28*) showed that increasing the number of *ortho-ortho* linkages in novolac improves its performance. Honda *et al.* concluded that "the content of the unhindered *ortho-ortho* bond in novolac is a good indicator of the propensity for dissolution inhibition" (*29*). "*Ortho-ortho*" means that the linkage between the cresol units is *ortho* to the hydroxyl groups on both rings. Other linkages include *ortho-para* and *para-para* linkages (Figure 3).

Templeton *et al.* (*30*) used semi-empirical molecular modeling calculations and x-ray crystal structures of novolac oligomeric residues to show that *ortho-ortho* linkages in novolac result in an extensive network of intramolecular hydrogen bonding. Templeton proposed that this hydrogen bonding causes the phenols to form hydrophilic clusters and the polymer backbone to form a hydrophobic barrier that decreases the dissolution rate. While Tempelton's work contributed greatly to the understanding of novolac dissolution phenomena, it still did not explain why some inhibitors are better than others.

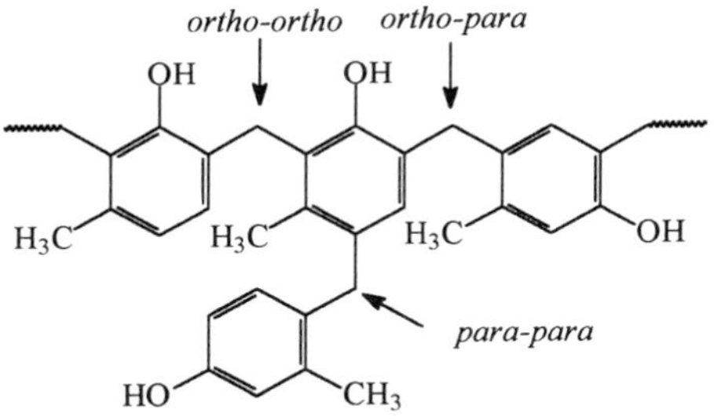

Figure 3. *Ortho-ortho*, *ortho-para*, and *para-para* linkages in novolac.

Inductive Cluster Polarization. Reiser (*21*) noted that the rate changes caused by dissolution inhibitors are disproportionate to the amount of inhibitor used. For example, the dissolution rate of novolac is decreased by several orders of magnitude with an inhibitor loading of only 10 % (wt/wt). The percolation model predicts that the inhibitor must block the active sites of the novolac to the point that p falls below p_c. Reiser concluded that for such small loadings of inhibitor to block so many active sites each inhibitor must block several phenolic sites in the percolation field.

Reiser proposed that the inhibitor induces the polarization of hydrogen bonding clusters of phenolic sites (Inductive Cluster Polarization - Figure 4) He indicated that the blocking effect of the inhibitor may propagate and influence up to 16 sites. Interestingly, Coleman *et al.* (*31*) showed that the magnitude of the extended hydrogen-bonded network in novolac is limited to clusters that are two to three monomer units in length.

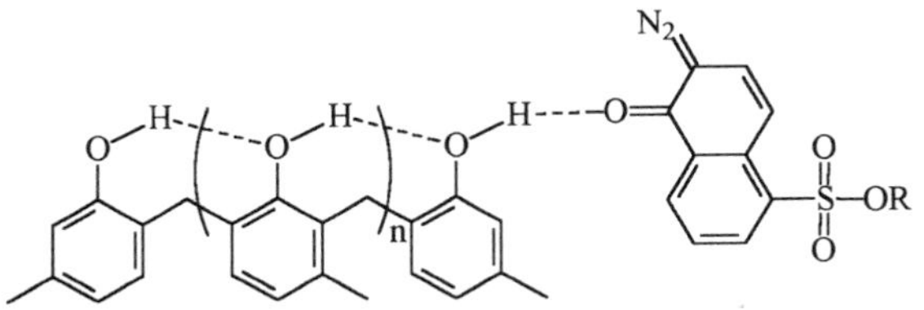

Figure 4. Inductive Cluster Polarization (Adapted from reference 22.)

Octopus-Pot Model. Honda *et al.* (*32*) proposed that the high *ortho-ortho* structure of novolac allows multiple phenols to bind to the DNQ carbonyl, thus, blocking developer diffusion into the resist. The lipophilic backside of the polymer forms a hydrophobic "pot" while the DNQ PAC rests atop the pot resembling an octopus (Figure 5). The hydrogen bonding of the novolac to the DNQ carbonyl was observed by infrared spectroscopy. In terms of percolation theory, the active sites of the polymer are blocked, which effectively lowers the percolation parameter, p.

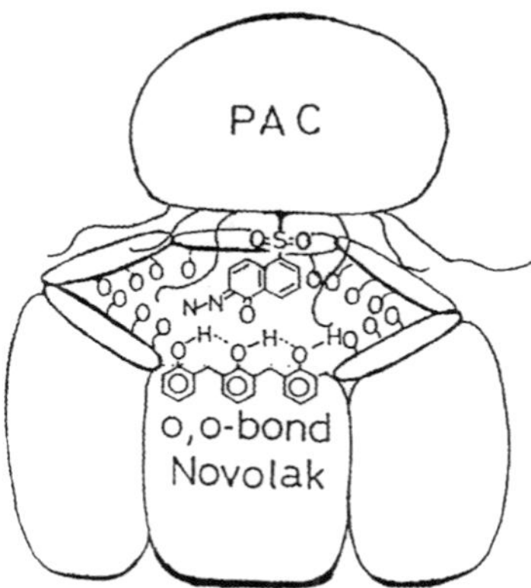

Figure 5. The Octopus-pot model (Reproduced with permission from reference 32. Copyright 1990 K. Honda)

The Octopus-pot model and Reiser's Inductive Cluster Polarization model have one severe limit: DNQ is not a good dissolution inhibitor. As shown earlier, Murata (*12*) provided evidence that the DNQ moiety does little to aid in the dissolution inhibition phenomenon (Figure 2). Also, Borzo *et al.* (*33*) showed using ^{15}N NMR that the diazo group was influenced only slightly by hydrogen bonding. Dammel suggested that the sulfonate ester group acts as the key binding site (*34*).

Probabilistic Approach. Recently, our group presented the Probabilistic Approach (*10*), which is based on the principle that the development of novolac is more accurately described as an etching process rather than a diffusional process. Since the dissolution of novolac requires that the polymer undergoes a chemical transformation (*i.e.* deprotonation) then "the average degree of ionization" dictates the dissolution rate "... rather than a diffusive, transport process." (*10*)

We proposed that the dissolution rate (*DR*) is proportional to the ratio of ionized sites *[A-]* to un-ionized sites *[HA]* on the surface of the dissolving film. This ratio is predicted by the apparent pK_a of the polymer and the pH.

$$DR \propto \frac{[A-]}{[HA]} \qquad (2)$$

The Probabilistic Approach allows for the explanation many previously unexplained dissolution phenomena: the effect of added salts (*7*), the MW dependence (*5*) and the effect of residual casting solvent (*35,36*) on the dissolution rate.

If this theory is correct, then the inhibitor must act to decrease the ratio of ionized to un-ionized sites (Equation 2). We propose that increased hydrogen bonding decreases this ratio by altering the "effective pK_a" of the novolac, and, as a result, hydrogen bonding is the primary mechanism of dissolution inhibition.

EXPERIMENTAL

Dissolution Measurements. Dissolution rate measurements of the benzophenone series were made on a Perkin-Elmer Dissolution Rate Monitor (DRM-5900) in puddle development mode with 0.25 N KOH developer. The DRM was computer-controlled using Perkin Elmer DRM analytical software (DREAMS 3.0). Measurements of the sulfonyl/carbonyl esters were made with 0.26 N tetramethyl-ammonium hydroxide (TMAH). Reported dissolution rates ($DR_{1/2}$) are at half film thickness.

Formulations. The benzophenone inhibitors were formulated to 0.4 *m* (molal = moles inhibitor/ kg polymer) in G-2 novolac resin (Shipley Company) in propylene glycol monomethyl ether acetate (PGMEA) (20-30 % solids wt/wt). The formulations were spin-coated on a silicon wafer at 5000 rpm and baked on a hot plate at 90 °C for 120 seconds. The sulfonyl and carbonyl ester inhibitors were formulated to 0.333 *m*. in G-2 novolac in PGMEA (20-30 % solids wt/wt). The formulations were spin-coated on a silicon wafer at 2500 rpm and baked on a hot plate at 90 °C for 90 seconds. Nominal film thicknesses were 1 - 2 μm.

Instrumentation. ^{1}H and ^{13}C NMR were performed on either a General Electric QE-300 or a Bruker AC-250 spectrometer. IR spectra were recorded on a Nicolet 550 Magna FTIR spectrometer with Omnic 1.2a analytical software.

Temperature controlled IR measurements of liquid samples (solution in CCl_4) were measured in a Wilmad temperature-controlled cell mount with calcium fluoride windows. The sample was surrounded by a heat sink that was filled with a heat transfer fluid (antifreeze), which was regulated by a Noah Precision temperature controller. The cell mount was sealed in the nitrogen-flushed IR compartment.

Materials. Compounds **1-8** and **10** and synthetic starting materials were obtained from Aldrich and used without purification unless otherwise stated. Compounds **9** and **11** were synthesized using the methods of Schlenk (*37*) *et al. via* the Freidel-Crafts acylation (*38*, *39*). All other inhibitors were prepared by standard methods. Elemental analysis was provided by Atlantic Microlabs.

4-(4-Methyl-benzoyl)biphenyl (9). Biphenyl (7.71 g, 50 mmol) and *p*-toluoyl chloride (7.73 g, 50 mmol) were dissolved in CS_2 (50 mL). $AlCl_3$ (6.67 g, 50 mmol) was added. The reaction refluxed at 46 °C for 5 hr. The solvent was removed by rotary evaporation. The residue was brought to basic pH with 10 % NaOH solution and extracted with ether. The solid precipitate was filtered off. The ether layer was separated, dried with $MgSO_4$ and filtered. The solvent was removed by rotary evaporation to give 8.3 g yield of a light-orange solid. The solid was chromatographed using 1:1 $CHCl_3$:cyclohexane to give **9**. Alternatively, the product was sublimed at 140 °C (< 250 mtorr). ^{1}H NMR ($CDCl_3$) δ 2.47 (s, 3H), 7.3 (d, *J* = 8Hz, 2H), 7.39 (tt, J = 7.1/2.3Hz, 1H), 7.47 (tt, J = 7.1/1.5Hz, 2H), 7.65 (dt, J = 7.4/1.9Hz, 4H), 7.75 (d, J = 8Hz, 2H), 7.85 (dd, J = 6.8/1.5Hz, 2H); ^{13}C NMR δ 21.6, 129.9, 127.2, 128.1, 128.9, 130.2, 130.6, 135.0, 136.5, 140.0, 143.1, 144.9, 196.6. IR (KBr) 3064, 3025, 2916, 2856, 1645, 1604, 1317, 1292, 739 cm^{-1}. MS

(CI) m/z = 273 (M+1), 301 (M+29), 119, 181. Anal. calcd. for $C_{20}H_{16}O$: C, 88.20; H, 5.92 Found: C, 88.29; H, 6.05.

4-(3-Trifluoromethyl-benzoyl)biphenyl (11). Biphenyl (1.54 g, 10 mmol) and 3-trifluoromethylbenzoyl chloride (2.08 g, 10 mmol) were dissolved in distilled nitroethane (20 mL), placed under N_2 and cooled to 0 °C. $AlCl_3$ (1.4 g, 10.5 mmol) dissolved in nitroethane (25 mL) was added dropwise over 30 min. The reaction turned orange immediately upon addition. The reaction stirred overnight at room temperature. The reaction mixture was brought to basic pH with 10 % NaOH solution. The solvent was removed by rotary evaporation. The mixture was extracted with CH_2Cl_2 and the organic layer was dried over $MgSO_4$. The mixture was filtered, and the solvent was removed by rotary evaporation to yield 1.721 g of gray solid (39 % pure by GC). The crude product sublimed at 140 °C (< 250 mtorr, 70 % pure by GC). The impure product was chromatographed in 1:1 $CHCl_3$:cyclohexane (R_f = 0.15) to give 0.449 g (14 % yield). ^{1}H NMR ($CDCl_3$) δ 7.4 (tt, 1H), 7.48 (tt, 2H), 7.64 (t, 3H), 7.72 (d, 2H), 7.85 (t, 3H), 8.0 (d, 1H), 8.08 (s, 1H); ^{13}C NMR δ 126.6, 126.7, 127.2, 127.3, 128.4, 128.8, 128.9, 129.0, 130.7, 133.0, 135.3, 135.3, 138.4, 139.7, 145.8, 194.8 cm^{-1}. MS (CI) m/z = 327 (M+1), 355 (M+29), 307 (M-F) 173, 181. Anal. calcd. for $C_{20}H_{13}F_3O$: C, 73.62; H, 4.02 Found: C, 73.07; H, 4.19.

***p*-Tolyl 2-naphthylsulfonate ester (12).** A solution of 2.54 g (23.5 mmol) of *p*-cresol in 20 mL of dry THF was added dropwise to 1.45 g (36.3 mmol) of NaH (60 % in oil) in 100 ml of dry THF in a 250-mL round-bottom flask with N_2. The reaction mixture was stirred for 10 min at room temperature. A solution of 5.08 g (22.4 mmol) of 2-naphthalenesulfonyl chloride in 20 mL of dry THF was added dropwise. After addition, the reaction mixture was refluxed for 18 hr. After cooling, 200 mL of water were added and the mixture was extracted with ethyl acetate (3 x 100 mL). The combined organic layers were dried over anhydrous $MgSO_4$. The solvent was removed by rotary evaporation. The residue was purified by recrystallization with ethanol to give 4.92 g of **12** (mp: 98 °C; yield: 73.8 %). ^{1}H NMR ($CDCl_3$) δ 8.33 (s, 1H), 7.78-7.96 (m, 4H), 7.55-7.68 (m, 2H), 7.00 (d, J = 8.7Hz, 2H), 6.83 (d, J = 8.6Hz, 2H), 2.40 (s, 3H). ^{13}C NMR δ 147.3, 136.9, 135.4, 132.3, 131.6, 130.3, 130.0, 129.4, 129.3, 129.2, 127.9, 127.7, 122.8, 121.9, 20.7. IR (KBr) ν_{max} 1595.5, 1504.1, 1379.1, 1181.9, 1080.9, 874.1, 835.6, 792.4 cm^{-1}. HRMS (CI): m/z calcd. for $C_{17}H_{15}O_3S$ (M+1) 299.0742 , found 299.0752.

***p*-Tolyl 2-naphthylcarboxylate (13).** A solution of 2.43 g (22.5 mmol) of *p*-cresol in 20 mL of dry THF was added dropwise to 1.48 g (37.0 mmol) of NaH (60 % in oil) in 100 ml of dry THF in a 250-mL round-bottom flask with N_2. The reaction mixture was stirred for 10 min at room temperature. To this reaction mixture a solution of 4.44 g (23.3 mmol) of 2-naphthoyl chloride in 20 mL of dry THF was added dropwise. After addition, the reaction mixture refluxed for 18 hr. After cooling, the reaction mixture was added to 200 mL of water and extracted with ethyl acetate (3 x 100 mL). The combined organic layers were dried over anhydrous $MgSO_4$. The solvent was removed by rotary evaporation. The residue was recrystallized with ethanol to give 5.28 g of **13** (mp: 122 °C; yield: 90 %). ^{1}H NMR ($CDCl_3$) δ 8.78 (s, 1H), 8.25 (d, J = 8.7, 1H), 7.88-8.00 (m, 3H), 7.59 (p, J = 6.8Hz, 2H), 7.20 (dd, J = 8.4Hz, 13.6, 4H), 2.38 (s, 3H). ^{13}C NMR δ 156.1, 148.7, 135.7,

135.4, 132.4, 131.7, 129.9, 129.4, 128.5, 128.2, 127.1, 126.8, 126.7, 125.3, 121.3, 20.8. IR (KBr) ν_{max} 1744.6, 1629.1, 1508.9, 1273.3, 1191.5, 1080.9, 946.3 cm^{-1}. HRMS (CI): *m/z* calcd. for $C_{18}H_{15}O_2$ (M+1) 263.1072 , found 263.1065.

***p*-Tolyl 1-naphthylsulfonate ester (14).** The compound was synthesized in a similar fashion to **12**. The residue was recrystallized with ethanol to give 4.88 g of **14** (mp: 81 °C; yield: 87 %). ^1H NMR ($CDCl_3$) δ 8.82 (d, J = 13.2Hz, 1H), 8.06 (t, J = 8.4Hz, 2H), 7.94 (d, J = 8.8Hz, 1H), 7.74 (t, J = 8.6Hz, 1H), 7.63 (t, J = 8.7Hz, 1H), 7.40 (t, J = 8.0Hz, 1H), 6.92 (d, J = 8.8Hz, 2H), 6.70 (d, J = 8.7Hz, 2H), 2.19 (s, 3H). ^{13}C NMR δ 147.4, 136.8, 135.5, 133.9, 131.1, 130.8, 129.9, 128.9, 128.8, 128.4, 127.2, 125.0, 123.8, 121.6, 20.7. IR (KBr) ν_{max} 1504.1, 1374.3, 1205.9, 1148.2, 869.3, 768.3 cm^{-1}. HRMS (CI): *m/z* calcd. for $C_{17}H_{15}O_3S$ (M+1) 299.0742, found 299.0741.

***p*-Tolyl 1-naphthylcarboxylate (15).** The compound was synthesized in a similar fashion to **13**. The residue was recrystallized with ethanol to give 4.02 g of **15** (mp: 60 °C; yield: 67 %). ^1H NMR ($CDCl_3$) δ 8.92 (d, J = 8.6, 1H), 8.32 (d, J = 8.5, 1H), 7.92 (d, J = 8.2Hz, 3H), 7.75 (d, J = 8.5Hz, 1H), 7.36-7.50 (m, 3H), 7.06 (dd, J = 8.5Hz, 4H), 2.38 (s, 3H). ^{13}C NMR δ 166.0, 148.7, 135.5, 134.1, 133.8, 131.6, 131.1, 130.0, 128.6, 128.0, 126.3, 125.9, 125.7, 124.4, 121.5, 20.9. IR (KBr) ν_{max} 1734.9, 1508.9, 1239.6, 1191.5, 1124.2, 984.7, 792.4 cm^{-1}. HRMS (CI): *m/z* calcd. for $C_{18}H_{15}O_2$ (M+1) 263.1072 , found 263.1062.

2-Naphthyl *p*-tosylate (16). The compound was synthesized in a similar fashion to **12**. The residue was recrystallized in ethanol to give 7.82 g of **16** (mp: 124 C; yield: 75 %). ^1H NMR ($CDCl_3$) δ 7.68-7.80 (m, 5H), 7.43-7.48 (m, 3H), 7.25 (d, J = 8.3Hz, 2H), 7.07 (dd, J = 2.3Hz, 4.5, 1H), 2.40 (s, 3H). ^{13}C NMR δ 147.1, 145.3, 133.3, 132.3, 131.7, 129.7, 129.6, 128.4, 127.7, 127.6, 126.7, 126.3, 121.0, 119.8, 21.6. IR (KBr) ν_{max} 1600.3, 1379.1, 1181.9, 912.6 cm^{-1}. HRMS (CI): *m/z* calcd. for $C_{17}H_{15}O_3S$ (M+1) 299.0742 , found 299.0736.

2-Naphthyl *p*-tolylcarboxylate (17). The compound was synthesized in a similar fashion to **13**. The residue was recrystallized in ethanol to give 7.82 g of **17** (mp: 142 °C; yield: 75 %). ^1H NMR ($CDCl_3$) δ 8.05 (d, J = 8.2Hz, 2H), 7.70-7.80 (m, 3H), 7.59 (d, J = 2.2Hz, 1H), 7.38 (p, 2H), 7.20-7.28 (m, 3H), 2.34 (s, 3H). ^{13}C NMR δ 165.6, 148.6, 144.4, 133.7, 131.4, 130.2, 129.3, 129.2, 127.7, 127.6, 126.7, 126.4, 125.6, 121.3, 118.6, 21.8. IR (KBr) ν_{max} 1725.1, 1607.7, 1270.1, 1245.7, 1085.2, 1015.8, 893.5, 810.3 cm^{-1}. HRMS (CI): *m/z* calcd. for $C_{18}H_{15}O_2$ ($M+H^+$) 263.1072 , found 263.1058.

RESULTS AND DISCUSSION

Linear Free-Energy Relationship. Substituted benzophenones (**4**) with electron donating and electron withdrawing characteristics (*i.e.* a Hammett Linear Free-Energy Relationship (*40*)) were used to determine the relationship between inhibitor basicity and inhibiton. The perturbation of a substituent on the free energy is defined in terms Hammett $\sigma_{TOT}{}^+$ values. Through-resonance stabilization of a positive charge (*40*) is usually best represented using σ^+ constants for the Hammett Plots. σ^+_{TOT} is the summation of all σ^+_{para} and σ^+_{meta} for substituents of that compound.

The Hammett plot (log($DR_{1/2}$) vs. σ^{+} – Figure 6) for these compounds shows that has a positive slope (i.e. electron donating groups decrease the dissolution rate). Electron donating groups ($\sigma^{+} < 0$) increase the dissolution inhibition efficiency while electron withdrawing groups ($\sigma^{+} > 0$) decrease it.

Table I: Benzophenone derivatives (Ph = phenyl)

	-A	-B	-C	-D	σ^{+}_{TOT}
1	$-OCH_3$	-H	-H	-H	-0.78
2	$-CH_3$	$-CH_3$	-H	-H	-0.38
3	$-CH_3$	-H	-H	-H	-0.31
4	-H	-H	-H	-H	0
5	-F	-H	-H	-H	-0.07
6	-H	$-CF_3$	-H	-H	0.57
7	$-CF_3$	-H	-H	-H	0.61
8	$-CF_3$	-H	-H	$-CF_3$	1.18
9	$-CH_3$	-H	-Ph	-H	-0.59
10	-H	-H	-Ph	-H	-0.18
11	-H	$-CF_3$	-Ph	-H	0.39

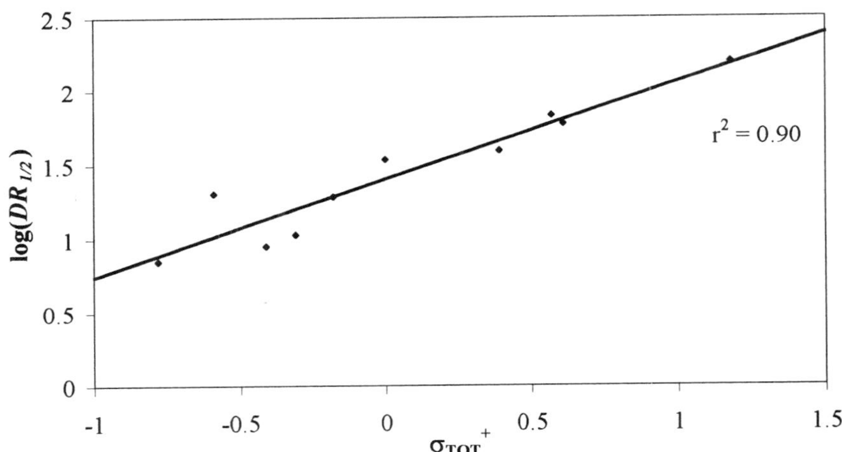

Figure 6: log($DR_{1/2}$) has a positive correlation with σ^{+}_{TOT}.

Hydrogen Bond Strength of Inhibitors with *m*-Cresol. The free energy, enthalpy, and entropy (ΔG, ΔH and ΔS, respectively) of the hydrogen bonding interaction (Figure 7) between each inhibitor (**B**) and *m*-cresol (**A**) were determined from the equilibrium binding constants (K) in CCl_4 (CCl_4 was chosen as a solvent to minimize the solvent effects on hydrogen bonding) solution (*41*). The monomer of novolac, *m*-cresol, was chosen as a model for the novolac, because measuring the binding constants of the inhibitor with the polymer was difficult.

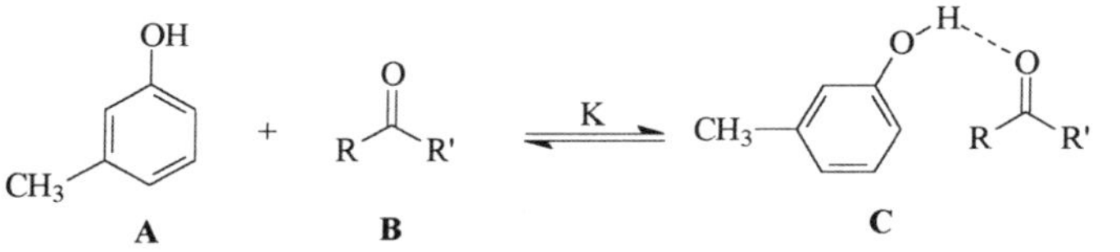

Figure 7. Inhibitor/*m*-cresol hydrogen bonding

K is calculated using equation 3 from the equilibrium concentrations: $[A]$, $[B]$ and $[C]$. These quantities are measured or derived when the initial concentrations of each additive, $[A]_o$ and $[B]_o$, are known. $[A]$ is measured directly by quantitative infrared spectroscopy as the area of the free OH peak at 3612 cm^{-1}.

$$K = \frac{[C]}{[A][B]} = \text{Binding Constant} \quad (3)$$

$$[A] = [A]_{meas} \quad (4)$$

The amount of associated **A** (or **B**) equals $[C]$ (equation 5), and $[B]$ is the portion of $[B]_o$ that is unassociated (equations 6 and 7)

$$[C] = [A]_{assoc.} = [B]_{assoc.} = [A]_o - [A]_{meas} \quad (5)$$

$$[B] = [B]_o - [B]_{assoc.} \quad (6)$$

$$[B] = [B]_o + [A]_{meas} - [A]_o \quad (7)$$

So, once $[A]$ is measured, K is easily derived. The change in free energy (ΔG) is determined from K in equation 8. Then, given the thermodynamic relationship in equation 9, the equations may be substituted and divided by T. Plotting equation 11 (-R ln K vs. 1/T) gives ΔH as the slope of the line and -ΔS as the intercept. The plot fro benzophenone is shown in Figure 8.

$$\Delta G = -RT \ln K \quad (8)$$

$$\Delta G = \Delta H - T\Delta S \quad (9)$$

$$-RT \ln K = \Delta H - T\Delta S. \quad (10)$$

$$-R \ln K = \frac{\Delta H}{T} - \Delta S \quad (11)$$

Four compounds from the benzophenone series (**1**, **3**, **4** and **6**) were chosen to make the measurement. When ΔH was plotted against $\log(DR_{1/2})$ (Figure 9), the benzophenone series showed a general negative trend. We suspect that the looseness of the correlation is probably due to the oversimplification of the model compounds.

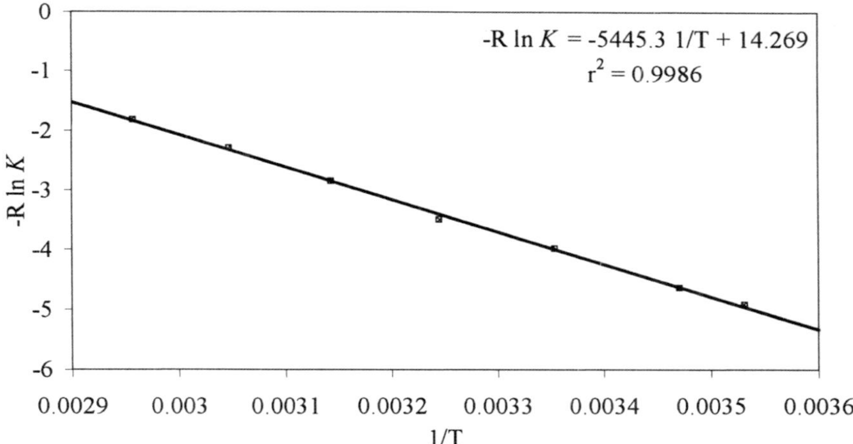

Figure 8. Change in equilibrium as a function of temperature for benzophenone and *m*-cresol.

Table II.

Inhibitor	σ_{TOT}^{+}	ΔH (kcal/mol)	ΔS (eu)	$DR_{1/2}$
4-methoxy benzophenone	-0.78	-5.67	13.9	10.1
4-methyl benzophenone	-0.41	-6.07	15.9	8.9
benzophenone	0	-5.45	14.3	22.8
3-trifluoromethyl benzophenone	0.57	-4.25	11.2	29.8

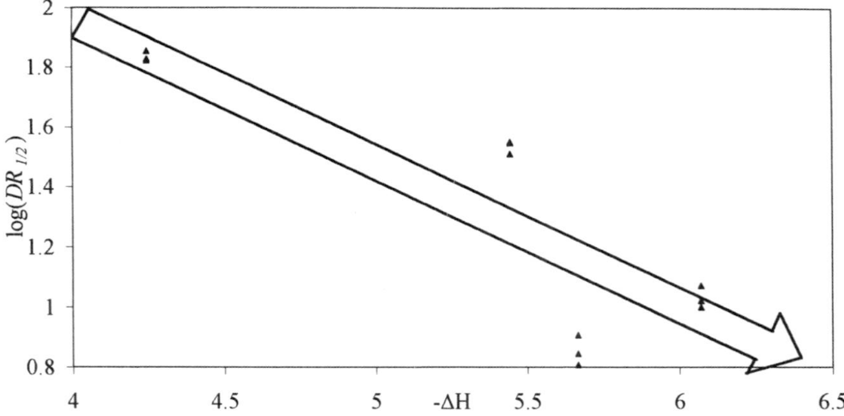

Figure 9. -ΔH vs. log($DR_{1/2}$) for substituted benzophenones.

Sulfonyl Esters vs. Carbonyl Esters. The presence of sulfonyl ester groups in the PAC has been shown by Murata to play a critical role in the inhibition of novolac dissolution (*12*). Below, we demonstrate that dramatic effect in a direct comparison of sulfonyl esters to their corresponding carbonyl esters for a series of phenyl naphthyl esters (Figure 10).

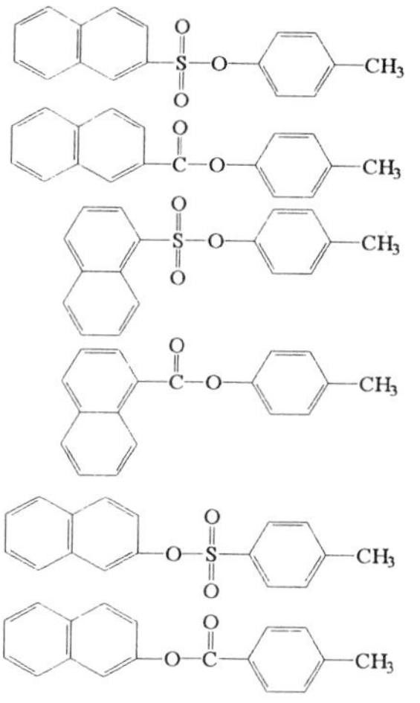

12	2-naphthyl–SO_2–O–C$_6$H$_4$–CH_3	5.2 Å/sec
13	2-naphthyl–C(=O)–O–C$_6$H$_4$–CH_3	19.7
14	1-naphthyl–SO_2–O–C$_6$H$_4$–CH_3	3.3
15	1-naphthyl–C(=O)–O–C$_6$H$_4$–CH_3	19.9
16	2-naphthyl–O–SO_2–C$_6$H$_4$–CH_3	2.6
17	2-naphthyl–O–C(=O)–C$_6$H$_4$–CH_3	22.5
	Uninhibited Novolac	285

Figure 10. Inhibitors 0.26 N TMAH

The sulfonyl esters inhibit the dissolution rate of novolac far more efficiently than the corresponding carbonyl esters. We suspect that this is due to the highly localized charge present on the sulfonyl oxygen as depicted in the commonly drawn resonance structure (Figure 11).

$$\left[\; R-\overset{O}{\underset{O}{\overset{\|}{\underset{\|}{S}}}}-O-R \;\longleftrightarrow\; R-\overset{O^-}{\underset{O^-}{\overset{|}{\underset{|}{S^{++}}}}}-O-R \;\right]$$

Figure 11. The resonance structure of sulfonyl esters shows the highly localized charge on the oxygen.

CONCLUSIONS

We conclude that a primary mode of dissolution inhibition is a strong polarization of the intramolecular hydrogen bonding clusters found in novolac. The polarization increases the effective pK_a of novolac and decreases the ratio of ionized to un-ionized phenolic sites. polymer (Equation 2). Inhibitors with stronger hydrogen bonds induce a larger polarization of the cluster increasing the "effective pK_a" of the polymer (*i.e.* decreasing the ratio of ionized to un-ionized sites).

We have demonstrated an excellent correlation between a structural characteristic (basicity) and dissolution inhibitor efficiency. We have shown two methods that help to strengthen the hydrogen bonding interaction between the inhibitor and the polymer: the introduction of resonance stabilizing electron donating groups and the substitution of sulfonyl esters in place of carbonyl esters in the inhibitor. Much more work of this sort is required to fully understand the interactions between dissolution inhibitors and novolac resin.

ACKNOWLEDGMENTS

This research was financially supported by the Semiconductor Research Corporation (SRC Contract #95-LP-409 -- Advanced Resist Materials Research) and a grant from the Texas Advanced Technology Program (TD&T-278). The authors would like to acknowledge the helpful discussions with Arnost Reiser and Ralph Dammel.

LITERATURE CITED

1. Dill, F. H. *IEEE Trans. Elect. Dev.* **1975**, *ED-22 (7)*, 440-444.
2. Rao, V.; Hinsburg, W. D.; Frank, C. W.; Pease, R. F. W. *Proc. SPIE Int. Soc. Opt. Eng.* **1993**, *1925*, 538-551.
3. Arcus, R. A. *Proc. SPIE Int. Soc. Opt. Eng.* **1986**, *631*, 124-134.
4. Templeton, M. K.; Szmanda, C. R.; Zampini, A. *Proc. SPIE Int. Soc. Opt. Eng.* **1987**, *771*, 136-147.
5. Tsiartas, P.C.; Simpson, L. L.; Qin, A.; Willson, C. G.; Allen, R. D.; Krukonis, V. J.; Gallagher-Wetmore, P. M. *Proc. SPIE Int. Soc. Opt. Eng.* **1995**, *2438*, 261-271.
6. Hinsberg, W. D.; Gutierrez, M. L. *Proc. SPIE Int. Soc. Opt. Eng.* **1984**, *469*, 57-64.
7. Henderson, C. L.; Tsiartas, P. C.; Simpson, L. L.; Clayton, K. D.; Pancholi, S.; Pawloski, A. R.; Willson, C. G. *Proc. SPIE Int. Soc. Opt. Eng.* **1996**, *2724*, 481-490.
8. Yeh, T.-F.; Shih, H.-Y.; Reiser, A.; Toukhy, M. A.; Beauchemin, B. T. Jr. *J. Vac. Sci. Technol.* **1992**, *10*, 3.
9. Hanabata, M.; Uetani, Y.; Furuta, A. *Proc. SPIE Int. Soc. Opt. Eng.* **1988**, *920*, 349-354.
10. Tsiartas, P. C.; Flanagin, L. W.; Henderson, C. L.; Hinsberg, W. D.; Sanchez, I. C.; Bonnecaze, R. T.; Willson, C. G. *Macromolecules* **1997**, *30*, 4656-4664.
11. Grant, B.D.; Clecak, N.J.; Tweig, R.J.; Willson, C.G. *IEEE Transactions on Electronic Devices* **1981**, *ED-28*, 1300-1305.
12. Murata, M.; Koshiba, M.; Harita, Y. *Proc. SPIE Int. Soc. Opt. Eng.* **1989**, *1086*, 48-55.
13. Huang, J.-P.; Kwei, T. K.; Reiser, A. *Macromolecules* **1989**, *22*, 4106-4112.
14. Yeh, T.-F.; Shih, H.-T.; Reiser, A. *Macromolecules* **1992**, *25*, 5345-5352.
15. Yeh, T.-F.; Shih, H.-T.; Reiser, A. *Proc. SPIE Int. Soc. Opt. Eng.* **1992**, *1672*, 204-213.
16. Reiser, A.; Lin, C.; Yeh, T. *Proc. SPIE Int. Soc. Opt. Eng.* **1993**, *1925*, 647-656.
17. Lin, C. C.; Yeh, T.-F.; Reiser, A.; Honda, K.; Beauchemin, B. T. *J. Photopolym. Sci. Technol.* **1993**, *6 (1)*, 147-160.

18. Yeh, T.-F.; Reiser, A.; Dammel, R. R.; Pawlowski, G.; Roeschert, H. *Proc. SPIE Int. Soc. Opt. Eng.* **1993**, *1925*, 570-581.
19. Shih, H.-T.; Yeh, T.-F.; Reiser, A.; Dammel, R. R.; Merrem, H. J.; Pawlowski, G. *Proc. SPIE Int. Soc. Opt. Eng.* **1994**, *2195*, 514-523.
20. Yeh, T.-F.; Huang, J.-P.; Reiser, A.; Honda, K.; Beachemin, B. T.; Hurditch, R. J. *Proc. SPIE Int. Soc. Opt. Eng.* **1994**, *2195*, 663-672.
21. Shih, H.-T.; Reiser, A. *Proc. SPIE Int. Soc. Opt. Eng.* **1995**, *2438*, 305-311.
22. Shih, H.-Y.; Reiser, A. *Macromolecules* **1995**, *28*, 5595-5600.
23. Shih, H.-Y.; Reiser, A. *Macromolecules* **1996**, *29*, 2082-2087.
24. Kim, M. S.; Reiser, A. *Macromolecules* **1997**, *30*, 4652-4655.
25. Essam, J. W. *Rep. Progr. Physics* **1980**, *43*, 833-912.
26. Stauffer, D. *Introduction to Percolation Theory;* Taylor & Francis, London, 1985.
27. Huang, J.-P.; Kwei, T. K.; Reiser, A. *Proc. SPIE Int. Soc. Opt. Eng.* **1989**, *1086*, 74-84.
28. Furata, A.; Hanabata, M.; Uemura, Y. *J. Vac. Sci. Technol.* **1986**, *B4(1)*, 430-436.
29. Honda, K.; Beauchemin, Jr., B. T.; Fitzgerald, E. A.; Jeffries, III, A. T.; Tadros, S. P.; Blakeney, A. J.; Hurditch, R. J.; Tan, S.; Sakaguchi, S. *Proc. SPIE Int. Soc. Opt. Eng.* **1991**, *1466*, 141-148.
30. Templeton, M. K.; Szmanda, C. R.; Zampini, A. *Proc. SPIE Int. Soc. Opt. Eng.* **1987**, *771*, 136-147.
31. Coleman, M. M.; Serman, C. J.; Bhagwagar, D. E.; Painter, P. C. *Polymer* **1990**, *30*, 1298.
32. Honda, K.; Beauchemin, Jr., B. T.; Hurditch, R. J.; Blakeney, A. J.; Kawabe, Y.; Kobuko, T. *Proc SPIE Int. Soc. Opt. Eng.* **1990**, *1262*, 493-500.
33. Borzo, M.; Rafalko, J. J.; Joe, M.; Dammel, R. R.; Rahman, M. D.; Ziliox, M. A. *Proc. SPIE Int. Soc. Opt. Eng.* **1995**, *2438*, 294-304.
34. Dammel, R. In *Diazonaphthoquinone-based Resisits;* O'Shea, D. C., Ed.; Tutorial Text in Optical Engineering; SPIE Optical Engineering Press, Bellingham, Washington, 1993; Vol. TT-11, p 80.
35. Mueller, K. E.; Koros, W. J.; Wang, Y. Y.; Willson,, C. G. *Proc. SPIE Int. Soc. Opt. Eng.* **1997**,*3049,* 871-873.
36. Gardiner, A. B.; Qin, A.; Henderson, C. L.; Pancholi, S.; Koros, W. J.; Willson, C. G.; Dammel, R. R.; Mack, C.; Hinsberg, W. D. *Proc. SPIE Int. Soc. Opt. Eng.* **1997**, *3049*, 850-860.
37. Schlenk. *Justus Liebigs Ann. Chem.* **1928** *463*, 120.
38. Arnold, R. T.; Murai, K.; Dodson, R. M. *J. Am. Chem. Soc.* **1950**, *72*, 4193-4195.
39. Orchin, M.; Woolfolk, E. O.; Reggel, L. *J. Am. Chem. Soc.* **1949**, *71*, 1126-1127.
40. Lowry, T. H.; Richardson, K. S. In *Mechanism and Theory in Organic Chemistry 3d;* Berger, L. S., Ed.; Harper & Row: New York, New York, 1987; pp 143-159.
41. Liddel, U.; Becker, E. D. *Spectrochim. Acta* **1957**, *10*, 70-84.

Chapter 23

Photoacid Generating Polymers for Surface Modification Resists

Masamitsu Shirai[1], Mitsuho Masuda[1], Masahiro Tsunooka[1], Masayuki Endo[2], and Takahiro Matsuo[3]

[1]**Department of Applied Chemistry, College of Engineering, Osaka Prefecture University, Sakai, Osaka 599-8531, Japan**
[2]**Association of Super-Advanced Electronics Technologies, Yokohama Research Center, 292 Yoshida-cho, Totsuka-ku, Yokohama, Kanagawa 244, Japan**
[3]**Semiconductor Research Center, Matsushita Electric Industrial Company, Ltd., 3-15 Yagumo-Nakamachi, Moriguchi, Osaka 570, Japan**

Terpolymers of methyl methacrylate, 1,2,3,4-tetrahydro-1-naphthylideneamino p-styrenesulfonate, and an amide-containing monomer or tetrahydrofurfuryl methacrylate were prepared. These polymers can be used as oxygen plasma-developable surface modification resists. When the irradiated polymer films were exposed to the vapor of alkoxysilanes at 30 °C, polysiloxane networks were formed at the film surface. The polysiloxane formation rate was enhanced by incorporation of amide or ether unit into polymers. Amide units were more effective than ether units. The film surface modified with polysiloxanes showed a good resistance to the etching with an oxygen plasma. Using the surface modification resist system, negative-tone images were generated by an ArF excimer laser lithography.

Photolithographic processes using surface modification techniques have been reported by several groups. The predominant approach involves the post-exposure silylation of organic polymer films (*1,2*) and the selective deposition of Si containing materials on the exposed or unexposed film surface (*3*). Previously we reported plasma-developable photoresists based on the photoinduced acid-catalyzed polysiloxane formation at the irradiated polymer surface by a chemical vapor deposition (CVD) method (*4-6*). The methodology of the system is as follows: Upon irradiation with UV light the surface of the polymers having photoacid generating units becomes hydrophilic because of the formation of acids. Water sorption from the atmosphere occurs at the surface. When the irradiated film is exposed to the vapor of alkoxysilanes, polysiloxane networks are formed at the surface of the polymer films (Figure 1). No polysiloxane networks are formed in unirradiated areas because the photochemically formed acids are necessary for the polysiloxane formation by hydrolysis and subsequent condensation of alkoxysilanes. This system gives a negative tone image by oxygen reactive ion etching (O_2 RIE).

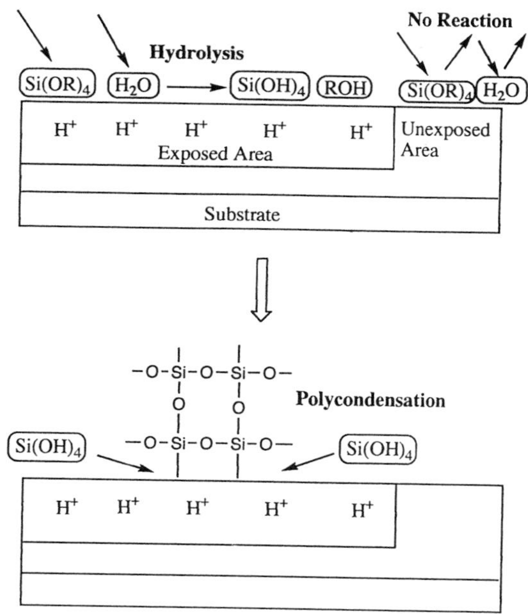

Figure 1. Selective deposition of polysiloxanes at the irradiated polymer surface.

In the previous work we obtained by a KrF (248 nm) excimer laser lithography 0.25 μm line and space images based on the surface modification process using a copolymer of methyl methacrylate and 1,2,3,4-tetrahydro-1-naphthylideneamino p-styrenesulfonate (NIS) as photoacid generating units (*7*). The present system has an advantage to allow imaging at both 248 and 193 nm wavelengths, without the need for introducing changes in the chemistry involved in the image forming process. However, the copolymer showed relatively low sensitivity toward ArF (193 nm) excimer laser exposure. To obtain a highly sensitive resist system, enhanced polysiloxane formation at the exposed film surface is required. The polysiloxane formation rate at the irradiated surface is reported to be proportional to the number of acid units generated photochemically and to the number of water molecules adsorbed (*6*). In this study we synthesized polymers containing both NIS units and amide or ether units. We also studied the surface modification of the polymer films by a chemical vapor deposition (CVD) method using alkoxysilane vapor at 30 °C. The incorporation of amide groups into polymers enhanced the sorption of water and thus increased the polysiloxane formation rate at the irradiated film surface. Pattern fabrication based on the present system was also investigated using an ArF excimer laser lithography.

Experimental

Materials. Synthesis of 1,2,3,4-tetrahydro-1-naphthylideneamino p-styrenesulfonate (NIS) was reported in detail elsewhere (*6*). Methyltriethoxysilane (MTEOS), tetramethoxysilane (TMOS), tetraethoxysilane (TEOS) and methyltrimethoxysilane (MTMOS) were of reagent grade and used without further purification. Methyl methacrylate (MMA), N,N-dimethylacrylamide (DMAA), acryloylmorpholine (AMO), and tetrahydrofurfuryl methacrylate (THFMA) were distilled before use.

Preparation of Polymers. Terpolymers (**2a-c** and **4**) and copolymers (**1**, **3**, **5**, and **6**) were prepared by the conventional radical copolymerization of corresponding monomers with 2,2'-azobis(isobutyronitrile) (AIBN) as an initiator at 55 °C. The concentrations of total monomer and AIBN in benzene or N,N-dimethylformamide were usually 4.5 and $1.6\text{X}10^{-2}$ mol/L, respectively. The sample solution was degassed under vacuum by repeating freeze-thaw cycles before polymerization. The content of the NIS units in the copolymers was determined by measuring the absorbances at 254 nm in CH_2Cl_2. The molar extinction coefficient of the NIS units in the polymers was estimated to be equal to that of the model compound, 1,2,3,4-tetrahydro-1-naphthylideneamino p-toluenesulfonate, ε being 15,300 L/mol·cm at 254 nm at room temperature. The composition of the terpolymers was determined by measuring absorbance at 254 nm in CH_2Cl_2 and from ^{1}H-NMR spectra. Polymerization conditions and characteristics of the polymers are shown in Table I. Structures of the polymers used in this study are shown in Scheme I.

Table I. Polymerization Conditions and Polymer Properties[a)]

Polymer[b)]	Monomer in Feed					Polymerization	Conversion	M_n	Composition (mol %)[b)]			T_g [c)]
	NIS (g)	MMA (g)	DMAA (g)	AMO (g)	THFMA (g)	Time (h)	(%)	X10-4	x	y	z	(°C)
1	1.3	3.7				7	40	5.6	0	83	17	96
2a	0.31	2.4	2.4			3	52	3.8	37	59	4	106
2b	1.3	1.8	1.8			2.5	34	4.1	39	37	24	- d)
2c	2.6	0.8	1.6			5	30	- e)	44	8	48	- d)
3	1.3		3.2			3.5	38	2.6	84	0	16	- d)
4	1.3	1.5		2.1		4	46	2.3	20	60	20	- d)
5	1.1			4.0		3.5	32	5.5	85	0	15	- d)
6	1.1				2.9	3	33	16.5	76	0	24	78

a) [Total monomer]=4.5 mol/L, [AIBN]=1.6X10-2 mol/L. N,N-Dimethylformamide (**2a-c**, **3**, **4** and **5**) and benzene (**1** and **6**) were used as a solvent. b) See Scheme I. c) Glass transition temperature from DSC. d) No distinct T_g was observed. e) Not measured.

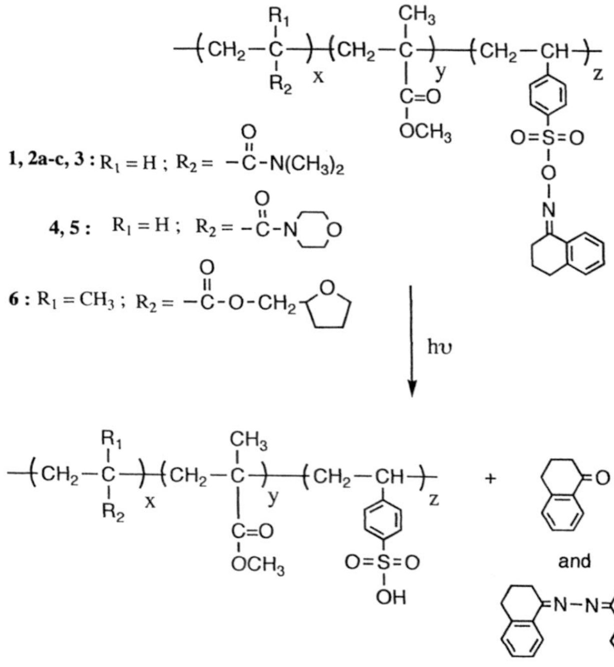

1, 2a-c, 3 : $R_1 = H$; $R_2 = -C(=O)-N(CH_3)_2$

4, 5 : $R_1 = H$; $R_2 = -C(=O)-N$(morpholino)

6 : $R_1 = CH_3$; $R_2 = -C(=O)-O-CH_2-$(tetrahydrofuranyl)

hυ

and

Scheme I. Structures and photochemical reaction of polymers

Water Sorption. A laboratory-constructed piezoelectric apparatus (quartz crystal microbalance) was used to measure water sorption in the polymer films. The AT-cut quartz crystal with gold electrodes (Hokuto Electronics) had a resonance frequency of 10.000 MHz. With this crystal, a frequency shift of 1 Hz corresponded to a mass change of 0.58 ng. The frequency change is linearly related to the mass sorbed on the quartz plate (*8,9*).

Polymers were deposited onto the quartz crystal (0.8 cm diameter) by casting from chloroform solution. The area coated with the polymer film was usually 0.13 cm^2. The quartz crystal was placed in the middle of the sealed glass vessel which had a quartz window for UV irradiation. 2M KNO_3 aqueous solution was placed at the bottom of the vessel to control its humidity (RH=95%) at 25 °C. Irradiation of polymer films on the quartz crystal through the quartz window of the vessel was carried out with 254-nm light using a 5-W low-pressure Hg lamp (Toshiba LP-11B). The intensity of the incident light determined with a chemical actinometer (potassium ferrioxalate) (*10*) was 0.1 $mJ/cm^2 \cdot sec$ at 254 nm.

Deposition of Polysiloxane. The polymer films (8.8 X 22 mm) were prepared on silicon wafers by spin-casting from diglyme solutions and drying under vacuum at room temperature. After exposure with 254-nm light using a low-pressure Hg lamp, the silicon wafer coated with polymer film was placed at the center of a 500 mL glass vessel which had gas-inlet and -outlet valves. Fifty mL of water was placed at the bottom of the vessel to adjust the relative humidity in the vessel and equilibrated for 10 min prior to introduction of the vapor of alkoxysilanes. During the polysiloxane network formation nitrogen gas (50 mL/min) flowed through a bubbler which contained liquid alkoxysilanes. The bubbler and reaction vessel were placed in a thermostatic oven at 30 °C. The amounts of polysiloxanes formed at the near surface of the polymer films were determined from the difference between the FT-IR absorbance at 1121 cm^{-1} of the sample plate before and after exposure to the vapor of alkoxysilanes.

Etching with Oxygen Plasma. Oxygen plasma etching was carried out at room temperature using a laboratory-constructed apparatus where the oxygen plasma was generated using two parallel electrodes and RF power supplies. The typical etching conditions were as follows: 20W power (13.56 MHz), power density of 1.0 W/cm^2, 125 mTorr, and oxygen flow of 1 sccm.

Lithographic Evaluation. Resists were spin-coated on Si substrate and baked at 90 °C for 90 sec. Diglyme was used as a solvent to make resist solutions. Exposure was carried out using a prototype ArF excimer laser exposure tool (NA=0.55).

Results and Discussion

The polymers **1**, **2a-c**, **3**, **4** and **6** were soluble in organic solvents such as diglyme, tetrahydrofuran, and dichloromethane. The polymer **5** was insoluble in diglyme.

Glass transition temperatures (T_g) of **1**, **2a** and **6** were observed to be 96, 106 and 78 °C, respectively. No distinct T_g was observed for the polymers other than **1**, **2a** and **6**. Polymers in CH_2Cl_2 solutions showed an absorption peak at about 200 nm and shoulder peaks at about 235 and 255 nm. The absorption peak at about 200 nm was due to the NIS units and amide units. The shoulder peaks were due to the NIS units. It has been reported that upon irradiation at 254 nm the cleavage of -O-N= bonds in the NIS units and subsequent abstraction of hydrogen atoms from residual solvent in the polymer film and/or from polymer molecules lead to the formation of p-styrenesulfonic acid units, tetralone, and tetralone azine (see Scheme I) (*11*). Change in the absorption spectrum of the **2a** film upon irradiation at 254 nm is shown in Figure 2. The absorbances at 200 and 254 nm decreased with irradiation time, and an isosbestic point was observed at 230 nm. In the present system photolytic decomposition of the NIS units was complete after irradiation of 207 mJ/cm^2. The quantum yield for the photolysis of the NIS units incorporated into poly(methyl methacrylate) was about 0.3 for 254-nm irradiation in air. A photolysis degree of the NIS units of **1**, **2a**, **2b** and **3** is plotted as a function of exposure dose in Figure 3. The photolysis rate was not strongly dependent on either structure of the amide units or the NIS unit fraction of the polymers.

$$n\ CH_3Si(OC_2H_5)_3 \quad + \quad 1.5n\ H_2O \xrightarrow{H^+} \text{Polysiloxanes} \quad + \quad 3n\ C_2H_5OH$$

In the presence of water and strong acids, the hydrolysis and subsequent polycondensation reactions of methyltriethoxysilane (MTEOS) and its homologues lead to the formation of polysiloxane networks, which is well known as the sol-gel process for the silica glass formation (*12*). When irradiated polymer films bearing the NIS units were exposed to the vapor of MTEOS at 30 °C, polysiloxane networks were formed in the near surface region of the film, which was confirmed by FT-IR analysis (Figure 4). The sample films irradiated and subsequently exposed to the MTEOS vapor showed new peaks at 3500 (Si-OH), 1272 ($Si-CH_3$) ,1000-1200 (Si-O-Si), and 790 ($Si-CH_3$) cm^{-1}. The presence of the peak due to Si-OH suggested the incomplete polycondensation reaction of the silanol moieties. No polysiloxane networks were formed in the unirradiated areas, because the p-styrenesulfonic acid units formed photochemically are essentially important for the acid-catalyzed hydrolysis of MTEOS molecules.

Figure 5 shows the effect of incorporation of hydrophilic units such as amide or ether units into the photoacid generating polymers on the polysiloxane formation rate at the irradiated film surface. NIS unit fractions of **1**, **2b**, **3** and **6** were 0.17, 0.24, 0.16 and 0.24, respectively. The amounts of acids photochemically formed were adjusted to be 4.5 mol% by controlling the irradiation time. The amounts of polysiloxanes formed at the film surface were measured by the enhanced absorbance at 1121 cm^{-1} due to Si-O-Si. The amounts of polysiloxane networks increased with CVD treatment time for all the polymers. The polysiloxane formation rate decreased in the order **3** > **2b** > **6** >**1** and the rate for **3** was about 18 times higher than that for **1**. On the other hand, the rate for **6** was about twice as that for **1**. Thus the incorporation of amide units in polymers effectively enhanced the polysiloxane

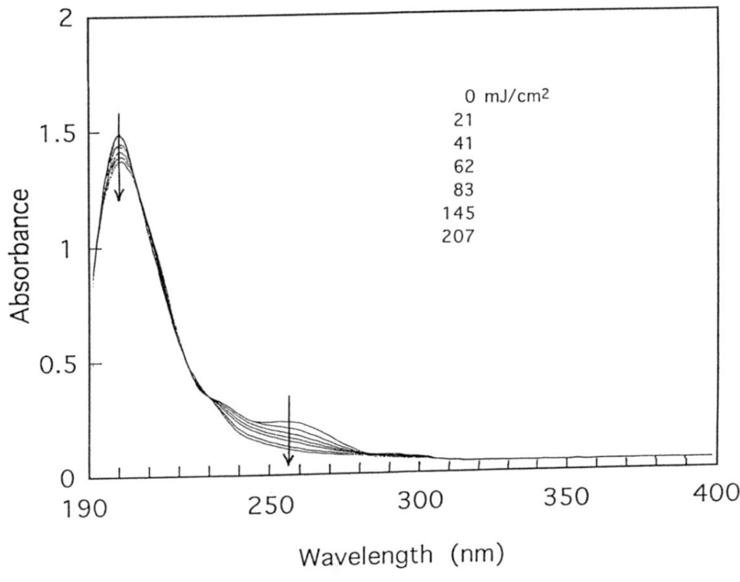

Figure 2. Spectral changes of the **2a** film upon irradiation with 254-nm light. Film thickness: 0.28 μm.

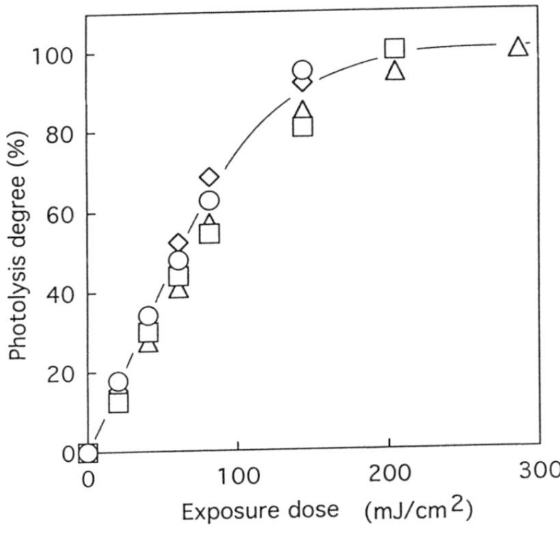

Figure 3. Relationship between photolysis degree of NIS units and exposure dose. Polymer: (○)**1**; (□) **2a**; (◇) **2b**; (△) **3**.

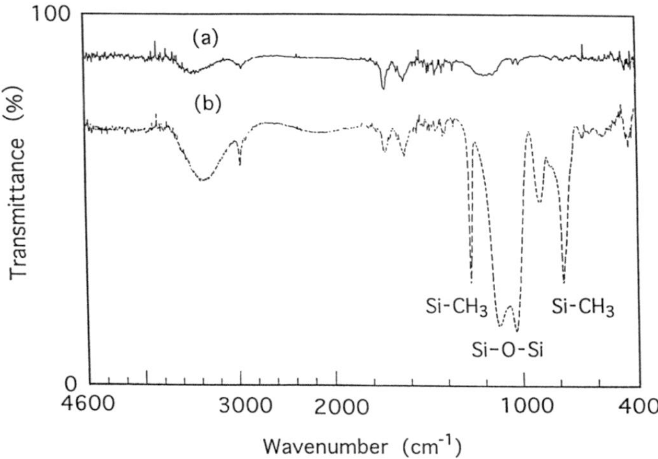

Figure 4. FT-IR spectra of the irradiated **2b** film (a) before and (b) after CVD treatment with MTEOS vapor at 30 °C for 20 min. Exposure dose: 166 mJ/cm^2.

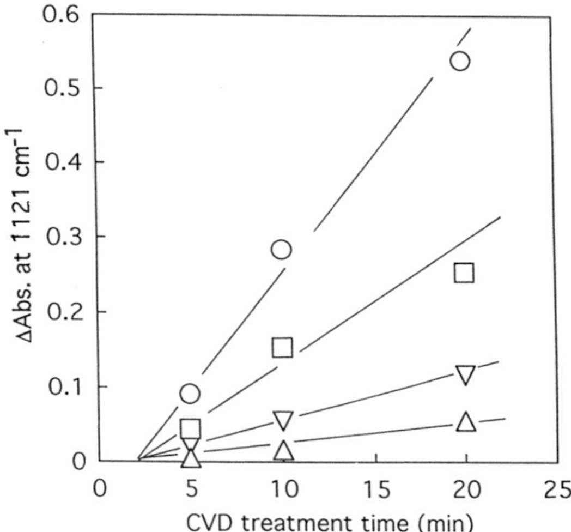

Figure 5. Effect of hydrophilic units in polymers on the polysiloxane formation at the irradiated film surface. CVD treatment was done with MTEOS vapor at 30 °C. Polymer: (△) **1**; (□) **2b**; (○) **3**; (▽) **6**.

formation rate. This may be due to the enhanced water sorption ability of the polymers because the amide units are generally hydrophilic. It was reported that the amounts of water sorbed onto the irradiated surface critically affected the hydrolysis rate of MTEOS molecules (*6*).

The enhanced water sorption ability of the amide-containing polymers was confirmed by measuring the amounts of water adsorbed using a quartz crystal microbalance. Figure 6 shows the relationship between the irradiation time and water sorbed into the films of **1**, **2b**, and **3** at relative humidity of 95%. The weight of polymer films cast on the quartz crystal was 2500 ng. If the polymer is assumed to have a density of 1 g/cm^3, the film thickness is 192 nm. Water sorption began the moment that the polymer film was irradiated with 254-nm light. It increased with the irradiation time and reached a constant value after irradiation. The water sorption ability of the polymers decreased in the order **3** > **2b** > **1**, suggesting that the amide units of the polymer enhanced the water sorption ability. The water sorption ability of **3** was about 5 times higher than that of **1**. It was found that the water sorbed onto the irradiated film surface was removed when the sample was placed under a dry nitrogen atmosphere.

Figure 7 shows the effect of acid generating unit fraction on the polysiloxane formation rate. The polysiloxane formation rate decreased in the order **2c** > **2b** > **2a**, if the photolysis degrees of the NIS units in the polymers were same (4.5 mol%). The higher the concentration of the photoinduced acids at the surface, the larger the polysiloxane formation rate.

The polysiloxane formation rate is also dependent on the structure of alkoxysilanes used. Although in the present system alkoxysilanes such as MTMOS, TMOS, MTEOS and TEOS could be used, the polysiloxane formation rate decreased in the order MTMOS > TMOS > MTEOS > TEOS. It was reported that the polysiloxane formation rate was determined by both hydrolytic reactivity of alkoxysilanes and vapor pressure of alkoxysilanes (boiling point of alkoxysilanes) (*6*).

As shown in Figure 8, the polysiloxane formation rate at the irradiated surface of **2b** and **4** films was almost the same. In this experiment, the photolysis degree of the NIS units of **2b** and **4** films was adjusted to be 4.5 mol%. Since the morpholine unit fraction of **4** was about half of the N,N-dimethylacrylamide (DMAA) units of **2b**, the morpholine units seem more effective for the enhancement of the water sorption ability. This was confirmed by the experiments on the water sorption of these polymers using a quartz crystal microbalance. The polymer **5** was not checked because the solubility of **5** in diglyme was too low to make sample films by spin-casting. Thus, polymer with high content of morpholine units has a disadvantage in terms of solubility in organic solvents.

Figure 9 shows the effect of surface modification of **2b** film on the oxygen plasma etching. The etching rate of **2b** film without modification was 0.08 μm/min under the present etching conditions. It was almost the same as that observed for poly(methyl methacrylate) film. The sample film was irradiated at 254 nm (10 mJ/cm^2) and subsequently exposed to the vapor of MTEOS for 5 min at 30 °C. The etch rate of the modified film was lower than 1/40 compared to that of the unmodified

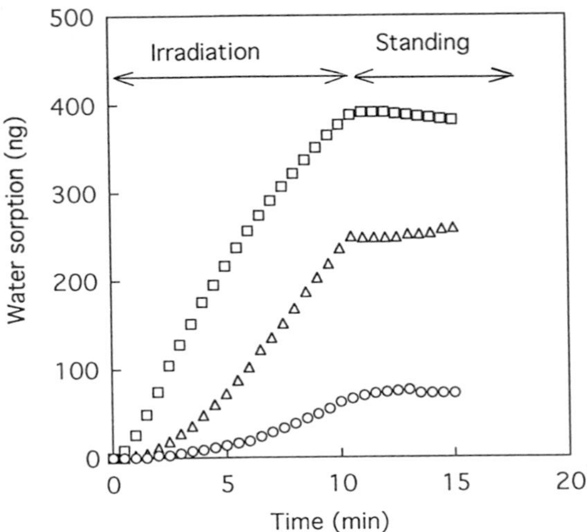

Figure 6. Relationship between irradiation time and water sorption into polymer films. Polymer weight: 2500 ng; relative humidity: 95%. Polymer: (△) **1**; (□) **2b**; (○) **3**.

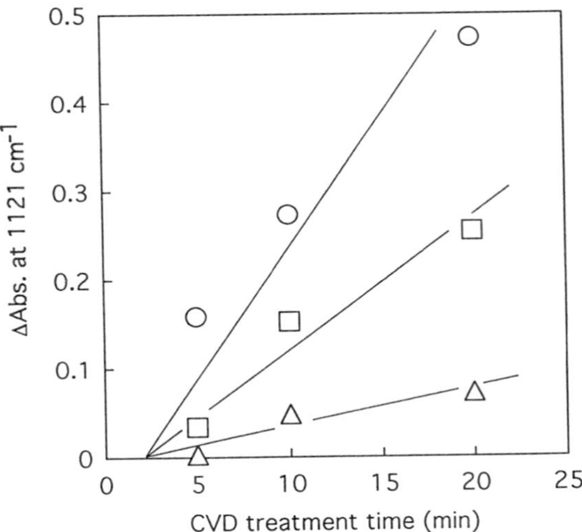

Figure 7. Effect of acid generating unit fraction of the polymers on the polysiloxane formation at the exposed areas. CVD treatment was done with MTEOS vapor at 30 °C. Polymer: (△) **2a**; (□) **2b**; (○) **2c**.

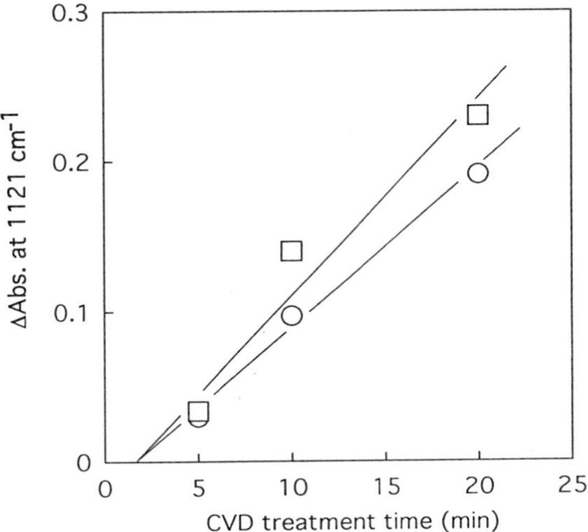

Figure 8. Effect of amide unit structure on the polysiloxane formation at the irradiated film surface. Polymer: (□) **2b** and (○) **4**.

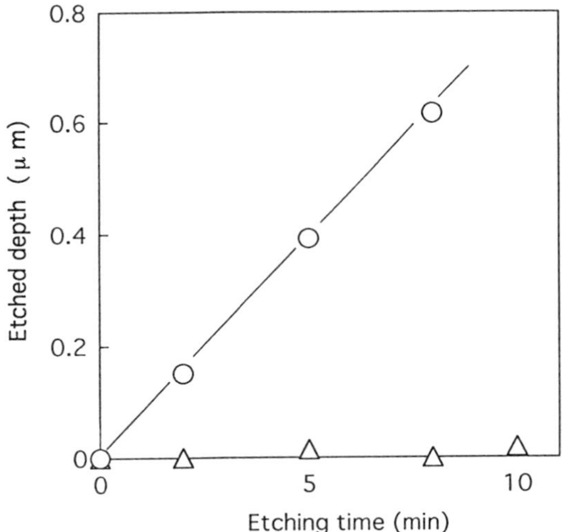

Figure 9. Effect of surface modification on the etching rate of **2b** film. CVD treatment was carried out with MTEOS vapor at 30 °C. Exposure dose: (○) 0 and (△) 10 mJ/cm^2.

2b film. The etching rate of the modified film decreased with increasing the number of Si element at the surface.

Preparation of line-space images using an ArF excimer laser stepper was studied. Figure 10(a) shows a SEM micrograph of 0.2 μm line-space images obtained using the polymer **1**. Exposure dose was 160 mJ/cm^2 and the CVD treatment was carried

(a)

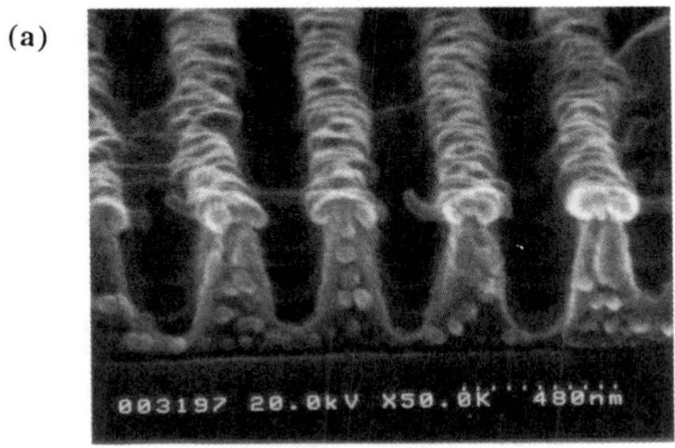

(b)

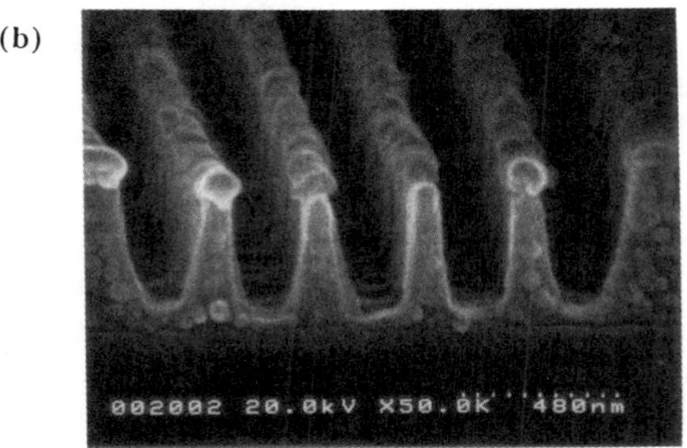

(c)

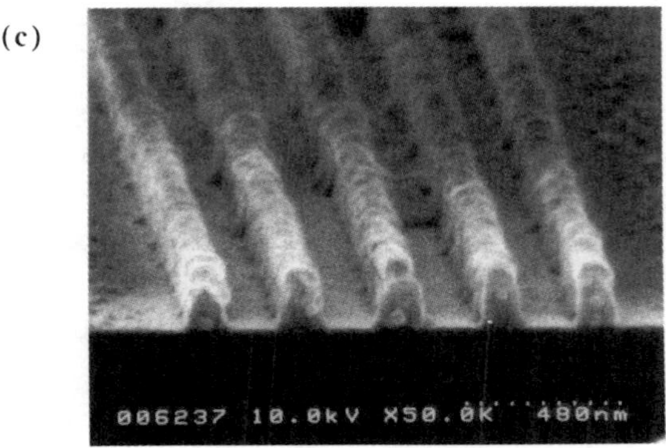

Figure 10. Scanning electron micrographs of negative images obtained by surface modification method using (a) **1**, (b) **2b** and (c) **4**.

out with MTEOS for 60 min at 30 °C. The resist sensitivity was relatively low. A SEM micrograph of 0.17 μm line-space features obtained by exposure (40 mJ/cm^2) of the polymer **2b** and subsequent CVD treatment with MTEOS vapor for 40 min is shown in Figure 10(b). Figure 10(c) shows a SEM micrograph of 0.16 μm line-space images obtained using polymer **4**. Exposure dose was 65 mJ/cm^2 and CVD treatment was carried out with MTMOS for 4 min. The resist sensitivity of the polymers **2b** and **4** was much improved compared to the polymer **1**.

Conclusions

Photoacid generating polymers bearing hydrophilic groups such as amide or ether groups were used as oxygen plasma-developable surface modification resists. When the irradiated polymer surface was exposed to the vapor of alkoxysilanes, polysiloxane networks were formed at the surface. The incorporation of amide units into the polymer chain strongly enhanced the rate of polysiloxane formation at the irradiated surface. This was due to the enhanced water sorption ability of the amide-containing polymers. Negative-tone images were obtained based on the present resist system by ArF excimer laser lithography and oxygen plasma etching.

Acknowledgment

A part of this work was performed under the management of ASET in the MITI's R&D Program supported by NEDO.

References

1. Maeda, K.; Ohfuji, T.; Aizaki, N.; Hasegawa, E. *Proc. SPIE*, **1995**, 2438, 465.
2. Palmateer, S. C.; Kunz, R. R.; Horn, M. W.; Forte, A. R.; Rothschild, M. *Proc. SPIE*, **1995**, 2438, 455.
3. *Microelectronics Technology-Polymers for Advanced Imaging and Packaging*, Reichmanis, E.; Ober, C. K.; MacDonald, S. A.; Iwayanagi, T.; Nishikubo, T. Eds., ACS Symposium Series No 614, American Chemical Society, Washington, DC, 1995.
4. Matsuo, T.; Endo, M.; Shirai, M.; Tsunooka, M. *J. Vac. Sci. Technol. B*, **1996**, 14, 4212.
5. Shirai, M.; Kinoshita, H.; Sumino, T.; Tsunooka, M. *Chem. Mater.*, **1993**, 5, 98.
6. Shirai, M.; Hayashi, M.; Tsunooka, M. *Macromolecules*, **1992**, 25, 195.
7. Shirai, M.; Tsunooka, M. *Proceedings of the 10th International Conference on Photopolymers*, New York; Mid-Hudson Section, SPE: Ellenville, New York, 1994; p. 128.
8. Sauerbrey, G. *Z. Phys.*, **1959**, 155, 206.
9. Alder, J. F.; McCallum, J. J. *Analyst*, **1983**, 108, 1291.
10. Murov, S. L. *Handbook of Photochemistry*; Dekker; New York, 1973.
11. Shirai, M.; Masuda, T.; Ishida, H.; Tsunooka, M.; Tanaka, M. *Eur. Polym. J.*, **1985**, 21, 781.
12. Hench, L. L.; West, J. K. *Chem. Rev.*, **1990**, 90, 33.

Chapter 24

Material Design and Development for Aqueous Base Compatible High-Performance Deep UV Negative-Tone Resists

Pushkara Rao Varanasi[1], Hiroshi Ito[2], Phil Brock[2], Gregory Breyta[2], William R. Brunsvold[1], and Ahmad D. Katnani[1]

[1]IBM Microelectronics, 1580 Route 52, Hopewell Junction, NY 12533
[2]IBM Almaden Research Center, 650 Harry Road, San Jose, CA 95120

The use of "acid catalyzed crosslinking" as a mechanism in the design of aqueous base compatible chemically amplified negative-tone resist materials for photolithography has been addressed (*1*). Although the potential of such a mechanism has been demonstrated in many negative-tone resist materials, a few studies have been directed toward understanding the correlation among the material properties, resist characteristics and lithographic features (*2,3*). In this study, we identified several "limiting factors" that affect the dense-line resolution of aqueous base compatible negative resists. We also analyzed various material systems against these limiting factors and provided an understanding of the material and resist characteristics.

Negative-Tone Resists: Performance Issues

The aqueous base compatible negative resists that perform based on acid induced crosslinking mechanism, generally suffer from two limitations: microbridging and line-collapsing. Microbridging, herein, means a series of tendrils or strands of unremoved resist between adjacent resist features in a developed photoresist. Line-collapsing could be viewed as adjacent developed resist features falling over one another. Both these factors limit the resolution capabilities of DUV negative resists. For any given feature size, increase in dose (photon-energy) tends to cause microbridging,while decrease in dose leads to line-collapsing. As the feature sizes become smaller and smaller (0.2 μm or below), both the microbridging and line-collapsing are the dominant failure modes for achieving resolution (Figure 1). The increase in resin matrix molecular weight (by several orders of magnitude) due to excessive crosslinking during the resist chemistry is believed to be responsible for the microbridging (*2*). On the other hand, the lack of structural integrity of the crosslinked features toward the developer is attributed to the line-collapsing phenomenon. With aqueous base developers, the developer normality is observed to play a vital role in dictating the dominance of these resolution limiting factors (*2,4*). The success in the design of high resolution DUV negative resists primarily depends on understanding the material properties, resist characteristics and process parameters that control the microbridging and line-collapsing.

Material Systems Studied

Two classes of negative resist materials have been studied to understand the structural features and resist characteristics that affect the microbridging and line-collapsing. The traditional classification is based on the number of components present in the resist formulation (e.g., two-component or three-component) excludingthe solvent used. The two-component systems consist of the polymer matrix and the photoacid generator. In such systems, the photoacid causes crosslinking reaction among the polymer chains, either intramolecularly or intermolecularly (*5*). The three-component systems are based on an external crosslinker, polymer matrix and photoacid generator. In this case, the photoacid induces a crosslinking reaction between the crosslinker and the polymer chain (*2-4,6,7*).

Two-Component Systems

To understand the lithographic potential or limitations of the two-component negative resist material systems, we reinvestigated the material system derived from copolymers of 4-vinylbenzyl acetate (VBA) and 4-hydroxystyrene (HS) and a photoacid generator, developed by Frechet and Willson et al. (*5*). The previous studies clearly demonstrated the mechanism of crosslinking reaction, the resist sensitivity and chemical contrast. The acid, generated from the photoacid generator during exposure, reacts with benzylacetate moieties and generates reactive electrophiles. The resultant benzylic cation undergo electrophilic substitution with the pendant phenolic units of the polymer and, thereby, generate the crosslinked matrix (Scheme 1). Despite the tidy understanding of this material system, virtually no lithographic data has been reported to examine its suitability to deep UV lithography.

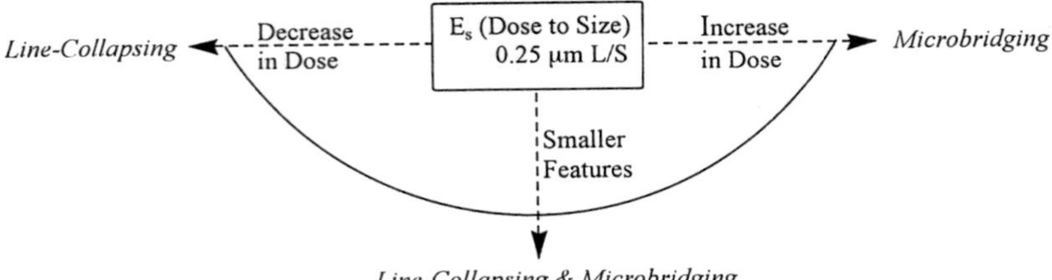

Figure 1. Crosslinking based DUV negative-tone resists: Resolution limiting factors.

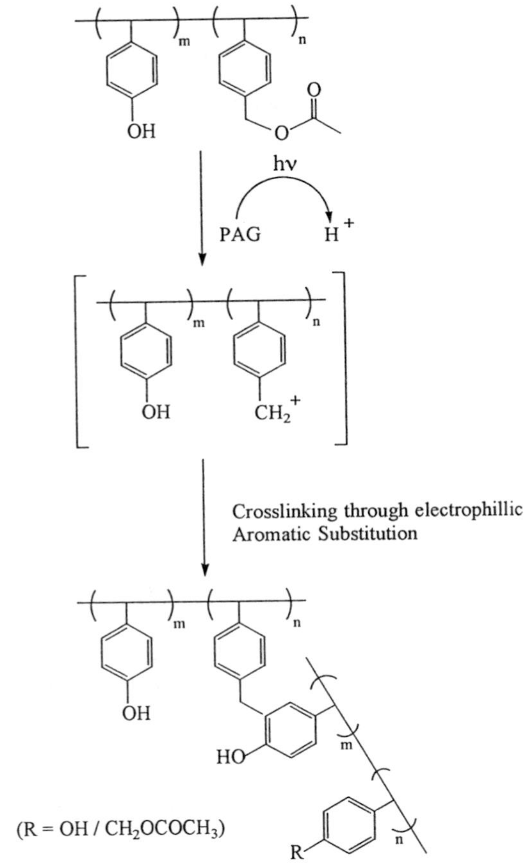

OH

hν

PAG H^+

OH CH_2^+

Crosslinking through electrophillic Aromatic Substitution

OH

HO

(R = OH / CH_2OCOCH_3)

Scheme 1. Crosslinking chemistry in a two-component system comprising 4-vinylbenzyl acetate (VBA) and 4-hydroxystyrene (HS) copolymer and photoacid generator (PAG).

The VBA-HS copolymers are highly soluble in commonly used organic casting solvents irrespective of the copolymer ratio. However, their solubility in the aqueous base developer (i.e., dissolution rate, DR) is strongly dependent on the VBA content in the copolymer. The dependence of VBA molar concentration on the dissolution rate of the VBA-HS copolymer in industry standard 0.263N tetramethylammonium hydroxide (TMAH) developer is shown in Figure 2. The data clearly indicate that the copolymer dissolution rate in the developer decreases with increase in VBA molar concentration. The copolymers containing 20% or above VBA content are virtually impossible to dissolve in the 0.263N TMAH developer. The copolymers containing 10% or lower VBA units possess acceptable dissolution rates and therefore, are suitable for lithographic studies.

Since benzylacetate moieties in the copolymer are the functionalities that react with the photoacid and induce the crosslinking reaction, there is a need to understand how much VBA concentration is essential to design a high resolution negative resist. And also, since the resin molecular weight is believed to influence the microbridging, it is imperative to understand the effect of copolymer molecular weight on the lithographic performance. For this reason, two sets of experiments were designed: for the first set of experiments, copolymers were made by maintaining the polymer molecular weight constant and varying the copolymer ratio (VBA:HS; 5:95 and 10:90). For the second set of experiments, copolymers of different molecular weights (Mn = 12,388 and Mn = 5,373) were prepared by keeping the VBA:HS ratio constant. Furthermore, since the copolymers possess reasonable dissolution inhibiting characteristics, the photoacid generator employed in the formulation must not cause any further dissolution inhibition which otherwise would require much longer development times (the industry practiced development rate range: ~ 30-120 seconds). For this reason, a noninhibiting N-hydroxyimide-triflate was used in preparing resist formulations. All resist formulations were made using propylene glycol monomethylether acetate (PGMEA) solvent. Spin coating with appropriate spin speed on a silicon wafer, followed by post apply bake (PAB) of 110 °C for 60 sec. resulted in 0.6 μm thin-films suitable for lithography. Exposures were first made with a 0.37NA deep UV stepper. Those samples that were determined to be suitable for high resolution negative resist were later exposed with a higher NA (0.5) deep UV stepper. Exposed resists were developed with aqueous tetramethylammonium hydroxide (TMAH) developers. The development time employed for each material system was maintained as 4 x t_c (t_c represents the time requires to develop or clear the unexposed resist film). Under these developing conditions, the resist systems studied in this work showed the optimized lithographic performance. The resist contrast, and representative dense-line profiles obtained from SEM analysis are shown to provide insight into the lithographic potential of these material systems.

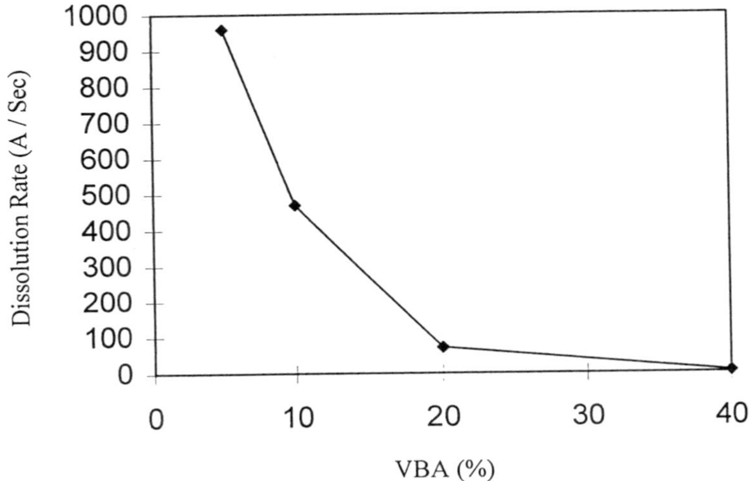

Figure 2. The effect of VBA concentration on the dissolution rate of VBA:HS copolymers in 0.263N TMAH developer.

The resist contrast behavior of the formulations containing copolymers with VBA:HS ratios of 5:95 and 10:90 is shown in Figure 3a. From the data, it is clear that the increase in VBA content makes the resist faster ($E_0 = 4$ mJ/cm^2 for VBA:HS = 10:90; $E_0 = 5$ mJ/cm^2 for VBA:HS = 5:95). However, the slope of the contrast curve is quite comparable for both the copolymers. Figure 3b represents the dense line resolution data obtained for these formulations. Since the NA of the stepper employed is 0.37, one would estimate the resolution limit of the tool as 0.35 μm. The data indicate that neither of these formulations seem to resolve dense lines of .35 μm. The line-collapsing seems to occur for features below 0.45 μm for the polymer containing lower VBA content. For the polymer containing higher VBA content, the line-collapsing may not be a serious limiting factor, but severe microbridging seems to be the failure mode. A general conclusion from this comparative study is that for the features to have structural integrity toward the developer, the polymer must possess certain minimum number of crosslinking sites. Below which the resist contrast may not be affected, but the resolution capabilities may be influenced.

The effect of resin molecular weight on the chemical contrast and lithographic performance is shown in Figures 4a and 4b. The copolymer ratio was maintained as 10:90 (VBA:HS) for this study. From the contrast curves, it is clear that lowering the resin molecular weight makes the resist slower, but the contrast is less sensitive to the resin molecular weight change. However, lowering the molecular weight is observed to reduce the microbridging between the dense lines and improve the resolution to 0.3 μm. This data is quite promising considering that the resolution limit of 0.37NA tool is 0.35 μm. To understand the suitability of this resist for 0.25 μm ground rules, the formulation containing the lower molecular weight copolymer was exposed with a higher NA (0.5) ASML tool. The SEM data is summarized in Figure 5. The data indicate that the resolution can be extended to below 0.25 μm dimensions without being affected by microbridging or line-collapsing. It is interesting to note that no microbridging is seen even for the lines of 0.22 μm.

The conclusion from this study is that the microbridgingcan be significantly eliminated by lowering the molecular weight, whereas the line-collapsing can be reduced by increasing the VBA content in the copolymer. The combination of both these factors is shown to make two component systems involving VBA-HS copolymers suitable for 0.25 μm ground rules. Further improvements in the masking linearity and the resolution below 0.2 μm can be sought by designing copolymers along these lines, and also by employing photoacid generators which produce acids of different strengths and different sizes.

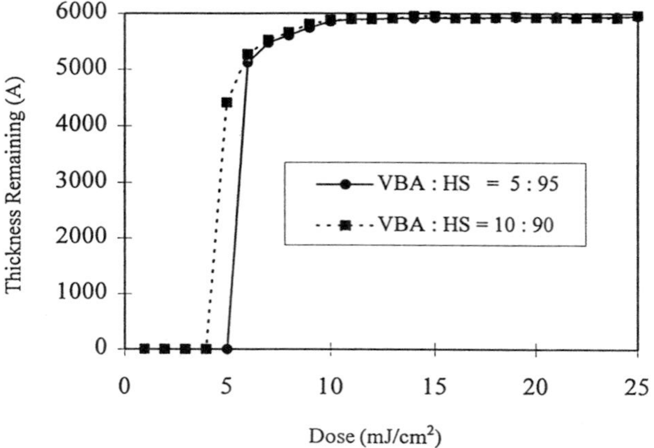

Figure 3a. The effect of VBA:HS copolymer ratio on the resist contrast; Resist composition: VBA:HS copolymer (Mn = 12388), N-hydroxyimide-triflate, PGMEA.

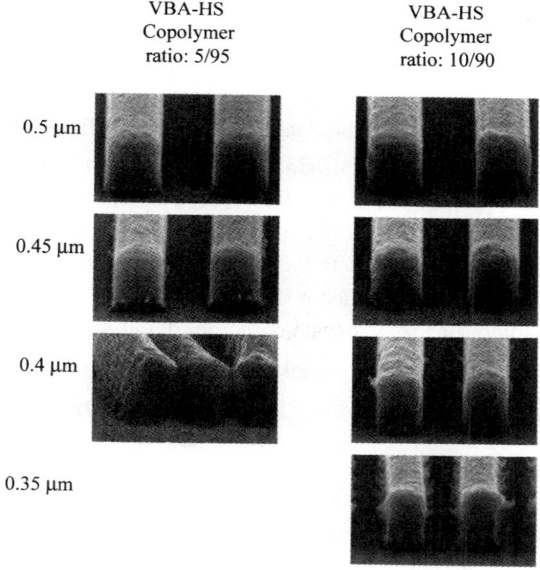

Figure 3b. The effect of VBA:HS copolymer ratio on the lithographic performance; Resist formulation: VBA:HS copolymer (Mn = 12388), N-hydroxyimide-triflate, PGMEA; Exposure system: Canon 0.37NA; Substrate: Si; PAB/PEB: 110°C / 110°C; Resolution of 1:1 L/S 0.5 μm and below.

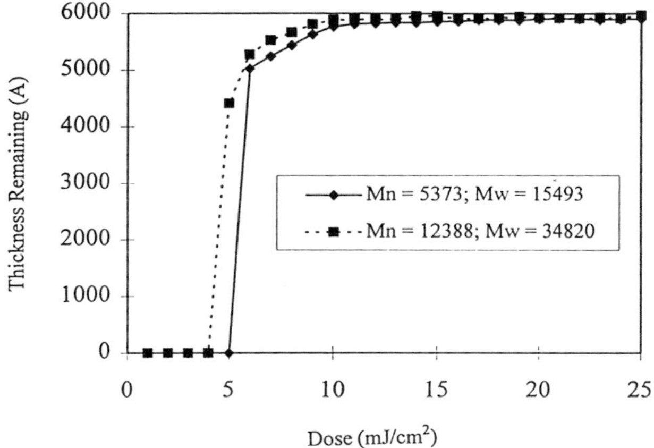

Figure 4a. Dependence of the resist contrast on the resin molecular weight for the formulation: Copolymer VBA:HS (10:90), N-hydroxyimide-triflate, PGMEA.

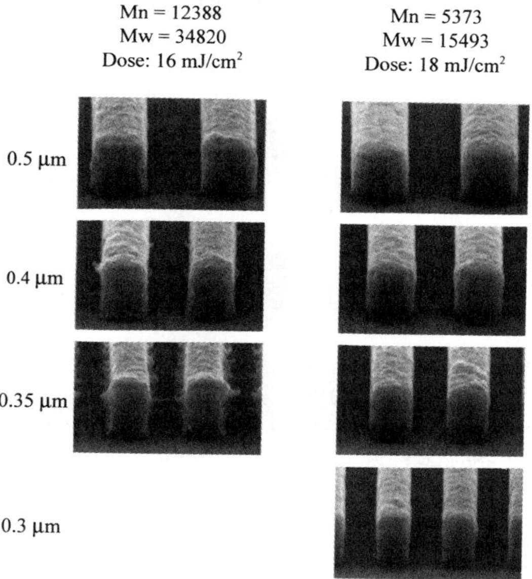

Figure 4b. The effect of resin molecular weight on the lithographic performance; Resist formulation: Copolymer VBA:HS (10:90); N-hydroxyimide-triflate, PGMEA; Exposure system: Canon 0.37NA; Substrate: Si; PAB/PEB: 110°C / 110°C; Resolution of 1:1 L/S 0.5 μm and below.

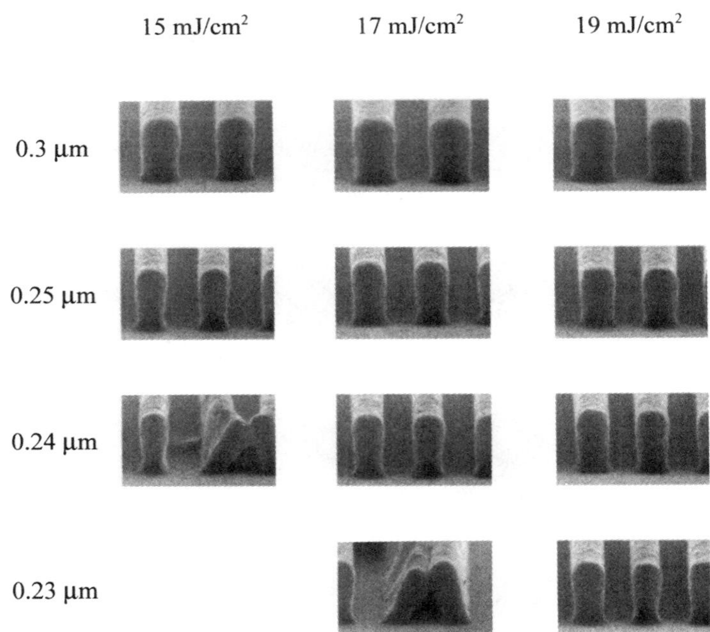

Figure 5. Lithographic performance of the resist formulation: Copolymer VBA:HS (10:90; Mn = 5373; Mw = 15493), N-hydroxyimide-triflate, PGMEA; Exposure system: ASML 0.5NA; Substrate: Barl 1200; PAB/PEB: 110°C / 110°C; Exposure latitude for 1:1 L/S 0.3 μm and below.

Three-Component Systems

In contrast to the two-component material systems discussed above, the crosslinking chemistry in the three-component systems is primarily due to the reaction between the resin matrix and an external crosslinker (*2-4,6,7*). In exploring this class of materials for DUV negative resists, we began our studies by using poly(4-hydroxystyrene), PHS, as the polymeric resin component and tetramethoxymethyl glycouril (TMMGU, trade name: Powderlink) as the external crosslinker. Similar to other aminoplast crosslinkers (*2,6a-c*), the multifunctional powderlink reacts with phenolic moieties in the presence of acid catalyst and forms a crosslinked matrix (Scheme 2). Some recent studies suggest that the crosslinking reaction between powderlink and PHS occurs exclusively through O-alkylation (*3,6a*). This observation is quite remarkable considering the fact that crosslinkers which yield multifunctional benzylic cations are known to react with phenolic resins through O-alkylation as well as C-alkylation (*6d-f*). Since the aqueous base solubility of phenolic resins is due to the phenolic hydroxyl, the crosslinking reaction that occurs through O-alkylation may have added advantages over that resulting from C-alkylation in providing structural integrity to the resolved dense-lines in aqueous base developed negative resists.

Using PGMEA as solvent, the resist formulations were made with PHS, powderlink and N-hydroxyimide-triflate (photoacid generator). Unlike the VBA-HS copolymers discussed above, PHS dissolve rapidly in the aqueous base developer. The dissolution rates of PHS in 0.263N and 0.14N TMAH developer are ~ 3000 A/sec. and ~ 300 A/sec. respectively. Since the dissolution characteristics of PHS are developer normality dependent, we studied the lithographic potential of this system using both 0.14N and 0.263N TMAH developers. The other components, powderlink and N-hydroxyimide-triflate barely contribute to the dissolution inhibition of the resist.

The resist contrast and lithographic features obtained for the resist formulation containing PHS, powderlink and N-hydroxyimide-triflate, are shown in Figure 6. From the Figure 6a, it is clear that the developer normality has effect on the photospeed: the lower the normality of the developer, the faster the photospeed. While the resist contrast appears to be comparable for both these normalities of the developer, the higher normality seems to provide marginally better contrast than the lower one. The dose to size (E_s) for 0.5 μm lines is 13 mJ/cm^2 for the 0.14N TMAH developer, whereas for the 0.263N TMAH developer the E_s is 17 mJ/cm^2. With 0.263N TMAH developer, the resolution is significantly retarded by line-collapsing: the features smaller than 0.4 μm either fall-over one another or do not resolve at all. On the other hand, with the 0.14N developer, resolution down to 0.3 μm is observed. Neither microbridging nor line collapsing

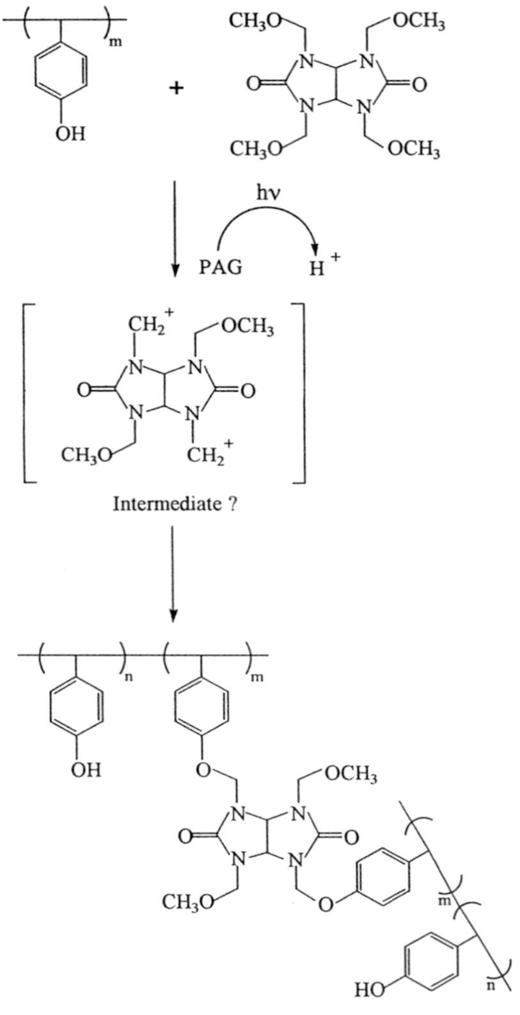

Scheme 2. Crosslinking chemistry in a three-component system comprising poly(4-hydroxystyrene), external crosslinker (powderlink) and photoacid generator (PAG).

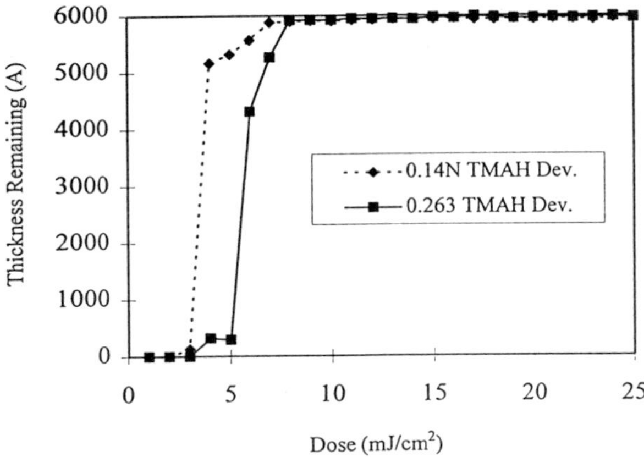

Figure 6a. The effect of developer normality on the resist contrast for the formulation: PHS, powderlink, N-hydroxyimide-triflate, PGMEA.

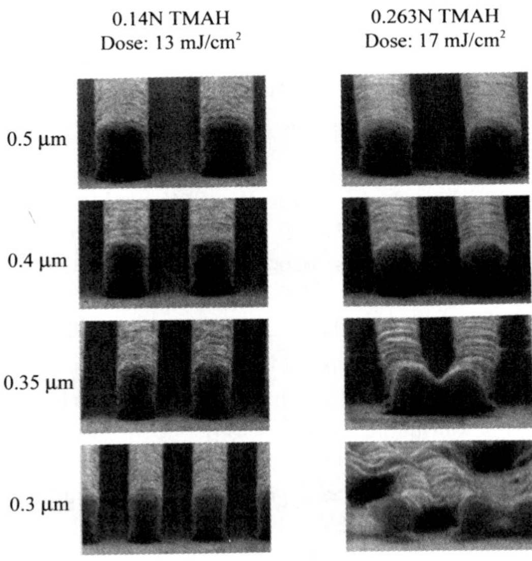

Figure 6b. The effect of developer normality on the lithographic performance; Resist formulation: PHS, powderlink, N-hydroxyimide-triflate, PGMEA; Exposure system: Canon 0.37NA; Substrate: Si; PAB/PEB: 110°C / 110°C; Resolution of 1:1 L/S 0.5 μm and below.

appears to affect the resolution with the weaker developer. The effects of developer normality may be better understood by experimentally monitoring the dissolution rate at each exposure dose and comparing the resist contrasts. In the absence of such data, with simple initial dissolution rate differences of PHS in 0.14N and 0.263N developers, one could predict that by simply creating dissolution inhibition on PHS through a functionalization of phenolic hydroxyl with hydrophobic groups, it would be possible to improve the resist resolution capabilities in 0.263N developer. Although the lithographic data clearly points out the high resolution for the lower developer normality, we focused the rest of the experimental design toward understanding the material characteristics that would be required to achieve high resolution for the 0.263N developer since this is the industry practiced developer normality.

The dissolution rate characteristics of PHS can be conveniently fine-tuned by substituting some of the phenolic hydroxyls with blocking groups (*7*). In this study, isopropyloxycarbonate was chosen as a functional group to block the phenolic hydroxyls. Any blocking group that can inhibit the dissolution rate, is thermally stable, and is less sensitive to the acid, would be suitable for this purpose. The isopropyloxycarbonate group satisfies all the requirements. Using 12% blocking level, the dissolution rate of PHS is controlled significantly (DR of PHS-IP is 542 A/sec. in 0.263N developer). The resist formulations were made exactly under the same conditions as the unprotected PHS material system. For these experiments, triphenylsulfonium triflate (TPS-TF) was employed as a photoacid generator. Unlike N-hydroxyimide-triflate, TPS-TF provides dissolution inhibition to the resist formulation.

The lithographic data obtained from an ASML 0.5NA exposure system for formulations containing PHS and PHS-IP are summarized in Figure 7. The SEM data clearly shows the advantages associated with the isopropyloxycarbonate substitution on PHS. With PHS based formulation, the microbridging limits the dense line resolution to 0.3 μm. Whereas, the formulation derived from PHS-IP shows the dense-line resolution up to 0.24 μm, the feature sizes smaller than that suffer from microbridging. Since lowering the resin molecular weight has been shown to reduce the microbridging in VBA-HS polymers discussed above, PHS-IP was prepared using a lower (Mw = 5000) molecular weight PHS. The formulations were made using TPS-TF as photoacid generator and their lithographic potential was studied using a Micrascan 0.5NA exposure system. The results from the exposure and focus latitude for 0.25 μm dense-lines and the masking linearity are summarized in Figure 8. The masking linearity data indicate the linearity down to 0.22 μm. No microbridging is seen. The observed 25% exposure latitude and 1.2 μm focus latitude for 0.25 μm features make this resist suitable for deep-UV lithography (0.25 μm ground rules). These studies unambiguously prove that three

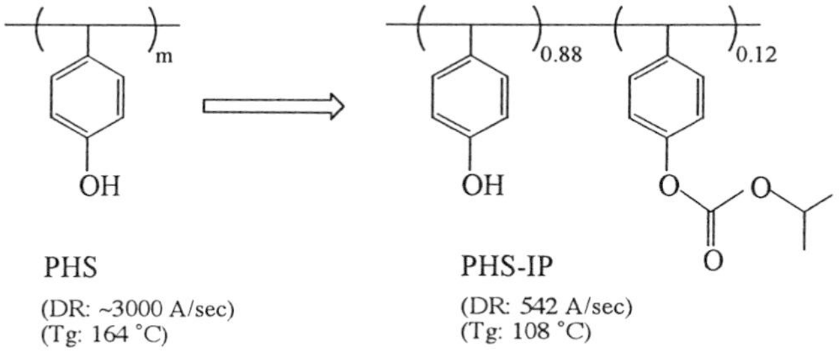

PHS

(DR: ~3000 A/sec)
(Tg: 164 °C)

PHS-IP

(DR: 542 A/sec)
(Tg: 108 °C)

Scheme 3. PHS and PHS-IP (isopropyloxy carbonate); DR represents dissolution rates in 0.263N TMAH; Tg represents glass transition temperature.

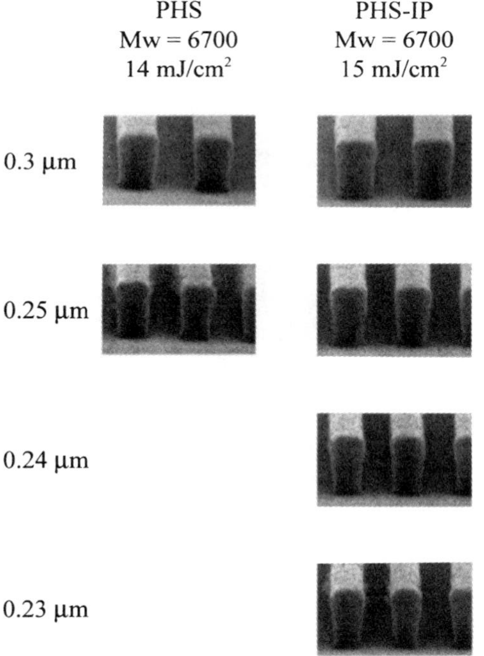

Figure 7. Influence of the blocking group on the lithographic performance; Polymers: PHS and PHS-IP (protection level: 12%); Photoacid generator: TPS-TF; Exposure system: ASML 0.5NA; Substrate: Si; PAB/PEB: 110°C / 110°C; Resolution for 1:1 L/S 0.3 μm and below.

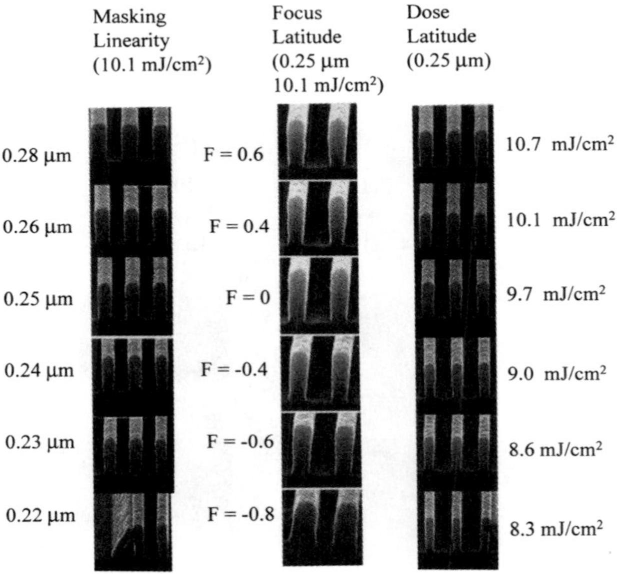

Figure 8. Lithographic performance of the resist formulation: PHS-IP (12%; Mw = 5000); Photoacid generator: TPS-TF; MSII (0.5NA) exposure data; Dose and focus latitude for 0.25 μm L/S and the masking linearity.

component systems based on PHS and external crosslinkers can be made suitable for 0.25 μm lithography by introducing inert blocking groups on PHS and fine-tuning the dissolution characteristics of PHS.

Summary

This study demonstrates that the dense-line resolution of aqueous base acid-catalyzed negative-tone resists is limited principally by line-collapsing and microbridging. Using two classes of materials, attempts have been made to understand the correlation among the material properties, resist characteristics and lithographic limiting factors. In the two-component systems containing 4-vinylbenzyl acetate and 4-hydroxystyrene copolymers, both the microbridging and line-collapsing can be reduced by lowering the resin molecular weight and increasing the 4-vinylbenzyl acetate content. In the three component systems containing poly(4-hydroxystyrene) and Powderlink, the developer normality plays a significant role in dictating the lithographic performance, with the resist performance getting better with lower developer normality. In poly(4-hydroxystyrene) based three component systems, inert blocking groups such as isopropyloxycarbonate will make the resist compatible for higher normality developers with improved lithographic performance. The effect of photo acid size, shape, and strength on the resolution limiting factors such as microbridging and line-collapsing will be the subject of our future publications.

Acknowledgements

The authors would like to thank Mr. W. Conley, Drs. N. Patel and M. Neisser for their technical contributions.

Literature Cited

1. a) Willson, C.G.; In *Introduction to Microlithography*, Thompson, L. F.; Willson, C. G.; Bowden, M. J., Eds.; American Chemical Society; Washington DC, 1994, Chapter 3; b) Iwayanagi, T.; Kohashi, T.; Nonogaki, S.; Matsuzawa, T.; Daita, K.; Yanazawa, H. *IEEE Trans. Electron Devices* **1981**, *ED-28*, 1306; c) Feely, W. E.; Imhof, J. C.; Stein, C. M.; Fisher, T. A.; Legenza, M. W. *Poly. Eng. Sci.* **1986**, *26*, 101.
2. Thackeray, J. W.; Orsula, G. W.; Denison, M. *Proc. SPIE* **1994**, *2195*, 152.
3. Lin, Q.; Katnani, A. D.; Willson, C. G. *Proc. SPIE* **1997**, *3049*, 974.

4. Linehan, L.; Smith, R.; Fahey, J.; Moreau, W.; Spinillo, G.; Puttlitz, E.; Collins, J. *Proc. SPIE* **1995**, *2438*, 211.
5. Frechet, J. M.; Matuszczak, S.; Reck, B.; Stover, H. D. H.; Willson, C. G. *Macromolecules* **1991**, *24*, 1746.
6. a) Thackeray, J. W.; Orsula, G. W.; Rajaratnam, M. M.; Sinta, R.; Herr, D.; Pavelchek, E. *Proc. SPIE* **1991**, *1466*, 39; b) Yamaguchi, A.; Kishimura, S.; Tsujita, K.; Morimoto, H.; Tsukamoto, K.; Nagata, H. *J. Vac. Sci. Technol.* **1993**, *B11*, 2867; c) Roschert, H.; Dammel, R.; Eches, Ch.; Kamiya, K.; Meier, W.; Przybilla, K-J.; Spiess, W.; Pawlowski, G. *Proc. SPIE* **1992**, *1672*, 157.; d) Lee, S. M.; Frechet, J. M. J.; Willson, C. G.; *Macromolecules* **1994**, *27*, 5154; e) Kajita, T.; Koboyashi, E.-I.; Ota, T.; Miura, T. *Proc. SPIE* **1993**, *1925*, 133; f) Lee, S. M.; Frechet, J. M. J. *Macromolecules* **1994**, *27*, 5160.
7. Brunsvold, W. R.; Conley, W.; Varanasi, P. R.; Khojasteh, M.; Patel, N. M.; Molless, A.; Neisser, M.; Breyta, G. *Proc. SPIE* **1997**, *3049*, 372.

Chapter 25

Advanced Chemically Amplified Resist Process Using Non-Ammonia Generating Adhesion Promoter

M. Endo and A. Katsuyama

ULSI Process Technology Development Center, Matsushita Electronics Corporation, 19 Nishikujo-Kasugacho, Minami-ku, Kyoto 601, Japan

We have developed a new non-ammonia generating adhesion promoter, 4-trimethylsiloxy-3-pentene-2-one. Its adhesion capability for a substrate is superior to a conventional adhesion promoter, hexamethyldisilazane, owing to its high reactivity. We obtained high aspect ratio and precise chemically amplified resist patterns on Si and TiN substrates using this new adhesion promoter.

Chemically amplified resist system is a promising technology to attain high resolution and high sensitivity for sub-quarter micron device fabrication. However, air-borne contamination (1-3), such as ammonia mainly generated from conventional adhesion promoter, hexamethyldisilazane (HMDS), severely affects this kind of resist. It causes surface insoluble layer of resist patterns, which results in failure of the pattern fabrication.

We developed a non-ammonia generating adhesion promoter, isopropenoxytrimethylsilane (IPTMS), in place of HMDS and its application to chemically amplified resist process was successful (4). However, the adhesion capability of the promoter has been an issue. The short treatment time of adhesion promoter in a gas phase is strongly necessary for actual device production in terms of throughput.

We have evaluated several adhesion promoters and finally have developed a new adhesion promoter, 4-trimethylsiloxy-3-pentene-2-one (TMSP). It effectively prevented air-borne contamination and substrate dependency (5). In this paper, we describe its adhesion capability.

Experimental

On a hydrophilic Si substrate (contact angle: 3°), an adhesion promoter is treated in a gas phase by N_2 bubbling of the promoter. We measured the contact angles at several treatment times and substrate temperatures for evaluation of adhesion capability. Semi-empirical molecular orbital calculation PM3 was performed to clarify the reactivity of an adhesion promoter.

We also evaluated contact angles on various device substrates (SiO_2, Al, poly Si, TiN) and measured minimum dot i-line resist pattern size fabricated on each substrate using each adhesion promoter.

For the pattern profile evaluation, KrF resist was coated to a thickness of 0.7 μ m on a Si or TiN substrate treated with an each adhesion promoter. After exposure, PEB and the alkaline development, we observed SEM of line-and-space pattern profiles. A 0.7 μ m thick KrF excimer laser positive chemically amplified resist with acid-labile protecting group was used.

Adhesion promoters evaluated are the newly developed TMSP, the formerly developed IPTMS and conventional HMDS (Fig.1). For exposure, a KrF excimer laser stepper of a numerical aperture (NA) 0.50 was used. The alkaline development was done with 2.38wt% tetramethylammonium hydroxide solution, NMD-3 (Tokyo Ohka) for 60 sec.

Results and Discussion

Comparison of Contact Angles. The contact angles using HMDS are shown in Fig. 2. The contact angle becomes higher as increasing treatment time and substrate temperature. The maximum contact angle is below 75° in the range of allowed throughput (within 30 sec. treatment).

The contact angles using IPTMS are below 60° (Fig. 3). The maximum contact angle is obtained at 140℃. Although the gas concentration of this promoter is high owing to its low boiling point (94.5℃), the low reactivity resulted in this phenomenon.

By using TMSP, the adhesion capability becomes better. The contact angle of greater than 75° is attained by treatment at 140℃ for 30 sec. (Fig. 4). This value is best in using any adhesion promoter. For TMSP, the Lowest-Unoccupied-Molecular-Orbital (LUMO) energy level is relatively small (0.77 eV) as compared with that of IPTMS (1.03 eV). As a smaller LUMO energy level represents a more electrophilic state of the molecule, this suggests the higher reactivity of $(CH_3)_3Si$ of TMSP to OH sites on a substrate, which leads to a higher contact angle.

For the use of TMSP, as the boiling point is relatively high (198.0℃), the preferred substrate temperature is over 100℃. Over 160℃, the contact angle

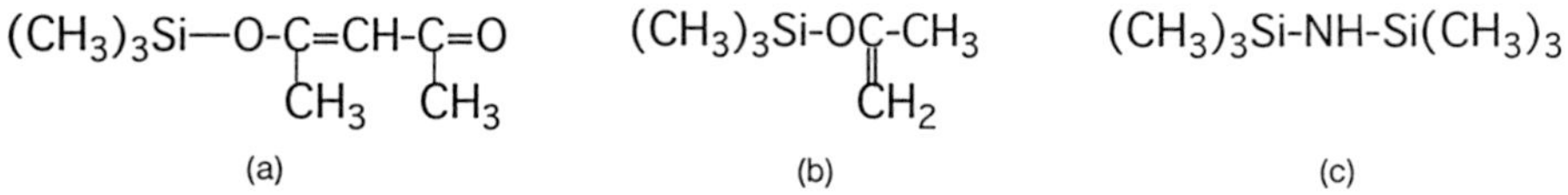

$(CH_3)_3Si{-}O{-}C(CH_3){=}CH{-}C(CH_3){=}O$ (a)

$(CH_3)_3Si{-}OC(=CH_2){-}CH_3$ (b)

$(CH_3)_3Si{-}NH{-}Si(CH_3)_3$ (c)

Figure 1. Chemical structures of (a) TMSP, (b) IPTMS and (c) HMDS.

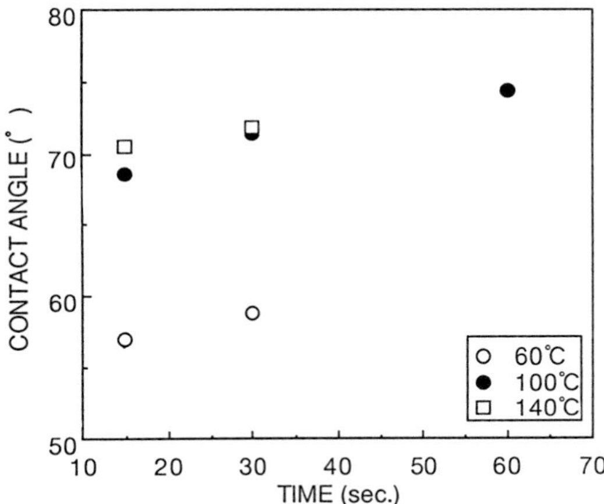

Figure 2. The relationship between treatment times and contact angles using HMDS.

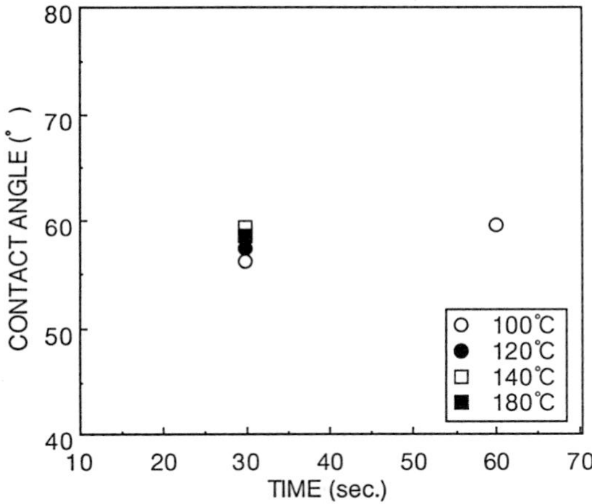

Figure 3. The relationship between treatment times and contact angles using IPTMS.

becomes lower. It is believed that unreacted trimethylsilyl group evaporated due to the excess heating.

The comparison of contact angles between TMSP, IPTMS and HMDS is shown in Fig. 5. The treatment time was 30 sec. It has been clarified that TMSP has a higher adhesion capability than IPTMS and HMDS.

Application to Various Substrates. Figure 6 summarizes contact angles on various substrates treated with TMSP or HMDS for 30 sec. at 100℃ in a gas phase by N_2 bubbling of the promoter. Under this actual use condition, TMSP showed equivalent adhesion capability as HMDS on the substrates tested.

Minimum dot patterns fabricated at the same exposure energy are similar with TMSP and HMDS (Fig.7), which agrees with the results of contact angles.

Resist Pattern Profiles. Figure 8 shows 0.24 μ m line-and-space patterns fabricated in a resist on a Si substrate treated with TMSP or HMDS for 30 sec. at 100℃. Figure 9 shows 0.3 μ m line-and-space patterns fabricated in the resist on a TiN substrate. For TMSP, as same as for HMDS, high aspect ratio and precise resist patterns were successfully achieved without any peeling off.

Conclusion

We have developed a new non-ammonia generating adhesion promoter, 4-trimethylsiloxy-3-pentene-2-one (TMSP). Its adhesion capability for a Si substrate is superior to IPTMS and HMDS owing to its high reactivity. It is applicable to various substrates. We obtained high aspect ratio and precise chemically amplified resist patterns on Si and TiN substrates using this new adhesion promoter without the anxiety about air-borne contamination and substrate dependency. This process is very promising for actual device production in terms of its simplicity and environmental safety.

Acknowledgments

We thank Dr. Ogura and Dr. Kubota for their continuous encouragement for this work.

Literature Cited

(1). MacDonald, S.A.; Clecak, N.J.; Wendt, H.R.; Willson, C.G.; Snyder, C.D.; Knors, C.J.; Deyoe, N.B.; Maltabes, J.G.; Morrow, J.R.; McGuire, A.E.; Holmes, S.J. In *Proc. SPIE*; Ito, H., Ed.; The International Society for Optical Engineering: Bellingham, WA, 1991, Vol. 1466; pp.2-12.

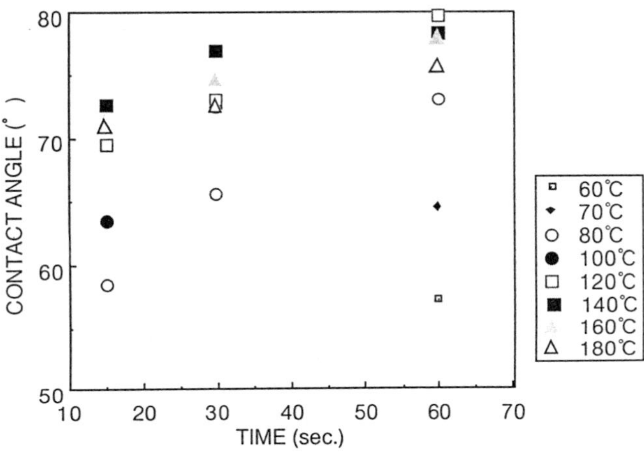

Figure 4. The relationship between treatment times and contact angles using TMSP.

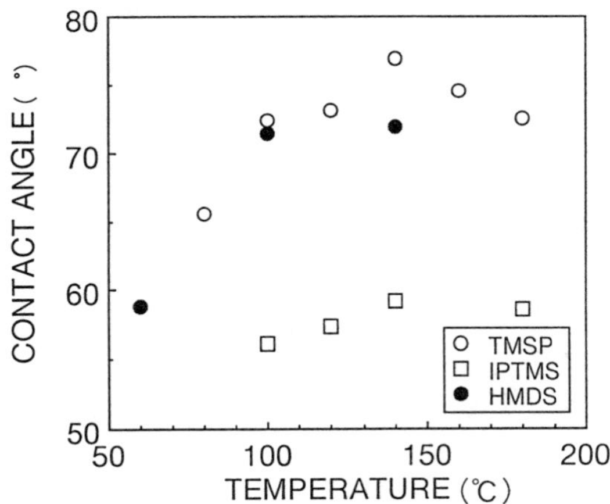

Figure 5. Comparison of contact angles between TMSP, IPTMS and HMDS (treatment time: 30 sec.).

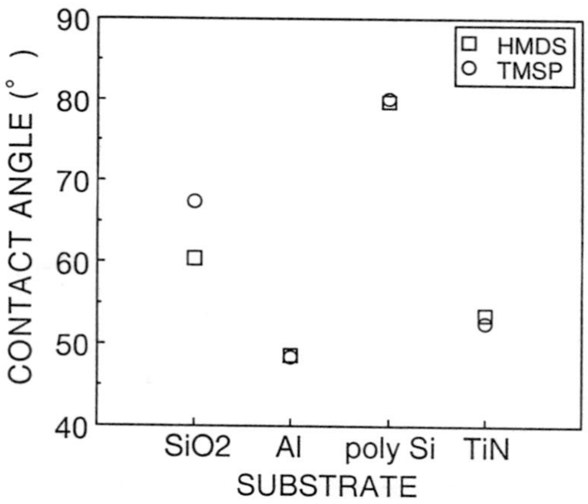

Figure 6. Comparison of contact angles between TMSP and HMDS on various substrates (treatment time: 30 sec., substrate temperature: 100℃).

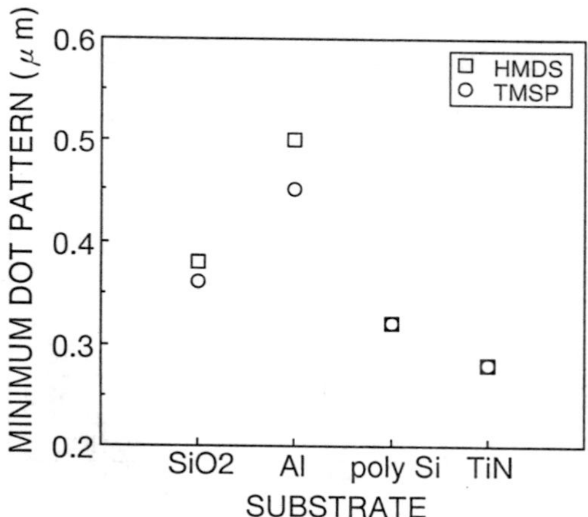

Figure 7. Comparison of minimum dot resist patterns fabricated using TMSP and HMDS on various substrates (treatment time: 30 sec., substrate temperature: 100℃).

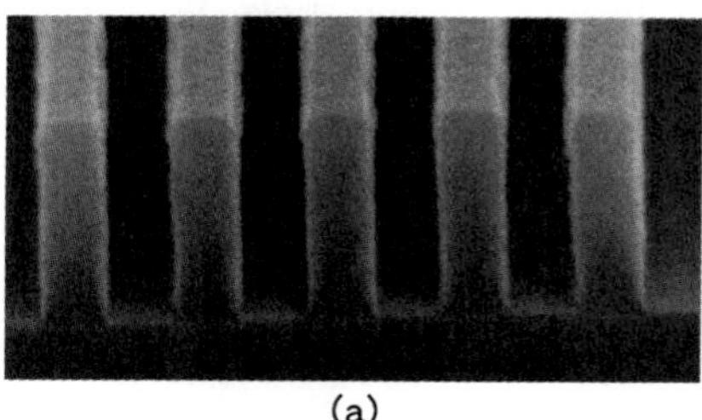

(a)

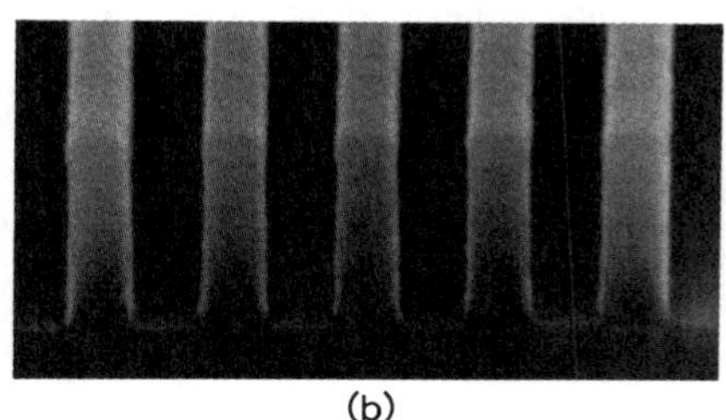

(b)

Figure 8. SEM photographs of 0.24 μm line-and-space resist patterns on a Si substrate treated with (a) TMSP or (b) HMDS (treatment time: 30 sec., substrate temperature: 100℃).

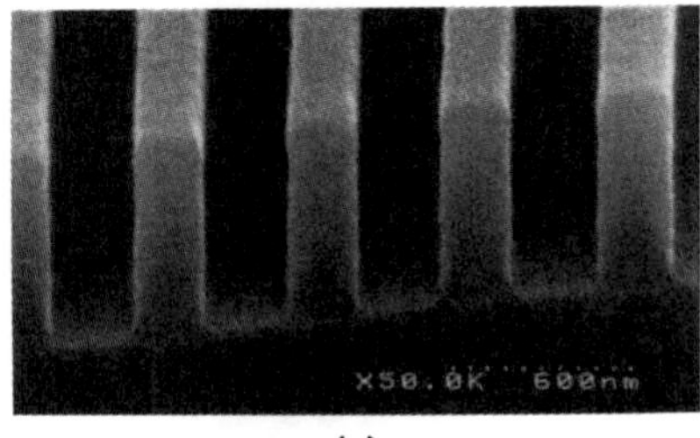

(a)

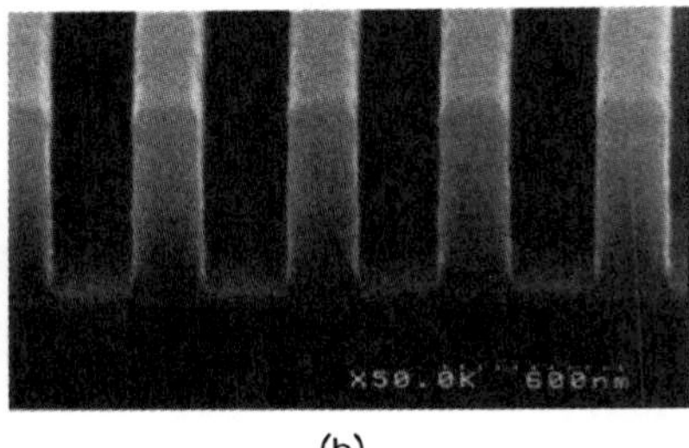

(b)

Figure 9. SEM photographs of 0.30 μm line-and-space resist patterns on a TiN substrate treated with (a) TMSP or (b) HMDS (treatment time: 30 sec., substrate temperature: 100℃).

(2). Oikawa, A.; Hatakenaka, Y.; Ikeda, Y.; Kokubo, Y.; Tanishima, M.; Santoh, N.; Abe, N. In *Proc. SPIE*; Allen, R., Ed.; The International Society for Optical Engineering: Bellingham, WA, 1995, Vol. 2438; pp.599-608.
(3). Dean, K.R.; Carpio, R.A.; Rich, G.K. In *Proc. SPIE*; Allen, R., Ed.; The International Society for Optical Engineering: Bellingham, WA, 1995, Vol. 2438; pp.514-528.
(4). Endo, M.; Kawasaki, S.; Katsuyama, A. In *Proc. SPIE*; Kunz, R., Ed.; The International Society for Optical Engineering: Bellingham, WA, 1996, Vol. 2724; pp.139-148.
(5). Endo, M.; Katsuyama, A. *Digest of Papers MicroProcesses and Nanotechnology '97*; The Japan Society of Applied Physics: Tokyo, Japan, 1997; pp.230-231.

Chapter 26

Post-Exposure Bake Kinetics in Epoxy Novolac-Based Chemically Amplified Resists

P. Argitis[1], S. Boyatzis[2], I. Raptis[1], N. Glezos[1], and M. Hatzakis[1]

[1]Institute of Microelectronics and [2]Institute of Physical Chemistry, NCSR Demokritos, 15310 Aghia Paraskevi, Attiki, Greece

The chemical mechanism and the kinetics of photoacid initiated crosslinking reactions in epoxy novolac based chemically amplified resists are consistent with the strong cage effect observed during post exposure bake (PEB) in these resists. FTIR has confirmed that the acid initiating the crosslinking reaction is bound to the polymer and suggests that the bulky carbonium intermediate is the actually moving species. Single pixel e-beam exposures used to measure the influence of acid diffusion in lithographic feature dimensions show limited diffusion with increasing PEB time. Microlithographic results from deep UV and e-beam contrast curve experiments reveal that the resist crosslinking is diffusion controlled at the PEB conditions of lithographic interest: the conditions that favour reactant diffusion facilitate reaction completion and increase resist contrast but can restrict resolution. Thermomechanical analysis and differential scanning calorimetry results are in agreement with the above observation.

The chemical amplification concept *(1)* is still the dominant strategy followed in the design of new resists for microlithographic applications including not only deep UV and 193 nm optical lithography but also XUV, x-ray, e-beam and lately micromachining. Since the chemical reactions inducing the solubility change in the chemically amplified resists occur during the post exposure bake the understanding and control of events occurring at this step is of great importance for the optimization of their formulations and processes.

Acid diffusion in particular determines to a significant extent the resist process windows, dose requirements, contrast and high resolution properties, as it controls the local reactant concentration and thus the chemical reaction inducing the solubility change in the resist. Thus the study of the phenomena related to the extent of diffusion has drawn great research effort. Since the acid diffusion related effects are independent from the source of radiation used for its generation there are opportunities for using different experimental conditions for their study *(2)*. In this context there has been developed by Raptis et al. a methodology based on single pixel e-beam exposures aiming at the study of post apply bake (PAB) and post exposure bake (PEB) effects of chemically amplified resists *(3)*. Based on this experimental technique diffusion coefficients can be calculated for different resists and processing conditions to be used in proximity correction.

On the other hand an insight into the specific resist chemistry should give further information on the expected behavior, rationalize the results obtained from microlithograpic experiments and help the resist process and formulation optimization. These goals motivated the study on the chemistry of epoxy novolac based chemically amplified resists presented here.

The concept of chemical amplification has been fruitfully applied to systems containing epoxy functionality giving resists with good performance in a number of applications: deep and near UV lithography *(4)*, micromachining *(5,6)* and fast high resolution e-beam lithography *(7,8)*. In this last case the resist formulations are mainly characterized by high sensitivity and capability for high aspect ratio microlithography down to the 0.1 μm resolution whereas they show very good process latitude regarding PEB conditions.

In the present study, the chemical mechanism and the kinetics of acid initiated crosslinking reactions of epoxy novolac based chemically amplified resists are examined. FTIR and thermal analysis have been used as the basic methods for elucidating chemical mechanism. Lithographic results obtained in a number of different processing conditions are interpreted in the context of the proposed mechanism.

Experimental

Materials. The epoxy novolac polymers used in this study were purchased from Dow (Quatrex 3710) and from Shell (Epikote 164, Epon). They were used either as received or after fractionation. Details on the fractionation procedure and material characterization have been published elsewhere *(8)*. Triphenylsulfonium hexafluoroantimonate was used as photosensitizer in most formulations and propylene glycol methyl ether acetate, as solvent. Concentration of resin in the formulation was varied in the 20%-35% range for the formation of films of desired thickness (thicker films are required in thermal analysis).

Lithographic Processing / Evaluation. Standard thermal processing and development were used in the microlithographic experiments *(8)*, unless otherwise specified. An EBMF 10cs/120 e-beam machine operated at 40 KV, and a prototype vector scan e-beam operated at 20-30 KV were used for e-beam exposures. An Oriel illuminator was used for deep UV exposures. The standard film thickness measurement and SEM techniques were employed for lithographic evaluation.

FTIR Study. The study was done on Si wafers of 500 μm thickness that are positioned in the FTIR instrument (Nicolet) at a specific angle with respect to the IR beam to minimize the noise produced by reflection.
Thermal analysis. A DuPont 943 TMA instrument was used for the measurement of mechanical properties of resist films of 2.0-0.5 μm thickness. A DuPont 910 DSC instrument was used for differential scanning calorimetry.

Results and Discussion

Resist Chemistry. The basic chemistry of epoxy novolac based chemically amplified resists has been proposed in the past by Stewart et al. *(9)*. According to this the Brønsted acid generated either photochemically or through electron beam exposure from the onium salt induces acid catalysed polymerization of the epoxy functionality. This mechanism implies that the proton generated by the exposure is actually bound to the polymer. Since the lithography consequences of this mechanism are obvious we decided to seek possible experimental evidence for the proton binding in the resist film under conditions of lithographic interest.

FTIR Results. FTIR spectroscopy has been applied to follow the chemical reactions occurring during PEB. The epoxy ring gives quite a few characteristic bands in the IR spectrum. In the first experiments the epoxy ring consumption has been followed through the bands assigned to epoxy ring vibrations (915.2 and 864.0 cm^{-1}) and the band assigned to epoxy ring C -- O stretching (1265.5 cm^{-1}). In Fig 1 spectra recorded after exposure (UV exposure was used in these experiments for practical reasons) and PEB corresponding to different exposure times are presented. As it can be seen all the epoxy related bands are reduced as the reaction proceeds but the degree of reduction of the two bands assigned to ring vibrations is distinctly higher than the corresponding one of the C -- O stretching band. Based on literature suggestions for other epoxy systems *(10)* the reaction has been also followed through a 3000 cm^{-1} band assigned to the epoxy ring CH_2 stretchnig. The results presented in Fig.1 are in good agreement with the ones obtained with the bending bands, suggesting that at the 1265.5 cm^{-1} a separate band must be superimposed to the epoxy C – O stretching. In Fig. 2 the degree of epoxy ring consumption vs the exposure time is presented as calculated from the 3000 cm^{-1} band decrease. It must be noticed that in the process conditions used in this specific case the exposure time to obtain the lithographically useful dose is 4 sec corresponding to an epoxy ring consumption of ~ 10%.

In the same IR spectrum series it is clearly shown that the OH band increases as the reaction proceeds. In Fig. 3 this increase is plotted vs the exposure time showing a kinetic behaviour in good correspondence with the consumption of epoxy ring presented in Fig.1. The increase can be attributed to the attachment of the photogenerated H^+ to the epoxy ring confirming that the H^+ in this specific reaction acts as a crosslinking initiator and not as a catalyst as is the case with most other chemically amplified resists. Thus, amplification in this case results from the diffusion of the intermediate of the epoxy ring opening reaction, ie the carbocation, and not of the H^+. This picture (elucidated in the mechanism at Scheme I) is in agreement with the slow reactant diffusion behaviour during PEB that has been reported before *(11)* : the diffusion of the bulky carbocation, which is part of a polymer chain, is expected to be much more difficult than the diffusion of the small H^+.

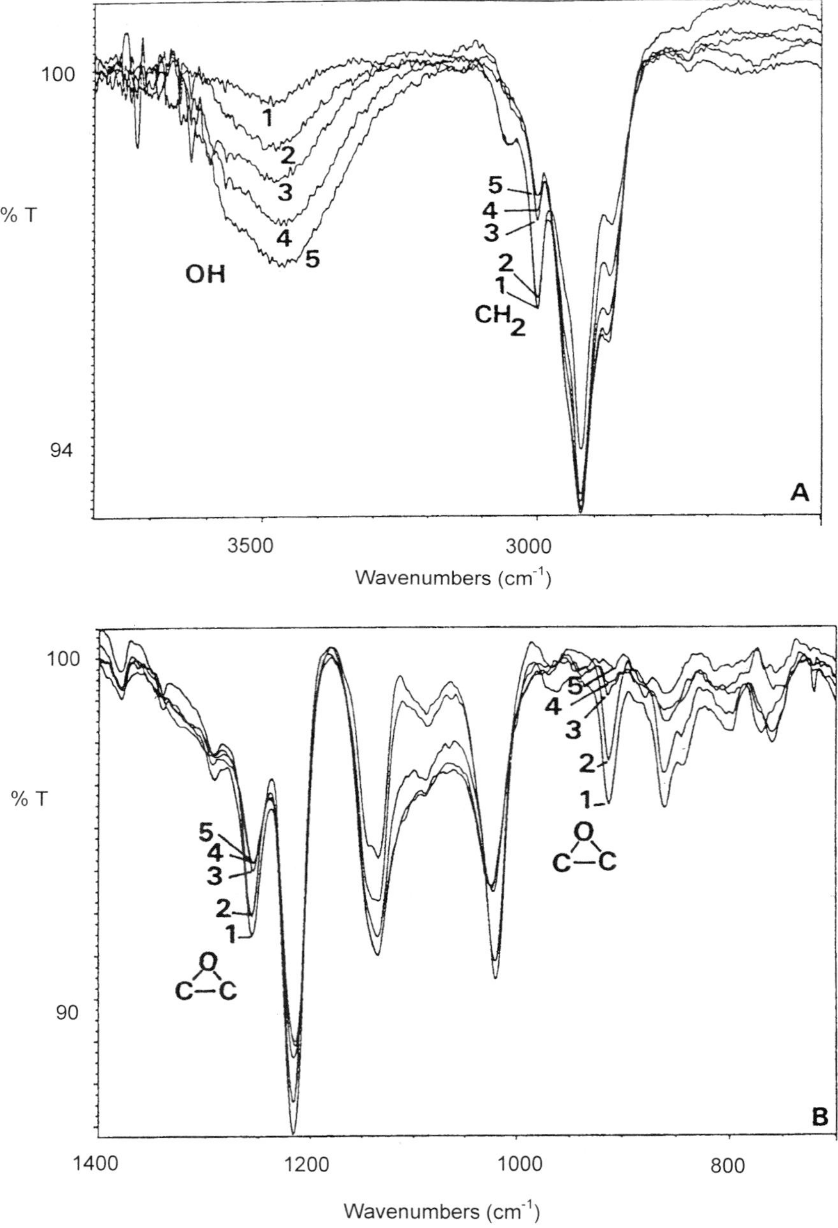

Figure 1. FTIR spectra of epoxy novolac resist on Si wafer recorded after PEB. Each spectrum corresponds to different exposure time: 1 to 0s, 2 to 4s, 3 to 30s, 4 to 120s and 5 to 240s. The lithographically useful dose is close to 4s in this case.

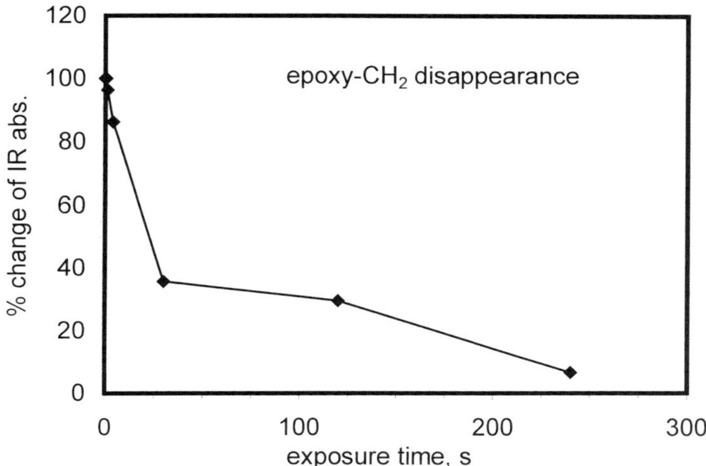

Figure 2. Decrease in the epoxy CH_2 band intensity during processing of the epoxy novolac resist monitored with FTIR. Absorbance values calculated from the areas under the 3000 cm^{-1} band of Fig. 1 spectra.

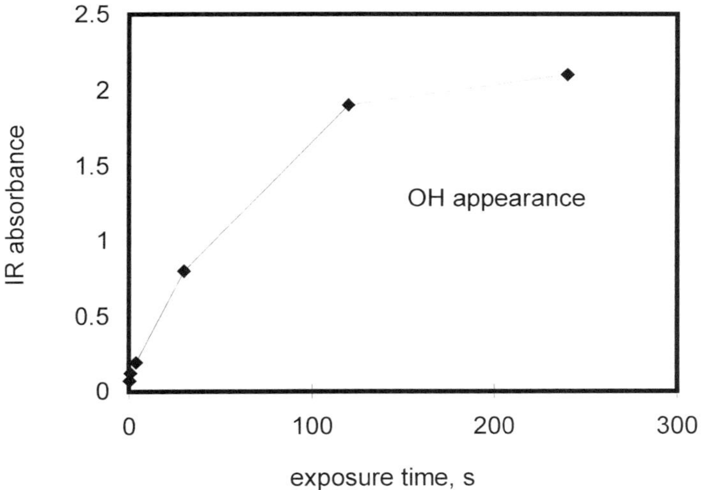

Figure 3. OH formation during processing of epoxy novolac resist. Absorbance values calculated from the areas under the 3450 cm^{-1} band of Fig. 1 spectra.

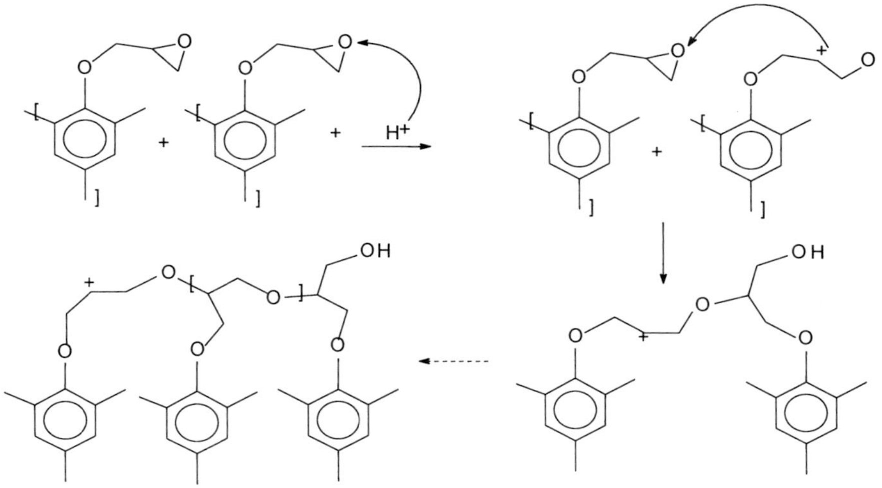

Scheme I: Proposed mechanism of crosslinking showing the H^+ attachment to the epoxy oxygen and the carbocation transfer. Arrows indicate the movement of the reacting species.

Thermal analysis studies. Thermomechanical analysis (TMA) has been used for the *in situ* determination of the resist thermal properties after processing steps *(12)*. In the present study, TMA showed the influence of the sample preparation, the prebake conditions and formulation parameters (MW distribution of the polymer) to the resist glass transition temperature (T_g) (Table 1).

Table 1. Tg values for epoxy novolac polymer and resist samples.

Sample	Polymer	Fractionated Polymer	Photoresist		
Processing	-	-	PAB 90 °C 4 min	PAB 90 °C 20 min	PAB 110 °C 4 min [(2)]
Tg (°C) [(1)]	40 (32)	70 (54)	45 (35)	70 (53)	70 (53)

[1]Measured in films with TMA (onset TMA values in parentheses)
[2] In this case Tg was 74 °C with DSC

Similar T_g values were obtained by differential scanning calorimetry (DSC) in samples prepared after scraping the resist films off the wafer. The last technique has also been used in the investigation of the resist chemical changes induced by light or temperature. It has been shown (in accordance with results presented in ref. 12) that the unexposed resist is stable up to ~170 °C, where an exothermic reaction leading to resist crosslinking starts. On the other hand, in exposed samples, where acid has been photogenerated, an exothermic reaction starts practically at temperatures above T_g (at scan rates 5 - 20 deg/min). During the PEB step this reaction proceeds up to a certain degree depending on the amount of acid produced, PEB temperature and time. If a limited amount of acid is produced (ie. at a lithographically useful dose or below), the available epoxy reaction sites are not fully consumed (cf. IR results) even at elevetated temperatures, and a separate thermal crosslinking reaction is observed at higher temperatures. This behaviour is reflected in the DSC curves presented in Fig. 4.

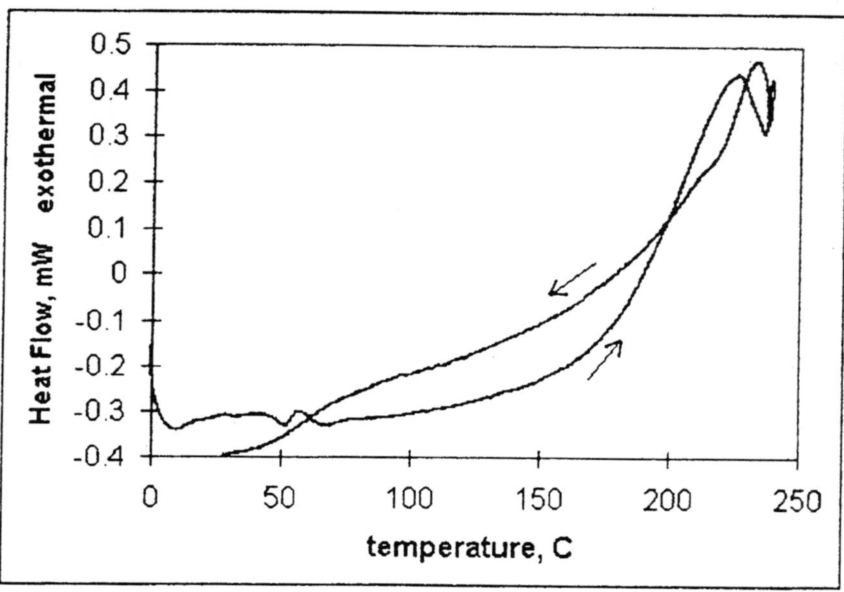

Figure 4. DSC curve of epoxy novolac resist after prebake and exposure at a low dose (0.2sec exposure time vs. 2 sec for the lithographically useful dose in the same conditions).

Microlithographic Results: PEB Dependence of Lithographic Features. Single pixel electron beam exposures have been introduced for the study of PEB acid diffusion and diffusion coefficient measurements. In Fig. 5 a representative column (single pixel exposure) and a representative isolated line (three pixel exposure) of comparable dimensions are shown. In Fig. 6 the dependence of column diameter on square root of PEB time is presented. This diagram shows limited acid diffusion effects. The diffusion coefficient concluded from a reaction-diffusion modeling scheme *(11)* based on single pixel exposure information is 10^{-6} μm^2/sec whereas for SAL-601 the corresponding value is 10^{-4} μm^2/sec. It is reasonable to attribute this difference to the different nature of the chemical mechanism presented above.

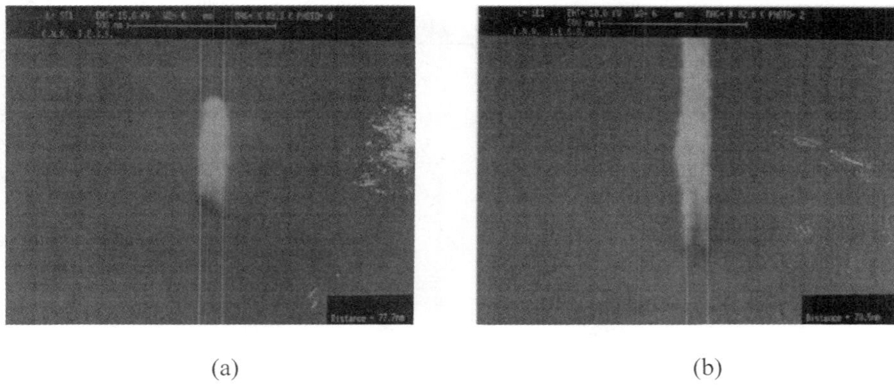

(a) (b)

Figure 5. Representative lithographic features of epoxy novolac resist obtained with e-beam lithography (40 KV). a) Resist column of 78 nm diameter and aspect ratio over 5. Exposure charge = 6 10^{-16} Cb, b) Isolated line of linewidth = 71 nm. Exposure dose = 4.25 $\mu C/cm^2$

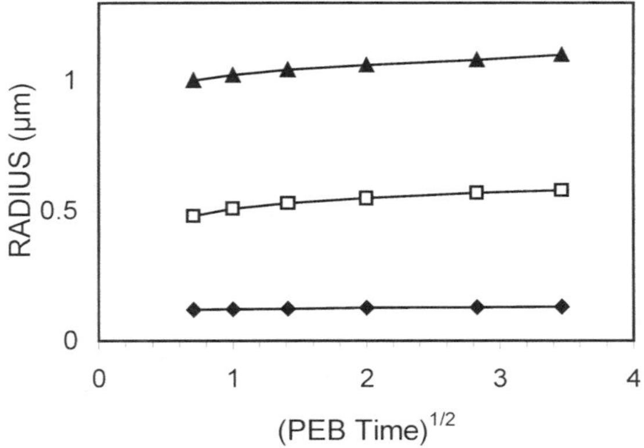

Figure 6. Column radius dependence on PEB time for 3 different column sizes.

Diffusion vs. Reaction Controlled Kinetics Contrast curves obtained from e-beam experiments at the lithographically useful doses at 90 °C and 110 °C for different PEB times show a very characteristic behavior for this epoxy chemistry based chemically amplified resist. The gel dose does not change in this PEB temperature and time range whereas the contrast increases in the higher temperature and time regime (Figures 7 and 8).

This behavior has been further examined in a series of similar simple lithographic experiments with UV exposure at different PEB times but at lower PEB temperatures (Figs. 9-12). These experiments show that the gel dose decreases constantly with PEB time at temperatures below T_g and practically levels at PEB temperatures above T_g which is approximately 70 °C for this material. Above T_g longer PEB times and higher PEB temperatures result in contrast increase in accordance with the e-beam experiments.

These experimental results can be rationalized on the basis of the complex reaction-diffusion behaviour encountered in chemically amplified resists (see also *15*). They are controlled by the relative rates of the chemical reaction resulting in crosslinking and of reactant diffusion at the different temperatures and crosslink densities. The rate of crosslink formation is characterized by diffusion controlled kinetics at temperatures below Tg and by reaction controlled kinetics above this temperature. When the system is above Tg the reaction proceeds too fast to be followed by lithographic experiments: gelation takes place at times shorter than 30 secs (compare with results on epoxy monomer polymerization in ref. 14). At temperatures below T_g the crosslink formation is controlled by the diffusion of reactants and thus there is a clear influence of the temperature and acid concentration on the initial reaction rate. At PEB temperatures of lithographic interest (above T_g) the parameter that changes with PEB conditions is contrast, because the rate of the reaction decreases as the crosslink density increases (cage effect). In other words the system enters again in diffusion controlled kinetics since the Tg of the material locally approaches or reaches the PEB temperature (vitrification). Therefore, the reaction proceeds to high conversion degrees at low doses (increased contrast) only under PEB conditions (temperature, time) that favour reactant diffusion. Nevertheless, increased diffusion can degrade lithographic performance and this is the reason for the fact that increased contrast is not always accompanied by better lithography in the EPR resist *(8)*.

Conclusions

The study of the epoxy novolac resist chemistry confirms that the photogenerated acid is consumed during the PEB. According to the proposed mechanism the propagating species is the bulky carbonium intermediate in compliance with the relatively small diffusion coefficient measured. The microlithographic results also suggest diffusion controlled kinetics in the lithographically useful PEB time and temperature regime. Optimized lithography is obtained under processing conditions resulting in optimized diffusion length.

Acknowledgements. The above work was supported by the European Union ESPRIT #2284 project "NANCAR" (Nanofabrication with Chemically Amplified Resists). The

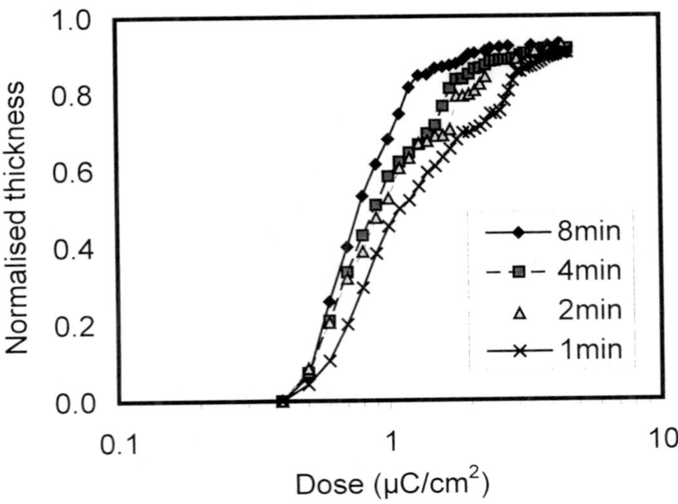

Figure 7: Contrast curves at 90 °C PEB temperature at different PEB times. E-beam exposure at 40 KeV.

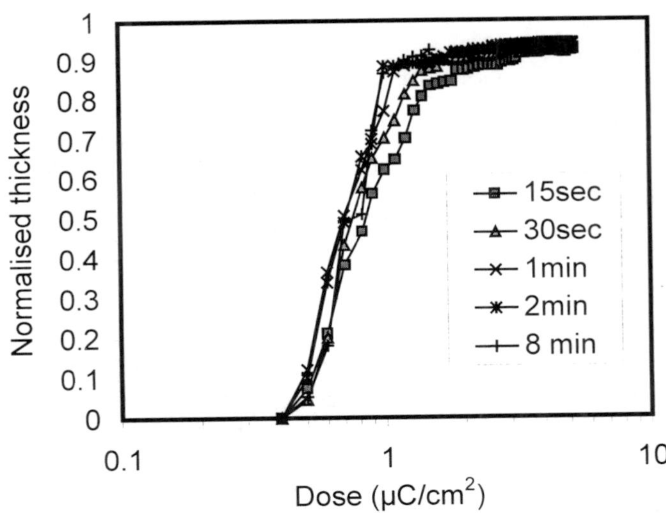

Figure 8: Contrast curves at 110 °C PEB temperature obtained for different PEB times. E-beam exposure at 40 KeV.

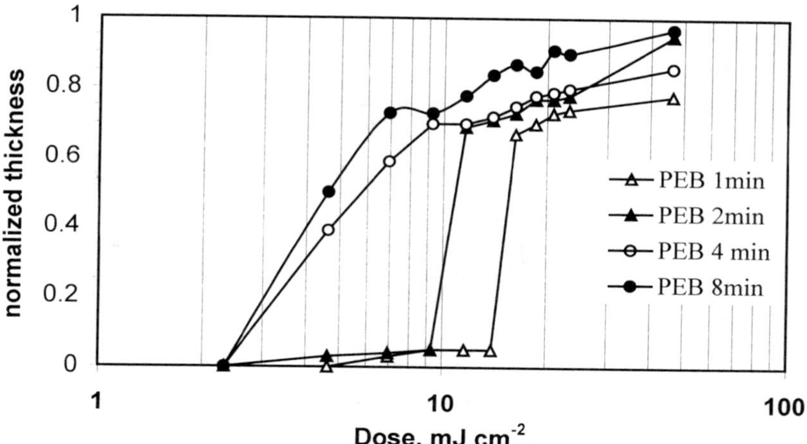

Figure 9: Contrast curves of epoxy novolac resists exposed to DUV and postexposure-baked at 40 °C for the times shown.

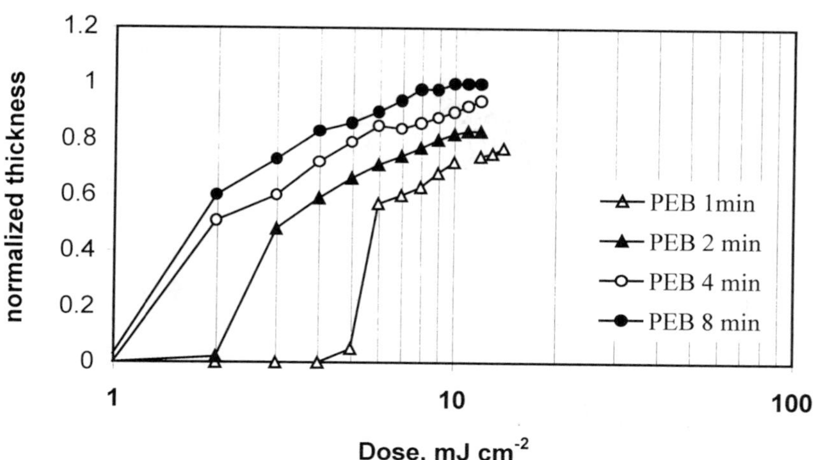

Figure 10: Contrast curves of epoxy novolac resists exposed to DUV and postexposure-baked at 52 °C for the times shown.

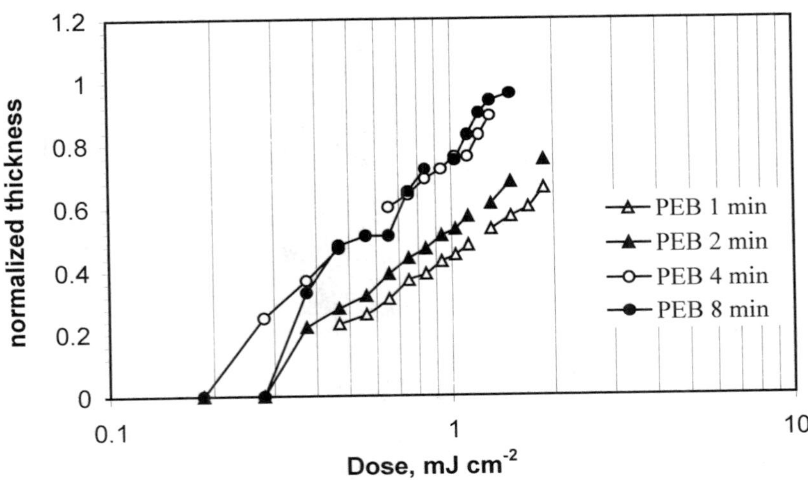

Figure 11: Contrast curves of epoxy novolac resists exposed to DUV and postexposure-baked at 75 °C for the times shown.

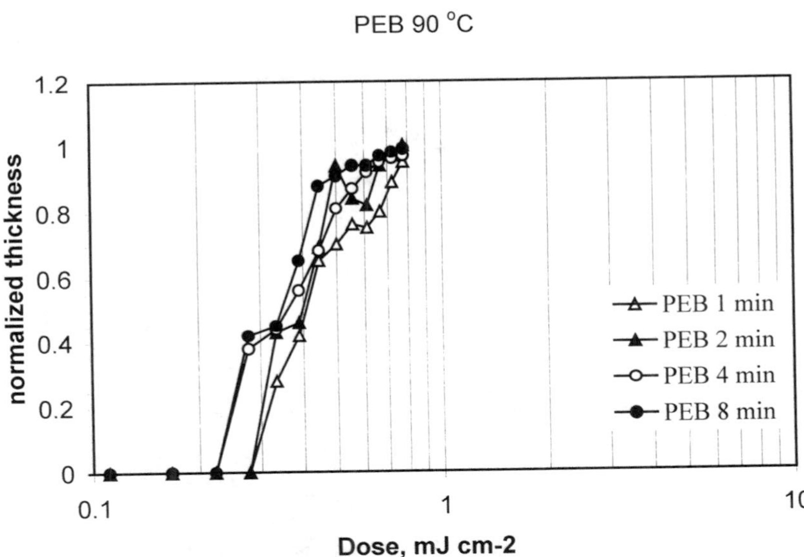

Figure 12: Contrast curves of epoxy novolac resists exposed to DUV and postexposure-baked at 90 °C for the times shown.

help of Mr. R.Maggiora, Mr. L.Scopa and Dr. M. Gentili of IESS-CNR Rome, Italy, with single pixel e-beam experiments is greatly acknowledged.

References

1. Ito, H.; Willson, C. G. *Polym. Eng. Sci.* **1983**, *23*, 1012.
2. a) Fedynyshyn, T.; Thakeray, J.; Georger, J.; Denison, M. *J. Vac. Sci. Technol. B* **1994**, *12(6)*, 3888, b) J. Nakamura, H. Ban, and A. Tanaka, *Polym. Mat. Sci. Eng.* **1995**, *72*, 155.
3. Raptis, I.; Grella, L.; Argitis, P.; Gentili, M.; Glezos, N.; Petrocco G. *Microelecton. Engineering* **1996**, *30*, 295-9.
4. Allen, R. D.; Conley, W.; Gelorme, J. D. *SPIE Proceedings* **1992**, *1672*, 513.
5. Lee, K. Y.; LaBianca, N.; Rishton, S.A.; Zolgharnain, S.; Gelorme, J.D.; Shaw J.; Chang,, T.H.P. *J. Vac. Sci. Technol. B* **1995**, *13(6)*, 3012.
6. Despont, M.; Lorenz, H.; Fahrni, N.; Brugger, J.; Renaud, P.; Vettiger, P. *Proc. of 10th IEEE Int'l, Workshop on Micro Electro Mechanical Systems (MEMS '97)* **1997**.
7. Hatzakis, M.; Stewart, K. J.; Shaw, J. M.; Rishton., S.A. *J. Electrochem. Soc.* **1992**, *138*, 1076.
8. a) Argitis, P.; Raptis, I.; Aidinis, C.J.; Glezos, N.; Baciocchi, M.; Everett J.; Hatzakis, M. *J. Vac. Sci. Technol.. B* **1995**, *13(6)*, 3030, b) Miller Tate, P. C.; Jones, R.G.; Murphy, J.; Everett, J. *Microelecton. Engineering* **1995**, *27*, 409.
9. a) Stewart, K.J.; Hatzakis, M.; Shaw, J. M.; Seeger, D. E. *J. Vac. Sci. Technol.. B.*, **7**, 1734 (1991), b) Haller, I.; Stewart K.J. *J. Vac. Sci. Technol.. B* **1991**, *9(6)*, 3370.
10. Crivello J. V.; Narayan, R. *Macromolecules* **1996**, *29*, 439.
11. Glezos, N.; Patsis, G.; Raptis, I.; Argitis, P.; Gentili M.; Grella L. *J. Vac. Sci. Technol. B* **1996**, *14(6)*, 4252.
12. Tegou, E; Gogolides E.; Hatzakis, M. *Microelectron. Engineering* **1997**, *35*, 141.
13. Patsis, G. ; Raptis, I. ; Glezos, N. ; Argitis, P. ; Hatzakis, M. ; Aidinis, C. J.; Gentili, M. ; Maggiora, R. *Microelectron. Engineering* **1997**, *35*, 157.
14. a) Decker, C. ; Moussa, K. *J. Polym. Sci. Polym. Chem. Ed.* **1990**, *28*, 3429, b) Decker C. *Progr. Polym. Sci.* **1996**, *21*, 593.
15. Mack, C. A. in "Microelectronics Technology", E. Reichmanis, C.K. Ober, S.A. McDonald,T. Iwayanagi and T. Nishikubo eds, ACS Symp. Ser. **1983**, *614*, 57.

Chapter 27

Alkali-Developable Positive-Photosensitive Polyimide Based on Diazonaphthoquinone Sensitizer

T. Ueno, Y. Okabe, T. Miwa, Y. Maekawa, and G. Rames-Langlade

Hitachi Research Laboratory, Hitachi Ltd., Hitachi-shi, Ibaraki-ken 319-12, Japan

We report on the positive alkali-developable photosensitive polyimides based on an alkali-soluble polyimide precursor as a base polymer and diazonaphthoquinone (DNQ) sensitizer to improve process stability and sensitivity. Polyamic acid ester with pendant carboxylic acid (PAE-COOH) showed good dissolution behavior in aqueous alkali developer. The dissolution rate of PAE-COOH was controlled by the content of pendant carboxylic acid. It was found that a photosensitive system composed of butyl ester of PAE-COOH and a DNQ compound can avoid the residue at the edge of hole patterns (footing) after development, while that of methyl ester of PAE-COOH showed the residue. A DNQ compound containing sulfonamide derived from diaminodiphenylether renders improved sensitivity compared with DNQ compounds derived from phenol derivatives.

Polyimide is widely used as insulating materials and protecting coating material in microelectronics. Progress of LSI has been supported by mass production technology such as chip shrink and decrease in number of process steps. The use of photosensitive polyimide as protecting chip coating can simplify the back end of the line process. Commercially available photosensitive polyimides are mainly negative and require organic solvent as developer. Since industrial LSI manufacturing is primarily based on positive photoresist processes using aqueous base developer, alkali-developable photosensitive polyimides have been received a great attention from the view point of process compatibility with photoresist process as well as environment issues. Several approaches for alkali-developable positive photosensitive materials for high temperature stable polymers have been reported including diazonaphthoquinone (DNQ) system [1-4] and chemical amplification systems [5,6]. Here we report on the positive alkali-developable photosensitive polyimides based on polyimides precursor with pendant carboxylic acid and DNQ sensitizer for improved process stability and sensitivity.

Experimental

Polymer Synthesis. Basic structures of polyamic acids (PAA), polyamic acid esters (PAE), copolymers of PAA and PAE, and polyamic acid esters with pendant carboxylic acid (PAE-COOH) used for this study are described in Figure 1. After dissolving a diamine into N-methylpyrrolidone, polyamic acids (PAA) were synthesized by adding tetracarboxylic dianhydrides slowly for about 1hr into the ice-cooled diamine solution as described in the literarure [7].

PAEs were synthesized in the following way [8]. Tetracarboxylic dianhydrides were esterified in the alcohol solutions by heating under reflux. After removing the excess alcohol, thionyl chloride was added to the solution and refluxed with stirring. Then diesterified dicarboxylic acid dichloride solution was obtained removing the excess thionyl chloride. PAE was obtained by adding this diesterified acid chloride solution to a diamine solution. When this diesterified dicarboxylic acid chloride solution was added to the polyamic acid solution, a copolymer of PAA/PAE was obtained. PAE was deposited in vigorously stirred water. After filtering and washing with water, the polymers were dried in air and in a vacuum oven at 40°C.

Polyamic acids esters with pendant carboxylic acid (PAE-COOH) were synthesized in the following way. 3,3',4,4'-benzophenonetetracarboxylic dianhydride (12.3 g, 38.3 mmol) dissolved in a mixture of butanol (14.8 mL, 162 mmol) and 10 mL of xylene was refluxed for 1 hour under nitrogen. After removing xylene and the excess amount of butanol under reduced pressure, a solution of thionyl chloride (11 g, 46 mmol) in 10 mL of toluene was added dropwise at room temperature. The reaction mixture was stirred for 1 hour at 80°C, followed by removing toluene and the excess amount of thionyl chloride under reduced pressure to obtain the corresponding 3,3',4,4'-benzophenonetetracarboxylic acid dibutylester dichloride (BTDBuCl) as a solid (27.2 g). 4,4'-diaminodiphenyl ether (2.68 g, 13.4 mmol), 3,5-diaminobenzoic acid (3.5 g, 23 mmol) was added to the obtained BTDBuCl in 50 mL of NMP at a rate to maintain the temperature below 5°C. The resulting mixture was then stirred at room temperature for 1 hour. The polymer was isolated by pouring the solution into water. The polymer was washed and dried under reduced pressure to give PAE-COOH. Molecular weight: Mw=12000, Mn=7000. IR (KBr cm^{-1}) 3340, 2960, 1730, 1670, 1500, 1280, 1230, 1070. ^1H NMR (DMSO-d6) δ 10.9 (s, 1.2H), 10.6 (d, 0.8H), 7.7-8.4 (m, 9.4H), 7.0 (bs, 1.6H), 4.1-4.3 (m, 4H), 0.6-1.7 (m, 14H).

Diazonaphthoquinone (DNQ) Compounds Preparation. 2.3,4,4'-tetrahydroxy-benzophenone-1,2-diazonaphthoquinone-5-sulfonate (average esterification ratio is 75%) (TBP-DNQ) was purchased from Tokyo Gosei Kogyo Co. Ltd. 4,4'-bis(1,2-diazonaphthoquinone-5-sulfonamino)diphenyl ether (DDE-Q): To a solution of 4,4'-

diaminodiphenyl ether 4.00 g (20.0 mmol) and 1,2-diazonaphthoquinone-5-sulfonyl chloride 10.8 g (40.0 mmol) in 50 mL of dioxane was added a solution of triethylamine (6.7 mL, 120 mmol) in 5 mL of dioxane dropwise to maintain the temperature below 10°C. The stirring was kept on about 3 hours and the resulting mixture was poured into 2 L of 1N hydrochloric acid. The precipitate was washed several times with water and dried at 40°C for 16 hours in vacuo to give yield 12.5 g (yield=94%) of DDE-Q as a solid. IR (KBr cm^{-1}) 2970, 2110, 1600, 1500, 1360, 1260, 1200, 880. ^{1}H NMR (DMSO-d6) δ 10.50 (s, 2H), 8.39 (d, 2H), 8.21 (dd, 2H), 7.5-7.6 (m, 4H), 6.96 (d, 4H), 6.73 (d, 4H).

Characterization. The polymers were characterized with IR, NMR spectroscopy. Molecular weight and molecular weight distribution were measured with GPC using tetrahydrofuran/formacetoamide and phosphoric acid/lithium bromide as an eluant, calibrated polystyrene standard.

Lithographic Evaluation. To prepare the photosensitive polyimide solution, a polyimide precursor and a diazonaphthoquinone compound were dissolved in the solvent and filtered through 5 μm membrane filter. After spin-coating onto silicone substrate, the film was prebaked on a hot plate. The film was exposed to the light by a contact printer with high-pressure mercury lamp and developed in a standard developer of aqueous base (2.38% tetramethylammonium hydroxide, TMAH). After the lithographic patterning the precursor was converted to the polyimides by curing on the hot plate.

Results and Discussion

Alkali-Developable Base Polymers. Since the dissolution rate of PAAs is too fast in 2.38% TMAH solution, the photosensitive system composed of PAA and a DNQ compound suffers from low contrast of the difference in dissolution rate between exposed and unexposed area. It implies that the optimization of the dissolution rate of base polyimide precursor in the developer is needed. PAA/PAE blends, PAA/PAE copolymer, and PAE with pendant carboxylic acid (PAE-COOH) were evaluated as base polymers for photosensitive polyimides.

PAA/PAE Blends. PAAs and PAEs were blended and dissolution rate of their films was measured. Some of the mixture films were peeled off from the substrate during the dissolution. The results for two examples of the mixtures are shown in Figure 2. It can be seen from Figure 2 that the dissolution rate of the films for blends increases with the content of PAA as expected. We selected a polymer of blend ratio of PAA/PAE 50/50 for PAA(BTDA/DDE)/PAE(BTD-Me/DDE) as a base polymer with moderate dissolution rate, where BTDA is benzophenone tetracarboxylic acid dianhaydride, DDE is diphenyldiaminoether, and BTD-Me is benzophenone tetracarboxylic acid dimethyl ester. The exposure characteristic for a photosensitive material composed of the blend as a base polymer and TBP-DNQ

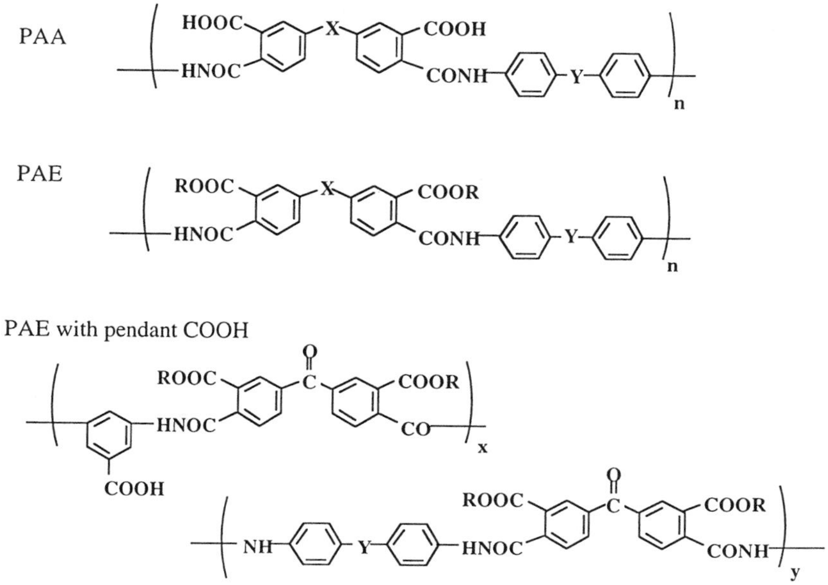

PAA

PAE

PAE with pendant COOH

Figure 1 Polymer structure of polyamic acid (PAA), polyamic acid ester (PAE) and polyamic acid ester with pendant carboxylic acid.

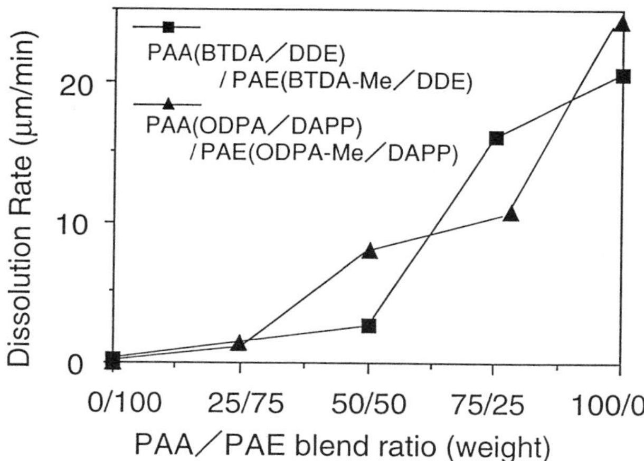

Figure 2 Relation between dissolution rate and PAA/PAE blend ratio. BTDA: 3,3',4,4'-benzophenonetetracarboxylic dianhydride, DDE: 4,4'-diaminodiphenyl ether, ODPA: 4,4'-oxydiphthalic anhydride, DAPP: 2,2-bis((3-aminophenoxy)phenyl)propane.

was evaluated. However, the contrast of this system is low of 2. In addition, the unexposed area was damaged and waved after development, which may be ascribed to partial dissolution and/or swelling due to incompatibility of the two polymers in the film.

PAA/PAE Copolymer. To avoid incompatibility and to obtain uniform dissolution in the developer, a copolymer of PAA(BTDA/DDE)/PAE(BTD-Me/DDE) was used as a base polymer. The dissolution rates of exposed (300mJ/cm^2) and unexposed area for a photosensitive system composed of the copolymer and TBP-DNQ is shown in Figure 3. The dissolution rate ratio of unexposed to exposed areas, about 2, is not enough for pattern formation.

PAEs with Pendant Carboxylic Acid. Next attempt for optimization of dissolution rate of base polymer for alkali-development is to use PAE with pendant carboxylic acid (PAE-COOH) as already shown in Figure 1. Figure 4 shows the relation between dissolution rate in a standard developer and the content of carboxylic acid. From this figure, alkali-solubility was controlled by the content of pendant carboxylic acid. It is noted that dissolution rate shows sharp change in the content range of 60 to 80% of x/(x+y) in Figure 4, while blends of polyamic acid and polyamic acid ester showed no such a sharp increase in dissolution rate with the content of carboxylic acid. We consider that the polymer which shows sharp change in dissolution rate is suitable for pattern forming capability, since the sharp increase in dissolution rate of exposed area can be expected when DNQ compound is converted to indenecarboxylic acid.

Figure 5 shows hole patterns obtained with the system composed of methyl esterified PAE-COOH and TBP-DNQ. In this photograph, there can be seen residue of film at the edge of the hole pattern after development (footing). The footing may be ascribed to imidization near the substrate during the prebaking, since the imidization of PAE leads to the decrease in dissolution rate of film in aqueous base. The effect of the structures of alkyl ester on curing processes (imidization temperature) is shown in Figure 6 as reported previously [8]. The imdization temperature was determined from peak temperature of endothermic reaction in the measurement of differential scanning calorimetry for PAEs. As shown in Figure 6, the imidization temperature increases with the carbon number of alkyl esters for polyamic acid ester. Therefore, it is expected that PAE of butyl ester renders much lower imidization degree than PAE of methyl ester during prebaking. SEM photograph of hole patterns obtained using PAEs of methyl ester and butyl ester are compared in Figure 5. It is clearly demonstrated that PAE of butyl ester shows no evidence of footing near the substrate.

DNQ Compounds. The use of the DNQ compounds derived from the phenol derivatives such as TBP-DNQ suffers from low sensitivity. Since dissolution inhibition of PAE containing pendant carboxylic acid by DNQ compounds is small compared to the novolak-DNQ system for a positive photoresist, we investigated the DNQ compounds which render higher dissolution rate in exposed area to improve

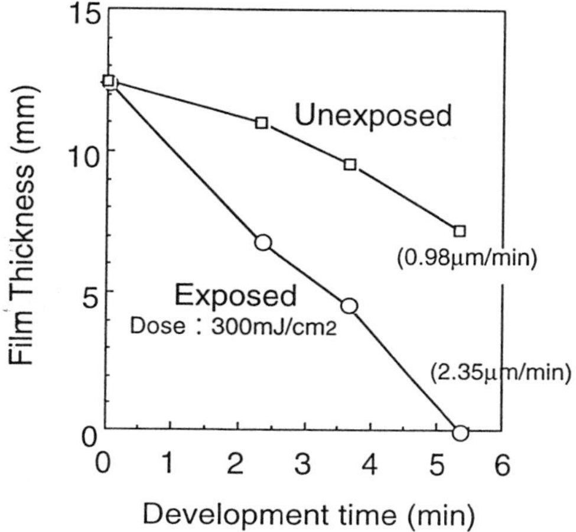

Figure 3 Change in film thickness with development time for exposed and unexposed area for a copolymer of PAE/PAA and THB-DNQ. Exposure dose is 300 mJ/cm^2.

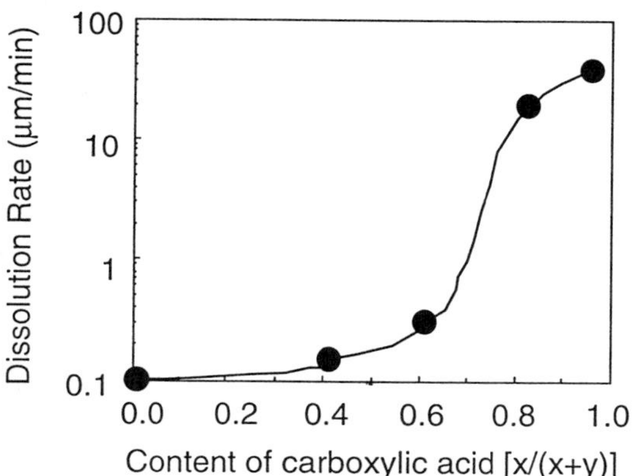

Figure 4 Relation between dissolution rate and content of carboxylic acid of PAE with pendant COOH.

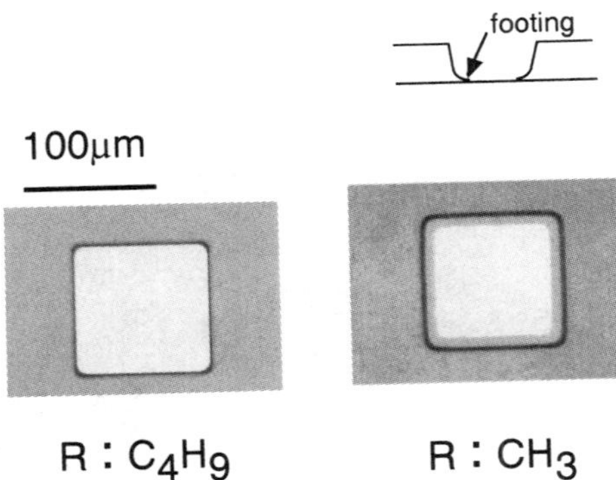

Figure 5 SEM photographs of hole patterns obtained from polyamic methyl- and butylesters containing pendant carboxylic acid and a DNQ compound.

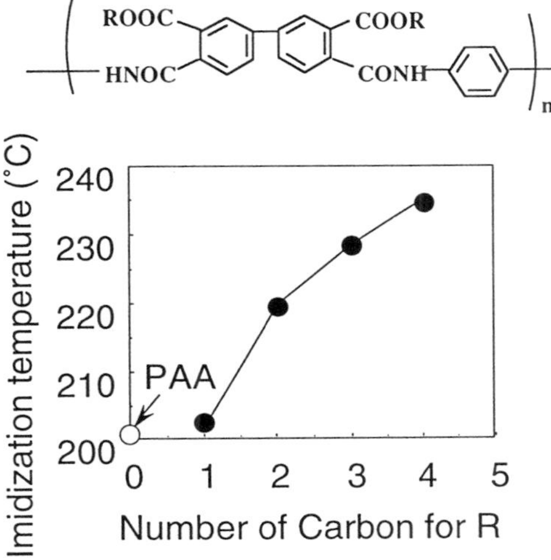

Figure 6 Dependence of imidization temperature on number of carbon for alkyl group (R) for polyamic acid esters. (Reproduced with permission from ref. 8. Copyright 1993 The Society of Polymer Science, Japan)

the sensitivity. It is known that sulfonamide is soluble in aqueous base. Therefore, we propose DNQ compounds containing sulfonamide group, since it is expected that indenecarboxylic acid containing sulfonamide formed with exposure promotes the dissolution in exposed area. DNQ compounds containing sulfonamide are easily synthesized from the reaction of diazonaphthoquinone sulfonyl chloride with an amine in the presence of base. The sensitivity of photosensitive system composed of PAE-COOH with DNQ derived from diaminodiphenylether is shown in Figure 7 compared with the sensitivity of PAE-COOH with TBP-DNQ. This figure clearly demonstrated that DNQ compound derived from the diamine shows higher sensitivity than DNQ compound from tetrahydroxybenzophenone. Dissolution behavior for these two DNQ compounds is shown in Figure 8. This figure indicates that the DNQ with sulfonamide gives higher dissolution rate than TBP-DNQ.

Lithographic Evaluation. The SEM photograph of patterns using PAE-COOH with a DNQ compoundderived from a diamine is shown in Figure 9. 10μ m hole patterns with 6 μm thickness were obtained. The 3.5 μm thick patterns after fully cured at 330℃ are also shown in Figure 9.

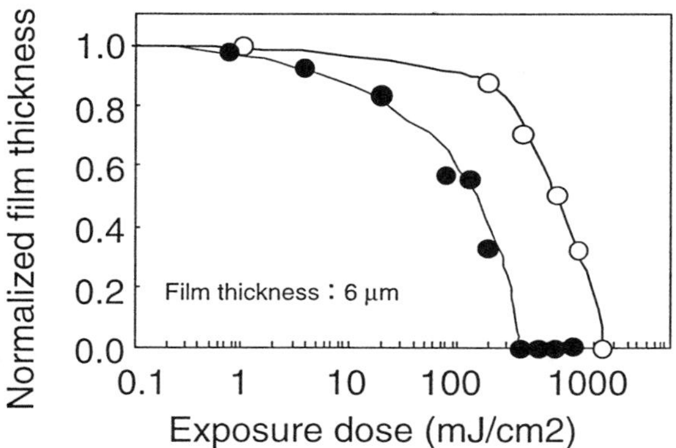

Figure 7 sensitivity curves for photosensitive materials composed of polyamic butylester containing pendant carboxylic acid and a DNQ compound derived from tetrahydroxybenzophenone (○) and that from diaminodiphenylether (●).

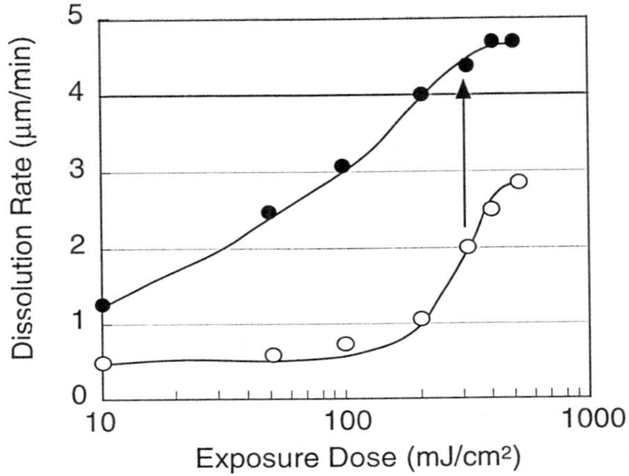

Figure 8 Change in dissolution rate with exposure dose for photosensitive systems composed of PAE-COOH using TBP-DNQ (○) and a DNQ compound derived from a diaminodiphenylether (●).

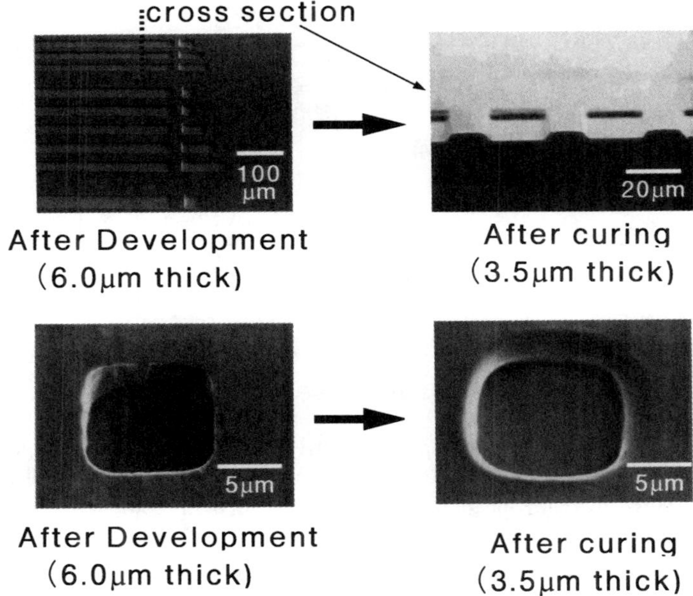

Figure 9 Scanning electron micrographs of patterns produced by a contact printer for PAE-COOH and a DNQ compound derived from a diamine and those after curing at 330 °C on a hot plate.

Acknowledgment

The authors thank to Mrs. Nobuaki Sato, Shyun-ichiro Uchimura, Masataka Nunomura, Mamoru Sasaki, Michiaki Hashimoto, Shigeru Koibuchi, and Makoto Kaji for their helpful discussion.

References

[1] D.N. Khanna and W. H. Mueller, *Polym. Eng. Sci.*, **1989**, 29, 954.
[2] T. Omote, H. Mochizuki, K. Koseki, and T. Yamaoka, *Polymer Commun.*, **1990**, 31(4), 131.
[3] R. Sezi, H. Ahne, R. Gestigkeit, E. Kühn, R. Leuschner, E. Rissel and E. Schmidt, *Proc. 10th international Conference of Photopolymers*, **1994**, p.444.
[4] A. Mochizuki, Y. Tamino, K. Yamada, and M. Ueda, *J. Photopolym. Sci. Technol.*, **1995**, 8(2), 333.
[5] H. Fuji-i, T. Omote, and S. Hayashi, *Polymer Preprint, Japan*, **1993**, 42(7), 2685.
[6] H. E. Simmons III, and D. R. Wipf, *ACS Polym. Mater. Sci. Eng.*, **1994**, 70, 235.
[7] S. Numata, S. Oohara, K. Fujisaki, J. Imaizumi, and N. Kinjo, *J. Appl. Polym. Sci.*, **1986**, 101.
[8] Y. Okabe, T. Miwa, A. Takahashi, and S. Numasta, *Jpn. J. Polym. Sci. Technol.*, **1993**, 50, 947.

Indexes

Author Index

Subject Index

R

S

X

Z

Highlights from ACS Books